Time-Varying Waveform Distortions in Power Systems

Time-Varying Waveform Distortions in Power Systems

Edited by

Paulo F. Ribeiro
Calvin College, Michigan, USA

Celebrating 125 Years
of Engineering the Future

A John Wiley & Sons, Ltd. Publication

This edition first published 2009
© 2009, John Wiley & Sons, Ltd

Registered office
John Wiley & Sons Ltd, The Atrium, Southern Gate, Chichester, West Sussex, PO19 8SQ, United Kingdom

For details of our global editorial offices, for customer services and for information about how to apply for permission to reuse the copyright material in this book please see our website at www.wiley.com.

MATLAB® MATLAB and any associated trademarks used in this book are the registered trademarks of The MathWorks, Inc.

For MATLAB® product information, please contact:

The MathWorks, Inc.
3 Apple Hill Drive
Natick, MA, 01760-2098 USA
Tel: 508-647-7000
Fax: 508-647-7001
E-mail: info@mathworks.com
Web: www.mathworks.com

Library of Congress Cataloging-in-Publication Data

Time-varying waveform distortions in power systems / edited by Paulo Ribeiro.
    p. cm.
  Includes bibliographical references and index.
  ISBN 978-0-470-71402-7 (cloth)
  1. Electric power system stability. 2. Phase distortion (Electronics) 3. Electric power
distribution–Alternating current. 4. Harmonics (Electric waves) 5. Time-series analysis.
I. Ribeiro, Paulo.
  TK1010.T56 2009
  621.319–dc22

                                                                              2009004133

A catalogue record for this book is available from the British Library.

ISBN: 978-0-470-71402-7 (H/B)

Set in 10/12 pt. Times by Thomson Digital, Noida, India.
Printed in Great Britain by CPI Antony Rowe, Chippenham, Wiltshire.

"No model is a catalogue of ultimate realities, and none is a mere fantasy. Each is a serious attempt to get in all the phenomena known in a given period, and each succeeds in getting in a great many. But also, no less surely, each reflects the prevalent psychology of its age almost as much as it reflects the state of that age's [scientific] knowledge."

C. S. Lewis, The Discarded Image, Cambridge University Press, 1964.

"These things are so delicate and numerous that it takes a sense of great delicacy and mathematical precision to perceive them and judge them correctly and accurately: Often it is not possible to set it out analytically, because the necessary principles are not ready to hand, and it would be an endless task to undertake. The thing must also be seen all at once, at a glance, intuitively, and not only as a result of progressive reasoning, at least up to a point."

Blaise Pascal, 1650

Contents

Contributors

Y. Baghzouz, University of Nevada, USA

T. Baldwin, Florida State University, USA

A. Capasso, University of Rome - La Sapienza, Italy

P. Caramia, Università degli Studi di Napoli Parthenope, Italy

G. Carpinelli, Università degli Studi di Napoli Federico II, Italy

D. Cartes, Florida State University, USA

J. R. Carvalho, Federal University of Juiz de Fora, Brazil

A. S. Cerqueira, Federal University of Juiz de Fora, Brazil

F. De Rosa, Second University of Naples, Italy

J. Driesen, Catholic University of Leuven, Belgium

C. A. Duque, Federal University of Juiz de Fora, Brazil

A. E. Emanuel, Worcester Polytechnique, USA

E. Gursoy, Drexel University, USA

B. R. Klingenberg, Calvin College, USA

R. Lamedica, University of Rome - La Sapienza, Italy

R. Langella, Second University of Naples, Italy

Y. Liu, Florida State University, USA

A. C. de F. Marotti, Federal University of Rio de Janeiro, Brazil

C. A. G. Marques, Federal University of Juiz de Fora, Brazil

P. J. Masson, Florida State University, USA

R. E. Morrison, Stafforshire University, UK

D. Niebur, Drexel University, USA

T. H. Ortmeyer, Clarkson University, USA

A. Prudenzi, University of l'Aquila, Italy

L. Qi, Florida State University, USA

L. Qian, Florida State University

M. V. Ribeiro, Federal University of Juiz de Fora, Brazil

P. F. Ribeiro, Calvin College, USA

A. Russo, Politecnico di Torino, Italy

C. Sandoval, Jr, Federal University of Rio de Janeiro, Brazil

N. Seroy, Indian Institute of Technology Delhi, India

P. M. Silveira, Federal University of Itajubá, Brazil

A. Sollazzo, Second University of Naples, Italy

M. Steurer, Florida State University, USA

S. Suryanarayanan, Colorado School of Mines, USA

A. Testa, Second University of Naples, Italy

J. Vanden Keybus, Catholic University of Leuven, Belgium

P. Varilone, Università degli Studi di Cassino, Italy

P. Verde, Università degli Studi di Cassino, Italy

J. Wikston, Hatch, Canada

S. Woodruff, Florida State University, USA

W. Xu, University of Alberta, Canada

G. Zhang, University of Alberta, Canada

Preface

The ever-present time-varying nature of waveform distortions in power systems requires a comprehensive and precise analytical basis that needs to be incorporated into the system studies and analyses. This time-varying behavior, which is due to continuous changes in system configurations and variations of linear and nonlinear load and equipment, presents conceptual and practical challenges. Figure 1 illustrates the nature of the problem by indicating the possible methods of analyzing waveform distortions; connecting the time domain to the frequency domain as a function of its time-varying condition. For example, for steady-state waveforms Fourier analysis is sufficient, whereas when the time-varying conditions prevail, then spectral, probabilistic, evolutionary spectrum and time-frequency techniques are required.

This publication has been in preparation for several years as part of an activity within the IEEE Task Force on Probabilistic Aspects of Harmonics (Harmonics Working Group) which I have had the privilege to convene, and many people have contributed to. During this process our understanding of the problem and the tools available has evolved and we have agreed on a more encompassing perspective of the subject. First we moved away from the strict steady-

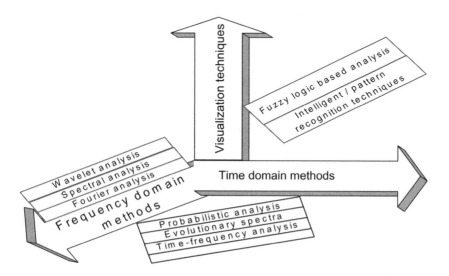

Figure 1   Time-varying distortions – connecting time and frequency domains

state "harmonic distortion" definition to a "waveform distortion" deviation where the time-varying nature (the main challenge of the problem) could be dealt with frequency/spectral, time-frequency, probabilistic, and computational intelligence methods. Second, several new techniques became available or were applied for the first time to power systems problems, and this prompted the editor to seek better understanding and additional contributions.

What seemed a settled issue, that is, that harmonics could only be dealt with as steady-state components and that the time-varying nature of waveforms could only be analyzed by probabilistic methods applied to short interval averages of individual harmonic components or decompositions by windowing techniques, was reconsidered. The new signal processing methods based on time-frequency decomposition such as wavelet transform and multi-rate filter methods presented in Section 6 have allowed us much more precise analyses of the behavior of time-varying waveform distortions and opened up new opportunities for monitoring and investigating power systems phenomena.

The text reviews the nature, analytical concepts, special situations and problems associated with the time-varying nature of waveform distortions and associated harmonics, and suggests solutions and ways to deal more effectively with the problem.

The text covers time-varying harmonics produced by different sources from single-phase appliances to Multi-Mega Watt power electronics converters. Also, analytical aspects related to background distortion, harmonic summation and harmonic impedance are discussed. The time-varying and time-frequency aspects are considered in the establishment of an integrated approach to deal with waveform distortions, which need to be carefully applied lest the new sophisticated techniques convolute rather than solve the difficulties.

Professor C.S. Lewis[1] once said:

"To use the microscope, yet not to focus or clean it, is foolishness. You are passing from uncorrected illusions to positively invited illusions. Here, as elsewhere, untrained eyes or a bad instrument produce both errors: they create phantasmal objects as well as miss real ones."

Paraphrasing this to signal processing applied to waveform distortions, one could say:

"To use advanced signal processing techniques, and yet not to tune them to the adequate phenomenon, scale, resolution, etc., is foolishness. You are passing from inconsequential information to affirmative mistakes. Here, as elsewhere, ignorant guessing or inattentive signal processing analysis produces both errors: they create illusional results as well as miss the real / desired information"

Thus, one needs to use these techniques with much engineering sensitivity and mathematical precision to avoid producing sophisticated but phantasmal results.

Figure 2 is an attempt to illustrate the big picture of how stationary, nonstationary and spatial nonstationary signals can be analyzed. The engineer or researcher needs to utilize them with both engineering intuition and analytical perceptiveness in order to make full use of the techniques' potential.

We expect that the information here will contribute to a better understanding of time-varying waveform distortions and will allow a better understanding of power systems behavior under time-varying conditions.

I would like to acknowledge the invaluable contributions and encouragement of all the authors and members of the IEEE TF on Probabilistic Aspects of Harmonics who provided the

---

[1]C.S. Lewis, Studies in Medieval and Renaissance Literature, Cambridge University Press, 1966.

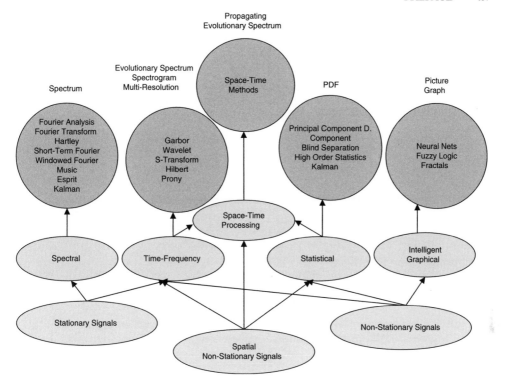

Figure 2   Overall perspective of the signal nature and corresponding analytical methods for analysis

motivation and insightful feedback, and in particular, Yahia Bagzouz, Alex Emanuel, Alfredo Testa, Roberto Langella, Tom Ortmeyer, Carlos Duque and Paulo Silveira, for helping to make this publication more intelligible and, hopefully, useful to the Power Sector. This text is intended to assist those who deal with harmonics and want to understand more clearly the time-varying nature and mechanisms of distortion generation, ways to analyze them, and design systems that are more cost/performance effective. I would also like to make a special mention to Dr. Robert Morrison who became one of the foremost influential researchers in this subject, and to Denis Howroyd, the developer of the first harmonic penetration program, who challenged me in the early eighties to seek more adequate tools for dealing with time-varying harmonics.

I would also like to thank Calvin College, Florida State University, New Mexico State University, USA, and the Federal University of Juiz de Fora, Brazil, for providing valuable time for preparing and editing this text.

Finally, I would like to thank my wife for her encouragement, support, and resignation for cancelling some of our kayak trips to work on this demanding but enjoyable effort.

Paulo F. Ribeiro
Grand Rapids, Michigan, USA

Website Information

Along with the publication of this book, a website has been created containing MATLAB® files for additional waveforms of typical non-linear loads which could be signal processed by different techniques for further understanding. Also two MATLAB based time-varying harmonic decomposition techniques will be available at the site for waveform processing. The website can be reached at:

www.laptel.ufjf.br

password: signals

Readers are welcome to send additional waveforms to the editor at pfribeiro@ieee.org to be included in the database.

Additional information and samples can be found in Appendix B: Sample of waveforms and decompositions.

Acknowledgments

We would very much like to acknowledge the following organizations for allowing us to include material taken from previously published papers.

Thanks to the IEEE for use of material within Chapter 1, taken from:

1) Baghzouz, Y.; Burch, R. F.; Capasso, A.; Cavallini, A.; Emanuel, A. E.; Halpin, M.; Imece, A.; Ludbrook, A.; Montanari, G.; Olejniczak, K. J.; Ribeiro, P. F.; Rios-Marcuello, S.; Tang, L.; Thaliam, R.; Verde, P., "Time-Varying Harmonics. I. Characterizing Measured Data," Power Delivery, IEEE Transactions on, Volume: 13 Issue: 3, July 1998, Page(s): 938–944.

2) Baghzouz, Y.; Burch, R. F.; Capasso, A.; Cavallini, A.; Emanuel, A. E.; Halpin, M.; Langella, R.; Montanari, G.; Olejniczak, K. J.; Ribeiro, P. F.; Rios-Marcuello, S.; Ruggiero, F.; Thallam, R.; Testa, A.; Verde, P., "Time-varying harmonics. II. Harmonic Summation and Propagation," Power Delivery, IEEE Transactions on, Volume: 17 Issue: 1, Jan. 2002, Page(s): 279–285.

Thanks to the IEEE for use of material within Chapter 2, taken from:

1) Ribeiro, P. F.; "A novel way for dealing with time-varying harmonic distortions: the concept of evolutionary spectra" Power Engineering Society General Meeting, 2003, IEEE, Volume: 2, 13–17 July 2003, Pages: 1153 Vol. 2.

Thanks to the IEEE for use of material within Chapter 4, taken from:

1) Emanuel, A. E., "The Randomness Power: An Other New Quantity to be Considered", IEEE Transactions on Power Delivery, July 2007, Vol. 22, No. 3, pp. 1304–08.

Thanks to the IEEE for use of material within Chapter 5, taken from:

1) Masson, P. J.; Silveira, P. M.; Duque, C.; Ribeiro, P. F.; "Fourier series: Visualizing Joseph Fourier's imaginative discovery via fea and time-frequency decomposition,"

Harmonics and Quality of Power, 2008. ICHQP 2008. 13th International Conference on Sept. 28 2008-Oct. 1 2008 Page(s): 1– 5.

Thanks to the IEEE for use of material within Chapter 6, taken from:

1) Marino, P. F. Ruggiero and A. Testa, "On the vectorial summation of independent random harmonic components", 7th ICHQP, Las Vegas (USA), 1996.

2) Cavallini, A., R. Langella, F. Ruggiero, A. Testa, "Gaussian Modeling of Harmonic Vectors in Power Systems", IEEE International Conference on Harmonics and Quality of Power, Athens, Greece, 14–16 October 1998, vol. 2 pp. 1010–1017.

3) Langella, R., A. Testa, "Interharmonics from a Probabilistic Perpective", IEEE PES Annual Meeting 2005, S. Francisco, USA, June 2005.

Thanks to the INESC Porto for use of material within Chapter 7, taken from:

1) Carbone, R., R. Langella, A. Testa: "Simplified Probabilistic Modeling of AC_DC_AC Power Converter Interharmonic Distortion", 6th PMAPS, Madeira (PG), Sept. 2000.

Thanks to the Associazione per gli Studi sulla Qualità dell'Energia Elettrica (AQEE) for use of material within Chapter 8, taken from:

1) Caramia, P., G. Carpinelli, T. Esposito, P. Varilone's 'Evaluation Methods and Accuracy in Probablistic Harmonic Power Flow' from the PMAPS Conference, Naples, September 2002.

Thanks to the IET for use of material within Chapter 8, taken from:

1) Carpinelli, G., A. Russo, M. Russo, P. Verde 'On the Inherent Structure Theory of Network in Presence of Harmonics', IEE Proc. Gen., Transm. and Distr. 1998, Vol. 145, No. 2, pp. 123–132.

2) Carpinelli, G., T. Esposito, P. Varilone, P. Verde 'First-order Probabilistic Harmonic Power Flow', IEE Proc. Gen., Transm. and Distr. 2001, Vol. 148, No. 6, pp. 541–548.

3) Caramia, P., G. Carpinelli, F. Rossi, P. Verde, 'Probabilistic Iterative Harmonic Analysis', IEE Proc. Gen., Transm. and Distr. 1994, Vol. 141, No. 4, pp. 329–338.

Thanks to the IEEE for use of material within Chapter 8, taken from:

1) Caramia, P., G. Carpinelli, A. Russo, P. Varilone, P. Verde, 'An Integrated Probabilistic Harmonic Index', IEEE PES Winter Meeting, New York, January 2002, pp. 1084–1089.

2) Carpinelli, G., T. Esposito, P. Varilone, P. Verde, 'Probabilistic Harmonic Power Flow for Percentile Evaluation', IEEE PES 2001 Canadian Conference on Electrical and Computer Engineering, Toronto, May 2001, pp. 831–838.

3) Esposito, T., A. Russo, P. Varilone, 'Probabilistic Modeling of Converters for Power Evaluation in Nonsinusoidal Conditions,' IEEE PES 2001Canadian Conference on Electrical and Computer Engineering, Toronto, May 2001, pp. 1059–1066.

4) Esposito, T., P. Varilone, 'Some Approaches to Approximate the Probability Density Function of Harmonics', IEEE proceedings of 10th ICHQP 2002, Rio de Janeiro, October 2002.

Thanks to the IEEE for use of material within Chapter 9, taken from:

1) Castaldo, D., R. Langella, A. Testa, "Probabilistic Aspects of Harmonic Impedances", invited paper at IEEE Winter Power Meeting 2002, New York (USA), 27–31 January 2002.

Thanks to the IEEE for use of material within Chapter 10, taken from:

1) Xu, W., Y. Mansour, C. Siggers, and M. B. Hughes. "Developing Utility Harmonic Regulations Based on IEEE STD 519- B. C. Hydro's Approach." IEEE Transactions on Power Delivery, Vol. 10, No. 3, July, 1995. pp. 1423–1431.

2) Xu, W., "Application of Steady-State Harmonic Distortion Limits to the Time-Varying Measured Harmonic Distortion." 2002 IEEE Power Engineering Society Summer Meeting, pp. 955–957.

Thanks to the IEEE for use of material within Chapter 11, taken from:

1) Caramia, P., G. Carpinelli, P. Verde, G. Mazzanti, A. Cavallini, G. C. Montanari, 'An Approach to Life Estimation of Electrical Plant Components in the Presence of Harmonic Distortions' 9th International Conference on Harmonics and Quality of Power, Orlando, October 2000, pp. 887–891.

2) Caramia, P., G. Carpinelli, A. Russo, P. Varilone, P. Verde, 'An Integrated Probabilistic Harmonic Index', IEEE PES Winter Meeting, New York, January 2002, pp. 1084–1089.

3) Caramia, P., G. Carpinelli, A. Russo, P. Verde, 'Some Considerations on Single Site and System Probabilistic Harmonic Indices for Distribution Networks,' IEEE PES General Meeting, Vol. 2, Toronto, July 2003, pp. 1160–1165.

Thanks to the IEEE for use of material within Chapter 13, taken from:

1) Wikston, J., "Harmonic Summation for Multiple Arc Furnaces", IEEE PES WM 2002 NY NY, Volume: 2, On page(s): 1072–1075 vol. 2.

Thanks to the International Power Systems Transients Conference for use of material within Chapter 14, taken from:

1) Carneiro Jr. S., and A. C. D. Marotti, "Treatment of Measured and Calculated Harmonic Currents in Filters of the Itaipu HVDC System", Proceedings of the International Power Systems Transients Conference- IPST'03, New Orleans, USA, Sept.28-Oct.3 2003.

Thanks to the CBQEE for use of material within Chapter 15, taken from:

1) Silveira, P. M., M. Steurer, P. F. Ribeiro, "Using Wavelet decomposition for Visualization and Understanding of Time-Varying Waveform Distortion in Power System," VII CBQEE, Aug. 2007, Brazil.

Thanks to the IEEE for use of material within Chapter 16, taken from:

1) Driesen J., R. Belmans, "Wavelet-based power quantification approaches," IEEE Transactions on Instrumentation and Measurement, vol. 52, no. 4, August, 2003; pp. 1232–1238.

Thanks to the IEEE for use of material within Chapter 17, taken from:

1) Klingenberg, B. R., and P. F. Ribeiro, "Fuzzy Logic for Harmonic Distortion Diagnosis in Power Systems," in Electro/information Technology, 2006 IEEE International Conference on, 2006, pp. 87–92.

Thanks to the IEEE for use of material within Chapter 18, taken from:

1) Liu, Y., M. Steurer, and P. F. Ribeiro, "A Novel Approach to Power Quality Assessment: Real Time Hardware-in-the-Loop Test Bed", IEEE Power Engineering Letter, IEEE Transactions on Power Delivery, Vol. 20, No. 2, April 2005, pp. 1200–1201.

Thanks to the IEEE for use of material within Chapter 19, taken from:

1) Gursoy, E., D. Niebur, "Blind Source Separation Techniques for Harmonic Current Source Estimation," Power Systems Conference and Exposition, 2006. PSCE '06. 2006 IEEE PES, Oct. 29 2006-Nov. 1 2006 Page(s): 252–255.

2) Gursoy, E., D. Niebur, "Harmonic Load Identification Using Complex Independent Component Analysis," Power Delivery, IEEE Transactions on Volume 24, Issue 1, Jan. 2009 Page(s): 285–292.

Thanks to the IEEE for use of material within Chapter 20, taken from:

1) Senroy, N., S. Suryanarayanan, P. F. Ribeiro, "An improved Hilbert-Huang method for analysis of timevarying waveforms in power quality," IEEE Transactions on Power Systems, vol. 22, no. 4, pp. 1843–1850, Nov. 2007.

2) Senroy, N., S. Suryanarayanan, "Two techniques to enhance empirical mode decomposition for power quality applications," in Proc. 2007 IEEE Power Engineering Society General Meeting, Tampa, FL, Jun 2007.

Thanks to the IEEE for use of material within Chapter 21, taken from:

1) Carbone, R., F. De Rosa, R. Langella, A. Sollazzo, and A. Testa: "Modelling of AC/DC/AC Conversion Systems with PWM Inverter", proc. of IEEE Summer Power Meeting 2002, Chicago, USA, July 2002.

2) Langella, R., A. Testa, "Interharmonics from a Probabilistic Perpective", invited paper at IEEE PES Annual Meeting 2005, S. Francisco (USA), June 2005.

3) Langella, R., A. Testa, "Interharmonics from a Probabilistic Perpective", IEEE PES Annual Meeting 2005, S. Francisco, USA, June 2005.

4) De Rosa, F., R. Langella, A. Sollazzo, and A. Testa: "On the Interharmonic Components Generated by Adjustable Speed Drives", IEEE Transaction on Power Delivery, Vol. 20, N. 4, Ottobre 2005, pp. 2535–2543.

Thanks to the IEEE for use of material within Chapter 22, taken from:

1) Duque, C., P. Silveira, T. Baldwin, Paulo Ribeiro: "Novel Method for Tracking Time-Varying Power Harmonic Distortions without Frequency Spillover", IEEE PES General Meeting 2008 Pittsburg, PA, USA.

Thanks to the IEEE for use of material within Chapter 23, taken from:

1) Siveira, P. M., C. A. Duque, T. Baldwin, P. F. Ribeiro, "Sliding Window Recursive DFT with Dyadic Downsampling - A New Strategy for Time-Varying Power Harmonic Decomposition," IEEE PES General Meeting 2009, Calgary, Canada.

Thanks to the IEEE for use of material within Chapter 24, taken from:

1) J. Carvalho, C. Duque, T. Baldwin, P. Ribeiro: "A DFT-Based Approach for Efficient Harmonic/Inter-Harmonic Analysis under Time-Varying Conditions", IEEE PES General Meeting 2008, Pittsburg, PA, USA.

Thanks to the IEEE for use of material within Chapter 25, taken from:

1) L. Qi, L. Qian, D. Cartes, S. Woodruff, "Prony analysis for power system transient harmonics," EURASIP (The European Association for Signal Processing) Journal on Applied Signal Processing, Volume 2007, Issue 1, January 2007, Pages: 170–170, ISSN:1110-8657.

Thanks to the IEEE for use of material within Appendix A, taken from:

1) Prudenzi A., "A novel procedure based on lab tests for predicting single-phase power electronics-based loads harmonic impact on distribution networks", Power Delivery, IEEE transactions on, Volume 19, Issue 2, April 2004 Page(s): 702–707.

2) Capasso A., Lamedica R., Prudenzi A., "Cellular phone battery chargers impact on voltage quality", Power Engineering Society Summer Meeting, 2000. IEEE Volume 3, 16–20 July 2000 Page(s) 1433–1438, vol. 3.

3) Lamedica R., Sorbillo C., Prudenzi A., "The continuous harmonic monitoring of single-phase electronic appliances: desktop pc and printers", Harmonics and Quality of Power, 2000. Proceedings. Ninth International Conference on, Volume 2, 1–4 Oct. 2000 Page(s): 697–702 vol. 2.

4) Capasso A., Lamedica R., Prudenzi A., "Estimation of net harmonic currents due to dispersed non-linear loads within residential areas", Harmonics and Quality of Power, 1998. Proceedings. 8th International Conference on, Volume 2, 14–16 Oct. 1998 Page(s): 700–705 vol. 2.

5) Prudenzi A., Grasselli U., Lamedica R., "IEC Std. 61000-3-2 harmonic current emission limits in practical systems: need of considering load and attenuation effects", Power Engineering Society Summer Meeting, 2001. IEEE Volume 1, 15–19 July 2001 Page(s) 277–282 vol. 1.

6) Grasselli U., Lamedica R., Prudenzi A., "Time-Varying Harmonics of Single-Phase Non-Linear Appliances", Power Engineering Society Winter Meeting, 2002. IEEE Volume 2, 27–31 Jan. 2002 Page(s) 941–946 vol. 2.

Part I
GENERAL CONCEPTS AND DEFINITIONS

This part covers general and introductory concepts. Chapter 1 presents an overview of probabilistic aspects of harmonics where initial models were restricted to the analysis of instantaneous values of voltages and currents. Direct analytical methods were originally applied with many simplifications. Attempts to use phasor representation of current also used direct mathematical analysis and simple distributions of amplitude and phase angle. Direct simulation was applied to test the assumptions used. When power systems measurement became more powerful the real distributions were measured for some loads and this enabled a significant increase in the accuracy of observations. The limitations to the application of harmonic analysis in general, and the issues that determine whether full spectral analysis should be used, are discussed. The chapter also reviews existing methods associated with harmonic measurement of nonstationary voltages and currents waveform, characterization of recorded data, harmonic summation and cancelation in systems with multiple nonlinear loads and probabilistic harmonic power flow.

Chapter 2 attempts to integrate spectral analysis and probability distribution concepts, for a better understanding of the nature of time-varying harmonics and possibly as a more precise way to treat time-varying harmonics and validate harmonic summation studies. Similarities between spectral analysis and probability distribution functions are considered and discussed.

Chapter 3 explores the basic definitions of harmonics (Fourier/spectral analysis) of periodic and nonperiodic functions, that is, discrete and continuous range of frequencies. Definitions of typical harmonics and transients phenomena are proposed.

Chapter 4 deals with the correct definitions that characterize the flow of electric power/ energy under probabilistic conditions.

Finally, Chapter 5 presents Joseph Fourier's heat transfer experiment through the use of finite element analysis. An iron ring is modeled and transient thermal analysis is performed to reproduce the data Fourier obtained experimentally. Simulated data give a clear view of how Fourier first thought of representing temperature distribution in a ring as a combination of sinusoidal functions and how this experiment gave information about how harmonics content is modified in time. The use of new signal processing methods, based on time–frequency decomposition, further illustrates Joseph Fourier's physical intuition to visualize the time varying components long before the mathematical foundation was developed.

1

Probabilistic aspects of time-varying harmonics

R. E. Morrison, Y. Baghzouz, P. F. Ribeiro and C. A. Duque

1.1 Introduction

This chapter presents an overview of the motivation, importance and previous development of the text. This chapter considers the early development of probabilistic methods to model power system harmonic distortion. Initial models were restricted to the analysis of instantaneous values of current. Direct analytical methods were originally applied with many simplifications. Initial attempts to use phasor representation of current also used direct mathematical analysis and simple distributions of amplitude and phase angle. Direct simulation was applied to test assumptions used. When power systems measurement systems became sufficiently powerful the real distributions were measured for some loads and this enabled a significant increase of accuracy. However, there is still a lack of knowledge of the distributions that might be used to model converter harmonic currents. The chapter concludes by considering the limitations to the application of harmonic analysis.

The application of probabilistic methods for analyzing power system harmonic distortion commenced in the late 1960s. Initially, direct mathematical analysis was applied based on instantaneous values of current from individual harmonic components [1]. Methods were devised to calculate the probability density function (pdf) of one total harmonic current generated by a number of loads, given the pdfs for the individual load currents. One of the first attempts to use phasor notation was applied by Rowe [2] in 1974. He considered the addition

Time-Varying Waveform Distortions in Power Systems   Edited by Paulo F. Ribeiro
© 2009 John Wiley & Sons, Ltd

of a series of currents modeled as phasors with random amplitude and random phase angle. Further, the assumption was made that the amplitude of each harmonic current was variable with uniform probability density from zero to a peak value and the phase angle of each current was variable from 0 to 2π. Rowe's analysis was limited to the derivation of the properties of the summation current from a group of distorted loads connected at one node.

Properties of the summation current were obtained by simplifying the analysis by means of the Rayleigh distribution. Unfortunately once such simplifications are applied, flexibility on modeling is not retained and the ability to model a bus bar containing a small number of loads is lost. However, Rowe was able to show that the highest expected value of current due to a group of loads could be predicted from the Equation (1.1):

$$I_{\mathrm{s}} = K(I1^2 + I2^2 + I3^2 + \cdots)^{1/2} \tag{1.1}$$

where I_{s} is the summation current from individual load currents ($I1, I2, I3$, etc.) and K is close to 1.5.

Equation (1.1) indicates that the highest expected value of summation current was not related to the arithmetic sum of all the individual harmonic current amplitudes. This factor was a major step forward. Also, by introducing the concept of the highest expected current it was noted that this would be less than the highest possible value of current, namely, the arithmetic sum. It was necessary to define the highest expected current as the lowest value which would be exceeded for a negligible part of the time. Negligible was taken to be 1%. To calculate this value, the 99th percentile was frequently referred to.

The early analysis depended on assumed probability distributions as well as variable ranges. Subsequently, some of the actual probability density functions were measured [3] and found to differ from the assumed pdfs stated above. Simulations were arranged to derive the cumulative distribution function (cdf) of the summation current for low-order harmonic current components. The estimated cdfs were then compared with the measured cdfs with reasonable agreement as shown in Figure 1.1.

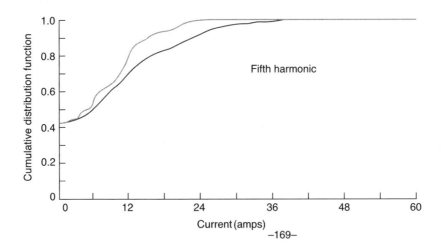

$-169-$

Figure 1.1   Measured (lower curve) and simulated (upper curve) cdf of fifth harmonic current at a 25kV AC traction substation

However, some shortcomings were noted in relation to the early methods of analysis:

- There was a lack of knowledge concerning the actual distributions for all but a limited number of loads.

- The interrelationship between the different harmonic currents for a single load was recognized as complex. However, there remained a lack of knowledge of the degree of independence between the different harmonic currents.

These problems remain and may be solved only following an extensive testing program.

For a period of time, there was little activity in the probabilistic modeling of harmonic currents since the interpretation of statistical parameters was difficult without extensive measurements. However, it was recognized by the engineers who were involved in the development of harmonic standards that some concepts from probability theory would have to be applied [5]. The concept of a compatibility level was introduced which corresponds to the 95th percentile of a parameter. To apply the standards, it was necessary to measure the 95th percentile of particular harmonic voltages to determine whether a location in a power system contained excessive distortion. Once this concept was introduced from a measurement point of view, it was important to be able to calculate the 95th percentile in order to effect the comparison between the calculated values and the maximum acceptable levels to determine whether a load would be acceptable at the planning stage of a project. This concept required knowledge of the true variation of the harmonic distortion, random or otherwise.

It was noted that the actual variation of power system distortion may not be totally random; that is, there is likely to be a degree of deterministic behavior [6]. Measurements made over 24 hours [4] clearly show that a good deal of variation is due to the normal daily load fluctuation. This factor complicates the analysis since it provokes a degree of deterministic behavior resulting in a nonstationary process. When considering a nonstationary process, it is known that the measured statistics are influenced by the starting time and the time window considered in the measurement. National power quality standards normally cover tests made over 24 hours, thus spanning a period within which it is not possible to assume that the variation of harmonic distortion is stationary. The so-called compatibility level is a 95th percentile which is intended to apply to the complete 24 hour period. Therefore, the methods used to evaluate harmonic levels at the planning stage must account for the nonstationary nature. To model the nonstationary effects, the variation of the mean harmonic current with time must be taken into account.

In order to model the complete nonstationary nature of power system distortion, it is necessary to gain additional knowledge from realistic systems. Measurements are needed to determine 24 hour trend values to enable suitable models to be found. The present harmonic audits may reveal such information [7].

1.2 Spectral analysis or harmonic analysis

Strictly speaking, harmonic analysis may be applied only when currents and voltages are perfectly steady. This is because the Fourier transform of a perfectly steady distorted waveform is a series of impulses suggesting that the signal energy is concentrated at a set of discrete frequencies. Thus, the transfer relationship between current and voltage

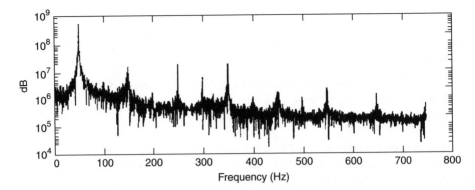

Figure 1.2   Spectrum of a time-varying current waveform

(impedance) is a single value at each component of frequency (harmonic), although different impedances at different frequencies have different values.

When there is variation of the distorted waveform, the Fourier transform of the waveform is no longer concentrated at discrete frequencies and the energy associated with each harmonic component occupies a particular region within the frequency band. This is illustrated in Figure 1.2 which shows the spectrum of an actual time-varying current waveform.

If the waveform variation is slow, the frequency range containing 80% of the energy associated with the signal variation may be restricted. Figure 1.3 shows a typical fifth harmonic voltage variation on a high voltage (230 kV) transmission bus during a world cup soccer event in Brazil [23].

Tests have also been carried out to determine the 'spread' of energy in the frequency domain for a limited set of loads in the past. An analysis carried out for an AC traction system demonstrated that the 80% energy bandwidth was less than 0.6 Hz for components up to the 19th harmonic [4].

For harmonic analysis to apply, there must be negligible variation of the system impedance within the frequency range covered by the 80% energy bandwidth. It was demonstrated [4] that

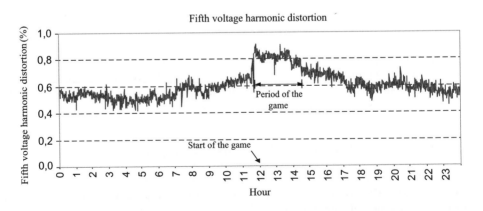

Figure 1.3   Fourier transform derived from fifth harmonic voltage variation

variation of system impedance over a range of 0.6 Hz is less than 2% even in unfavorable system circumstances. Thus, harmonic analysis is justified for applications using some types of locomotive load.

Clearly, further measurement should be carried out to demonstrate that harmonic analysis is also applicable to other types of loads when changes in current waveform properties may be more rapid than found in locomotive loads.

1.3   Observations

Probabilistic techniques may be applied to the analysis of harmonic currents from several sources. However, to generalize the analysis, there is a need to measure the pdfs describing harmonic current variation for a variety of loads. There is also a need to understand the nonstationary nature of the current variation in order to predict the compatibility levels. It is probable that the harmonic audits (currently in progress) could yield the appropriate information. To determine the compatibility level by calculation, it will be necessary to determine the nonstationary trends within the natural variation of power system harmonic distortion.

There are circumstances where harmonic analysis does not apply because the rate of change of current variation is too fast. It is possible to determine the limit to which harmonic distortion should be applied by considering information transferred into the frequency domain. A formal approach to understanding this problem might present new insight into the limit to which harmonic analysis is appropriate.

1.4   Typical harmonic variation signals

To show typical variation of harmonic signals, recorded data at two different industrial sites, denoted by Sites A and B, are presented. Site A represents a customer's 13.8 kV bus having a rolling mill that is equipped with solid-state 12-pulse DC drives and tuned harmonic filters. Figure 1.4 shows the variation of the current and voltage total harmonic distortion (THD) of one phase over a 6 hour period. The time interval between readings is 1 min, and each data point represents the average FFT for a window size of 16 cycles. It is known that the rolling mill was in operation only during the first 2.5 hours of the total recording time interval.

Note the reduction in current and voltage harmonic levels at Site A after 2.5 hours of recording. After this time, the rolling mill was shut down for maintenance and only secondary loads are left operating. The resulting low distortion in current and voltage are caused by background harmonics.

Site B is another customer's bus loaded with a 66 MW DC arc furnace that is also equipped with passive harmonic filters. Figure 1.5 shows the changes in current and voltage THDs during a period of 1 hour, but with a time interval of 1 s between readings and a window size of 60 cycles. The sampling rate of the voltage and current signals is 128 times per cycle at both sites. Finally, Figure 1.6 shows a polar plot of the current variation for a typical six-pulse converter with a firing angle variation.

While previous examples show long-term harmonic variation some kinds of load and equipment present a short-term harmonic variation. The illustrative case portrayed in Figure 1.7 is the well-known inrush current during transformer energization. The voltage and current variation presents a short-term nonstationary behavior and special digital signal processing tools should be used to analyze the frequency behavior. The waveforms

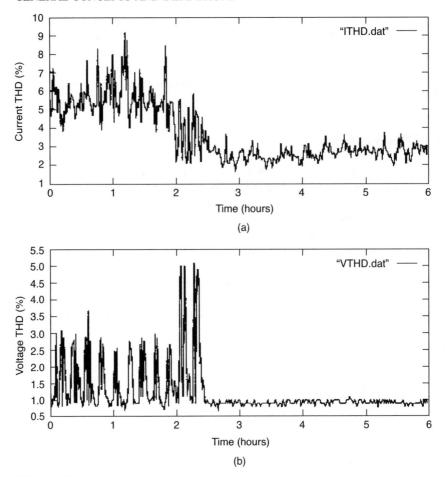

Figure 1.4   Variation of (a) current and (b) voltage THD at Site A

of the odd harmonic components are shown in Figure 1.8. These waveforms were obtained using the digital signal tools described in Chapter 22.

Figure 1.5 indicates that the voltage THD is quite low although it is known that the arc furnace load was operating during the 1 hour time span. This is due to the fact that the system supplying such a load is quite stiff. Note that the voltage and current THD drop simultaneously during two periods (4–10 min and 27–30 min) when the furnace was being charged. The two bursts of current THD occurring at 10 min and 30 min represent furnace transformer energization after charging.

It is of interest to analyze the effect of current distortion produced by a large nonlinear load on the distortion of the voltage supplying this load. One graphical way to check for correlation between these two variables is to plot one as a function of the other, or to display a scatter plot.

Figure 1.9 shows such a plot for Site B. In this particular case, it is clear that there is no simple relationship between the two THDs. In fact, the correlation coefficient which measures the strength of a linear relationship is found to be only 0.32.

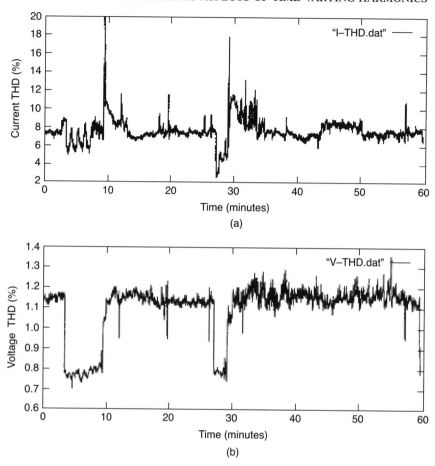

Figure 1.5   Variation of (a) current and (b) voltage THD at Site B

1.5   Harmonic measurement of time-varying signals

Harmonics are a steady-state concept where the waveform to be analyzed is assumed to repeat itself forever. The most common techniques used in harmonic calculations are based on the fast Fourier transform (FFT) – a computationally efficient implementation of the discrete Fourier transform (DFT). This algorithm gives accurate results under the following conditions: (i) the signal is stationary, (ii) the sampling frequency is greater than two times the highest frequency within the signal, (iii) the number of periods sampled is an integer and (iv) the waveform does not contain frequencies that are noninteger multiples (i.e. interharmonics) of the fundamental frequency.

   If the above conditions are satisfied, the FFT algorithm provides accurate results. In such a case, only a single measurement or 'snapshot' is needed. On the other hand, if interharmonics are present in the signal, multiple periods need to be sampled in order to obtain accurate harmonic magnitudes.

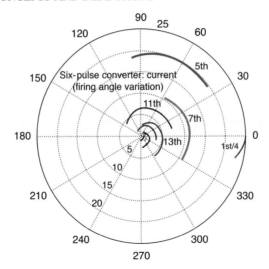

Figure 1.6   Polar plot of harmonic currents of six-pulse converter with varying load (firing angle variation)

In practical situations, however, the voltage and current distortion levels as well as their fundamental components are continually changing in time. Time-variation of each individual current and voltage harmonic is analyzed by a windowed Fourier transformation (short-time Fourier transform) and each harmonic spectrum corresponds to each window section of the continuous signal. However, because deviations exist within the smallest selected window

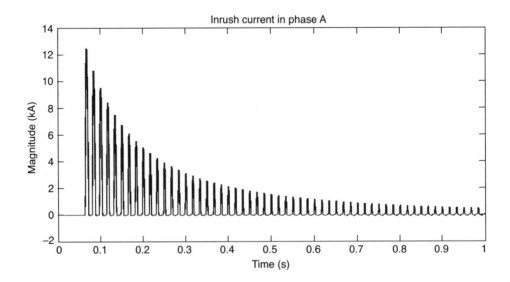

Figure 1.7   Inrush current during transformer energization, a typical short-term time-varying harmonic

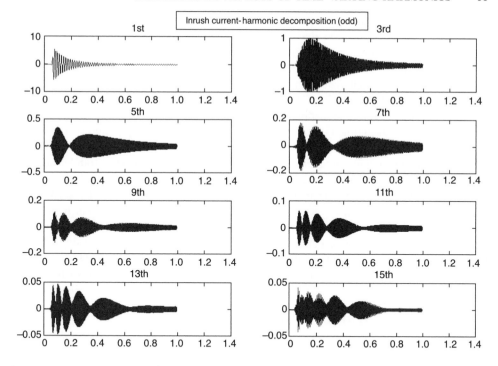

Figure 1.8 Harmonic decomposition for the inrush current during transformer energization, a typical short-term time-varying harmonic

length, different sized windows (i.e. the number of fundamental cycles included in the FFT) give different harmonic spectra. Furthermore, adequate window function length is a complex issue that is still being debated [13]. Besides hardware-induced errors, for example analog-to-digital converters and nonlinearity of potential and current transformers [14], several

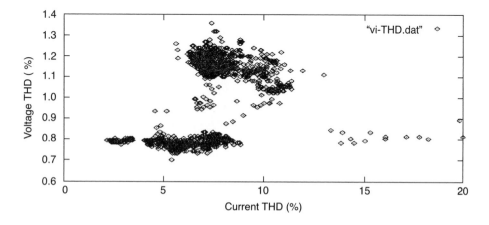

Figure 1.9 Scatter plot of voltage THD as a function of current THD at Site B

software-induced errors also occur when calculating harmonic levels by direct application of windowed FFTs. These include aliasing, leakage and the picket-fence effect [15].

Aliasing is a consequence of under-sampling, and the problem can be alleviated by use of anti-aliasing filters or by increasing the sampling frequency to a value greater than twice the highest frequency of components in the signal of interest. However, most modern measurements are relatively fast and a fourfold increase of frequency is considered more appropriate. Leakage refers to apparent spreading of energy from one frequency into adjacent ones if the number of periods sampled is not an integer multiple of the signals of interest. The picket-fence effect occurs if the analyzed waveform includes a frequency which is not an integer multiple of the fundamental frequency (reciprocal of the window length). Both leakage and picket-fence effects can be mitigated by spectral windows.

Several approaches have been proposed in recent years to improve the accuracy of harmonic magnitudes in time-varying conditions. These include the Kalman filter based analyzer [15, 18], the self-synchronizing Kalman filter approach [17], a scheme based on Parseval's relation and energy concept [11] and a Fourier linear combiner using adaptive neural networks [19]. Each one of these methods has advantages and disadvantages, and the search for better methods continues to be an active research area in signal processing.

1.6   Characterization of measured data

When considering charts of harmonic component variations with time, one often finds that the variables contain a large number of irregularities which fail to conform to coherent patterns. The physical processes which produce these irregularities involve a large number of factors whose individual effects on harmonic levels cannot be predicted. Due to these elements of uncertainty, the variations generally have a random character and the only way one can describe the behavior of such characteristics is in statistical terms which transform large volumes of data into compressed and interpretable forms [20].

At times, however, some general patterns can be noticed when examining some of the charts, thus indicating if a deterministic component exists in the recorded signal. In such cases, a more accurate description is to express the signal as the sum of a deterministic component and a random component. These descriptions are addressed below with illustrations using the recorded data (THD) shown in the previous section. These techniques can also be applied to individual harmonics, but such data were not recorded at the sites.

1.6.1   Statistical measures

Numerical descriptive measures are the simplest forms of representing a set of measurements. These measures include minimum value, maximum value, average or mean value and standard deviation which measures the spread and enables one to construct an approximate image of the relative distribution of the data set.

Mathematically, consider a set of n measurements X_i, $i = 1, \ldots, n$, with minimum value $X_{\min}$ and maximum value $X_{\max}$. The average value X_{avg} and standard deviation σ_x are calculated by

$$X_{\text{avg}} = \frac{\sum_{i=1}^{n} X_i}{n}, \tag{1.2}$$

Table 1.1   Statistical measures of signals displayed in Figures 1.5 and 1.6

Site	X	X_{min}	X_{max}	X_{avg}	σ_{min}
A	VTHD (%)	0.70	5.01	1.23	0.78
B	VTHD (%)	0.70	1.36	1.09	0.13
A	ITHD (%)	1.70	9.20	3.69	1.53
B	ITHD (%)	2.15	19.95	7.37	1.20

and

$$\sigma_X = \sqrt{\frac{\sum_{i=1}^{n} (X_i - X_{avg})^2}{n-1}}. \tag{1.3}$$

The statistical measures for the recorded data at Sites A and B are listed in Table 1.1 below. Given X_{avg} and σ_x one might think that the data is spread according to the Gaussian distribution,

$$p(x) = \frac{1}{\sqrt{2\pi}\sigma} \exp\left(\frac{-(X - X_{avg})}{2\sigma^2}\right). \tag{1.4}$$

The accuracy of this proposition depends on the level of randomness of the signal and whether it contains a deterministic component. If the signal is completely random, then the assumption of a Gaussian distribution is accurate. On the other hand, if the statistical measures are derived directly from a signal with a significant deterministic component such as Figure 1.5(a), then the actual probability distribution is expected to deviate significantly from a Gaussian distribution as will be seen in the next section. Furthermore, the time factor is completely lost with such statistical measures. For example, one cannot tell when the maximum distortion took place within the recording time interval an important element in trouble shooting.

1.6.2   Histograms or probability density functions

Because it is often difficult to determine a priori the best distribution to describe a set of measurements, a more accurate method is a graph that provides the relative frequency of occurrence. This type of graph, known as a histogram, shows the portions of the total set of measurements that fall in various intervals. When scaled such that the total area covered in each histogram is equal to one, the histogram becomes the density function (pdf) of the signal.

Figures 1.10 and 1.11 are the probability density functions corresponding to the recorded data in Figures 1.4 and 1.5, respectively. Note that pdfs of the current at Site A and voltage at Site B contain multiple peaks and cannot be described in terms of the common distribution functions. These irregularities are due to the presence of a deterministic component in both signals [21].

While histograms represent the measured data in a compressed form and provide more information than the statistical measures above, they hide some information, for example when an event takes place in time. Although they provide the total time duration during which

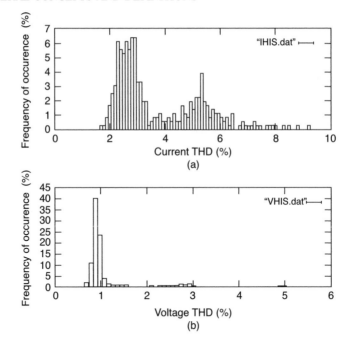

Figure 1.10   Probability density function of (a) current and (b) voltage at Site A

a harmonic component or distortion level is exceeded, one cannot determine whether such a level occurred in a continuous or in a pulsed fashion. Such knowledge is crucial when studying the thermal effects of harmonics on equipment.

1.6.3   Probability distribution functions

A probability distribution function $P_X(x)$ represents the same information as a pdf, $p_x(x)$, in a different form: $P_X(x)$ gives the summation of all the time intervals in which the variable exceeds a certain level. Mathematically, a distribution function represents the integral of a density function. One can also use it to express the probability of the event that the observed variable X is greater than a certain value x. Since this event is simply the complement of the event having probability $P_X(z)$, it follows that

$$Pr(X > x) = 1 - P_X(x) = 1 - \int_{-\infty}^{x} p_X(\tau)d\tau. \qquad (1.5)$$

Such probability curves are shown in Figure 1.12 for the recorded signals in Figures 1.4 and 1.5. The same advantages and disadvantages as previously mentioned for histograms apply to the statistical description of data by means of probability distribution functions.

1.6.4   Statistical description at sub-time intervals

One way to simplify complex probability density functions or histograms with multiple peaks and provide more accurate descriptive values, such as those in Figures 1.10(a) and 1.11(b), is to

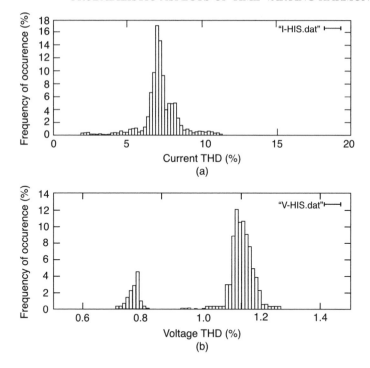

Figure 1.11   Probability density function of (a) current and (b) voltage at Site B

examine the recorded strip-charts and search for distinct variations at specific sub-intervals of recording time. Then, one can calculate the statistical variables for each of these intervals. To illustrate the point, it can be clearly seen that the variation in current THD at site A changes dramatically from one mode to another at $t = 2.5$ hours. Hence, two separate histograms should be derived: one for the time period between 0 and 2.5 hours and the other for the rest of the time, that is 2.5 to 6 hours.

One can visualize that this procedure decomposes the histogram in Figure 1.11(b) into two distinct ones, each having a single peak, thus making it easier to describe analytically. The same procedure can be applied to the voltage THD at Site B. Once again, two distinct modes are noticed: one corresponds to VTHD less than 0.9% (between 4–10 min and 27–30 min), and the other corresponds to a THD greater than 0.9% for the rest of the time. It should be clear that the two resulting histograms are equivalent to the two distinct ones shown in Figure 1.11(b). This method preserves the time factor to some extent while providing simple shapes of probability density functions, but an extra effort in examining the strip-charts is necessary.

1.6.5   Combined deterministic/statistical description

The recorded signals shown in Figures 1.4 and 1.5 are obviously nonstationary, their probability distribution functions change with the time, due to changes in load conditions (shutting down the rolling mill at Site A, and charging the arc furnace at Site B). Multiple peaks on histograms may also show signs of nonstationary signals, or the presence of a deterministic component within the signal. Theoretical analysis of nonstationary signals is complex, some

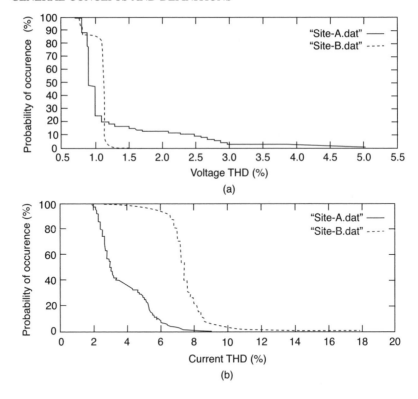

Figure 1.12   Probability distribution functions of (a) voltages and (b) currents

of which can be found in reference [22]. This section covers an alternative method to improve the accuracy of the above descriptions by treating the recorded signal as the sum of a deterministic component X_D and a random component X_R.

The values of X_D can be extracted by fitting a polynomial function of a certain degree to the recorded measurements using the method of least squared error. X_R is then defined as the difference between the actual signal and the deterministic component. It can be shown that the distribution of X_R approaches a normal distribution as the degree of the polynomial function representing X_D increases. In practice, however, it is desirable to work with the simplest model possible: either a linear or a quadratic function.

1.7   Conclusions

This chapter reviewed basic concepts associated with harmonic component variation and measurement of time-varying waveforms, and various ways of describing recorded data graphically and statistically. Statistical measures and histograms are the most commonly used methods. The shapes of probability densities are often found to have multiple peaks and cannot be represented by common probability functions. It is further known that these descriptions completely eliminate the time of occurrence of a certain event.

References

[1] W. G. Sherman, 'Summation of harmonics with random phase angles', *Proc. IEE*, 1972, **119**, 1643–1648.

[2] N. B. Rowe, 'The summation of randomly varying phasors or vectors with particular reference to harmonic levels', IEE Conference Publication 110, 1974, pp. 177–181.

[3] R. E. Morrison and A. D. Clark, 'A probabilistic representation of harmonic currents in AC traction systems' *IEE Proc. B*, 1984, **131**, 181–189.

[4] R. E. Morrison, 'Measurement analysis and mathematical modelling of harmonic currents in AC traction systems', Ph.D. thesis, Staffordshire University, November, 1981.

[5] R. E. Morrison, R. Carbone, A. Testa, G. Carpinelli, P. Verde, M. Fracchia and L. Pierrat, 'A review of probabilistic methods for the analysis of low frequency harmonic distortion', IEE Conference on EMC, Publication Number 396, September 1994.

[6] R. E. Morrison and E. Duggan, 'Prediction of harmonic voltage distortion when a non-linear load is connected to an already distorted network', *IEE Proc. C*, **140**, 1993, 161–166.

[7] V. Gosbell, D. Mannix, D. Robinson and S. Perera, 'Harmonic survey of an MV distribution system', Proceedings of AUPEC, Curtin University, Perth, Australia, September 2001.

[8] T. Shuter, H. Vollkommer and T. Kirkpatric, 'Survey of Harmonic Levels on the American Electric Power Distribution System', *IEEE Trans. Power Delivery*, **4**, 1989, 2204–12.

[9] A. Emanuel, J. Orr, D. Cyganski and E. Gulachenski, 'Survey of Harmonic Voltages and Currents at Distribution Substations', *IEEE Trans. Power Delivery*, **6**, 1991, 1883–90.

[10] J. Arrillaga, A. J. V. Miller, J. Blanco and L. I. Eguiluz, 'Real Time Harmonic Processing of an Arc Furnace Installation', Proceedings of the Sixth IEEE/lCHQP, Bologna, Italy, September 1994, pp. 408–414.

[11] IEEE Standard 519, *IEEE Recommended Practices and Requirements for Harmonic Control in Electric Power Systems*, IEEE Press, 1991.

[12] W. Xu, Y. Mansour, C. Siggers and M. El. Hughes, 'Developing Utility Harmonic Regulations Based on IEEE STD 519 - B.C. Hydro's Approach', *IEEE Trans. on Power Delivery*, **10**, 1995, 137–143.

[13] IEC Sub-Committee 77A Report, 'Disturbances Caused by Equipment Connected to the Public Low-Voltage Supply System: Part 2 – Harmonics', 1990 (revised Draft of IEC 555-2).

[14] G. T. Heydt, *Power Quality*, Star in a Circle Publications, 1991.

[15] A. A. Girgis, W. B. Chang and E. B. Makram, 'A Digital Recursive Measurement Scheme for On-Line Tracking of Power System Harmonics', *IEEE Trans. Power Delivery*, **6**, 1991, 1153–1160.

[16] H. Ma and A. A. Girgis, 'Identification and Tracking of Harmonic Sources in a Power System Using a Kalman Filter', EE/PES Winter Meeting, Baltimore, Maryland, USA, 1996, paper No. 96 WM 086-9-PWRD.

[17] I. Kamwa, R. Grondin and D. McNabb, 'On-Line Tracking of Changing Harmonics in Stressed Power Transmission Systems - Part 11: Application to Hydro-Quebec Network', 1996 IEEE/PES Winter Meeting, Baltimore, Maryland, USA, 1996, paper No. 96 WM 126-3-PWRD.

[18] C. S. Moo, Y. N. Chang and P. P. Mok, 'A digital Measurement Scheme for Time-Varying Transient Harmonics', IEEE/PES Summer Meeting, 1994, paper No. 94 SM 490-3 PWRD.

[19] P. K. Dash, S. K. Patnaik, A. C. Liew and S. Rahman, 'An Adaptive Linear Combiner for On-line Tracking of Power System Harmonics', IEEE/PES Winter Meeting, Baltimore, Maryland, USA, 1996, paper No. 96 WM 181-8-PWRS.

[20] G. T. Heydt and E. Gunter, 'Post-Measurement Processing of Electric Power Quality Data', IEEE/PES Winter Meeting, Baltimore, Maryland, USA, 1996, paper No. 96 WM 063-8-PWRD.

[21] G. R. Cooper and C. D. McGillem, *Probabilistic Methods of Signal and System Analysis*, Holt, Rinehart and Winston, Inc., Place, 1971.

[22] A. Cavallini, G. C. Montanari and M. Cacciari, 'Stochastic Evaluation of Harmonics at Network Buses', *IEEE Trans. on Power Delivery*, **10**, 1995, 1606–1613.

[23] J. J. A. Leitao, L. C. A. Fonseca, M. M. S. Lira, L. R. Soaresand P. F. Ribeiro, 'Harmonic Distortion on a Transmission System During Games of the Brazilian National Team in the 2006 World Cup', IEEE Power Engineering Society General Meeting, June 2007.

2

Probability distribution and spectral analysis of nonstationary random processes

P. F. Ribeiro and C. A. Duque

2.1 Introduction

The concept of harmonic decomposition can only be applied to steady-state, periodic waveform functions. In real life, however, voltages and currents are constantly changing with time. The pragmatic way of dealing with time-varying functions is to extend the time window and capture the variations along the time. However, this method can have some significant limitations and loss of physical meaning depending on the time-varying nature of the waveform being analyzed. This chapter suggests the application of spectral analysis and probability distribution concepts, in an integrated way, for a better understanding of the nature of time-varying harmonics and possibly as a more precise way to treat time-varying harmonics and validate harmonic summation studies.

Harmonic decomposition is a very convenient, but artificial way to manipulate nonsinusoidal waveforms. Therefore, they only 'exist', so to speak, and have physical meaning when associated with a periodic waveform or an instant in time, and they are characterized by a certain magnitude and phase angle. Thus, for example, changes in the waveform can be attributed to the phase angles only. As a consequence, analyzing only the magnitudes of the harmonic components for time-varying distortion may suffer from severe limitations and lead

Time-Varying Waveform Distortions in Power Systems   Edited by Paulo F. Ribeiro
© 2009 John Wiley & Sons, Ltd

to misinterpretation of the phenomena. This is particularly relevant when dealing with waveform sensitive harmonic distortion problems rather than heating effects for which the magnitudes alone are sufficient.

Traditionally we have overlooked the problem and applied statistical and probabilistic methods indiscriminately to time-varying harmonic distortion; but how can time-varying harmonic distortion be understood and analyzed more precisely? This chapter draws from some previous mathematical derivations and analysis [1] and attempts to apply them to power quality issues associated with harmonic distortion under time-varying conditions [2].

First, the similarity between spectral analysis and probability distribution functions are observed. Secondly, the concept of generalized frequency is presented and finally the concept of evolutionary spectra is discussed. A simple application is made to illustrate the usefulness of the concept/approach.

2.2   Concept 1 – similarities between spectral analysis and probability distribution functions

A curious engineer may have already noticed the similarity in shape between the graphs used to describe spectral analysis and the graphs of probability distribution functions as illustrated below in Figures 2.1 and 2.2.

This similarity is no coincidence. Indeed it can be demonstrated [1] that the normalized integrated spectrum has the same properties as a probability distribution function. This concept gives us the confidence necessary to work with spectral analysis of nonperiodic time-varying functions.

However, on the other hand, depending on the shape of continuous spectra (probability distribution function) one may conclude that probabilistic methods applied to harmonics may or may not be mathematically possible. This evolutionary concept, therefore, can be used to screen and validate harmonic analysis of time-varying waveforms.

2.3   Concept 2 – the generalized concept of frequency

In order to illustrate the generalized concept of frequency as presented in Equation (2.1) suppose that $X(t)$ is a deterministic function which has the form of a dampened sine wave

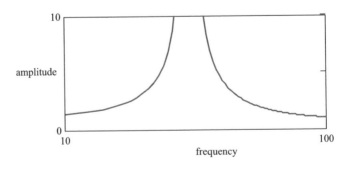

Figure 2.1   Continuous spectrum of a nonperiodic waveform

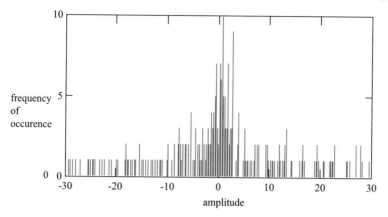

Figure 2.2   Probability distribution function of a nonstationary process

below and illustrated in Figure 2.3:

$$X(t) = A \cdot e^{\frac{-t^2}{\alpha^2}} \cdot \cos(\omega_0 \cdot t + \phi) \tag{2.1}$$

If one carries out a Fourier analysis of $X(t)$ one sees that it contains all frequencies as illustrated in Figure 2.4.

In fact, the Fourier transform of $X(t)$, as seen from Figure 2.4, consists of two Gaussian functions, one centered on w_0 and other on $(-w_0)$, the width of these functions being inversely proportional to the parameter α.

In other words, if we represent $X(t)$ as a sum of sine and cosine functions with constant amplitudes, we need to include components at all frequencies. However, we can equally well describe $X(t)$ by saying that it consists of two 'frequency' components, each having a time varying amplitude according to Equation (2.2)

$$A \cdot e^{\frac{-t^2}{\alpha^2}}. \tag{2.2}$$

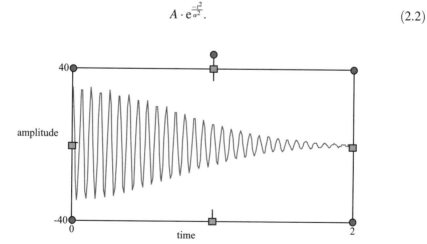

Figure 2.3   Dampened sine wave function

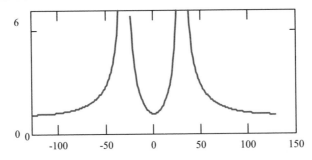

Figure 2.4   Fourier analysis of a dampened sinusoidal (including negative frequencies)

Indeed, if we were to examine the local behavior of $X(t)$ in the neighborhood of the time t_0, this is precisely what we would observe. For example, if the interval of the observation was small compared with α, $X(t)$ would appear simply as a cosine function with a frequency w_0 and amplitude as indicated before.

Caution: for the term frequency to be meaningful the function $X(t)$ must possess what we can loosely describe as an oscillatory form, and we can characterize this property by saying that the Fourier transform of such a function will be concentrated around a particular point w_0.

In conclusion, if we have a nonperiodic function $X(t)$ whose Fourier transform has an absolute maximum at a point w_0, we may define w_0 as 'the frequency' of this function, the argument being that locally $X(t)$ behaves like a sine wave with conventional frequency w_0, modulated by a 'smoothly varying' amplitude frequency.

2.4   Concept 3 – the evolutionary spectra

The evolutionary spectra have essentially the same physical interpretation as the spectra of stationary processes. The main distinction being that whereas the spectrum of a stationary process describes the power–frequency distribution for a whole process (over all time), the evolutionary spectrum is time dependent and describes the local power–frequency distribution at each instant of time.

The theory of evolutionary spectra is the only one which can preserve the physical interpretation for nonstationary processes. The evolutionary spectrum is a continuously changing spectrum or in other words, a time-dependent spectrum.

It is not practical to estimate the spectrum at every instant of time. However, if we assume that the spectrum is changing smoothly over time then, by using estimates which involve only local functions of the data, we may attempt to estimate some form of 'average' spectrum of the process in the neighborhood of any particular time instant.

Thus, in order to guarantee the confidence of probabilistic methods applied to harmonics the concept of evolutionary spectra can be applied when the FFT shows separate dominant frequencies. In this case the frequencies behave as sine waves at a particular point in time. When these conditions are not satisfied, probabilistic methods applied to harmonic summations will not result in accurate estimations.

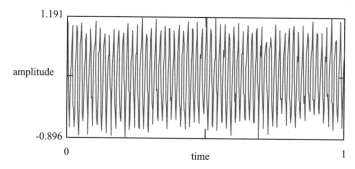

Figure 2.5   Waveform of voltage with time-varying harmonics

2.5   Considerations and applications

The concept of evolutionary spectra can then be applied to investigate and characterize the behavior of time-varying harmonic loads such as arc furnaces and electronic drives as well as aggregate harmonic loads. This approach, which needs to be investigated further and applied to a number of loads and systems conditions, may help to determine the behavior of time-varying harmonic distortion more precisely.

As a simple example let us consider the waveform in Figure 2.5 which may characterize the voltage at the PCC of an arc furnace.

The FFT of the previous waveform is illustrated in Figure 2.6 and clearly shows the dominance of two frequencies: the 60 Hz and the 180 Hz.

From Figure 2.6 and using the concept of evolutionary spectra one can confidently say that the behavior around these two frequencies at a certain instant in time is sinusoidal and linear summation methods could be applied. At other frequencies the level of confidence and meaning of the results are not to be trusted.

2.6   Conclusions

The concept of evolutionary spectra applied to time-varying harmonic distortion seems like a useful approach, and may help the power quality engineer to better understand the nature of

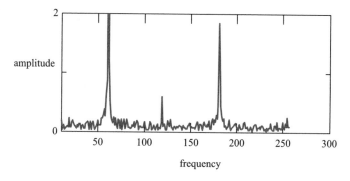

Figure 2.6   FFT of voltage with time-varying harmonics

such variations and properly utilize analytical tools to predict their behavior. Further investigations and applications using practical waveforms together with field measurements are necessary to validate this approach.

References

[1] M. B. Priestly, *Spectral Analysis and Time Series*, Academic Press, 1981.

[2] P. F. Ribeiro, 'Evolutionary Spectral For Dealing with Time-Varying Harmonic Distortion, Presentation for the Task Force on Probabilistic Aspects of Harmonics', IEEE PES Summer Meeting, Chicago, July 23, 2002.

3

Transients and harmonics

T. H. Ortmeyer

3.1 Introduction

This chapter will explore the basic definitions of harmonics (Fourier/spectral analysis) of periodic and nonperiodic functions, that is, discrete and continuous range of frequencies. Definitions of typical harmonics and transients phenomena are to be proposed.

The Comprehensive Dictionary of Electrical Engineering [1] includes the following definition:

Transient:

(a) The behavior exhibited by a linear system that is operating in steady state, in moving from one steady state to another.

(b) Any signal or condition that exists only for a short time.

(c) An electrical disturbance, usually on a power line.

(d) Refers to momentary overvoltages or voltage reductions in an electric power system due to lightning, line switching, motor starting and other temporary phenomena.

There is no definition of time-varying harmonics, but it can be expected that the voltage and current waveforms will vary from cycle to cycle to some extent.

Transient definitions (a) and (b) are limited, in that the power system has no true steady-state point, in the presence of time-varying loads and harmonics. The system does, however, exhibit movement from one steady-state point to another if these small changes are ignored. Definition (c) refers to an electrical disturbance. This definition avoids some of the

---

Time-Varying Waveform Distortions in Power Systems   Edited by Paulo F. Ribeiro
© 2009 John Wiley & Sons, Ltd

issues with the previous two definitions, by not referring to the steady state. The term 'disturbance' is defined as

> A sudden change or a sequence of changes in the components or the formation of a power system. Also called a fault.

This definition seems reasonable, apart from the final sentence – certainly, the normal switching of system components causes transients.

Definition (d) also refers particularly to power systems, but includes one important qualifier – that the event causes either momentary overvoltages or voltage reductions. On the other hand, the terms 'overvoltage' and 'voltage reduction' are typically taken to describe multiple cycle events, and certainly the (typically subcycle) fast electrical transients qualify as transients.

3.2   A proposed new definition

Possibly some new definitions could be considered.

Transient event: a power system event that has significant short-term impacts on voltages and currents in the system.
Transient: the short-term voltages and currents resulting from a transient event, which are superimposed on the longer-term system voltages and currents.

There are certain events that clearly qualify as transient events:

- system faults,

- capacitor, transformer, generator and line switchings,

- switching of large loads, including large motors.

On the other hand, there are certain events that do not cause transients:

- switching on and off of small loads,

- load and generation operating point variations,

- normal operating point variations of system devices including HVDC terminals, SVCs and other static compensators and converters and mechanical tap changers

Note that certain loads may challenge the notion that load variations do not cause transients – these loads could include arc furnaces and welders, traction loads and large impact loads.

Those events that do not cause transients will result in time-varying harmonic levels on the power system. Transient events will typically cause significant transients, most particularly, at and near the site of the event.

Harmonics measurements are taken at a specific site on the power system. A given set of measurements will provide a set of harmonic levels that vary with time. These measurements will include both time-varying harmonics and harmonic levels which include the effect of transients. It can be expected that some nearby transient events will result in harmonic levels significantly above the normal time-varying harmonic levels, while remote transient events will have little effect on the harmonic levels.

One solution to this issue would be to examine the time-varying nature of harmonics at a given site. Clearly, the transient events would be relatively rare, so that the majority of harmonics measurements will consist only of the time-varying harmonics. One definition of a 'remote' event would be an event that does not cause harmonic levels above the level of the normal variation of harmonics. This level could be defined as a certain number of standard deviations above the measurement – three standard deviations would typically limit the occurrence of transients to less than 1% of the measurements. Then, 'nearby' events would be defined as events which cause harmonic levels which are higher than the short-term harmonic mean, by at least three standard deviations.

The final caveat is that the measured harmonic levels are very much a function of the measurement process, for both transients and time-varying harmonic levels. While this is a separate topic, it must be pointed out here that the measurement process must be considered in defining the difference between a transient harmonic level and a time-varying harmonic.

Reference

[1] P. A. Laplante (ed.) *Comprehensive Dictionary of Electrical Engineering*, CRC Press/IEEE Press, 1999.

4

Electric power definitions under random conditions

A. E. Emanuel

4.1 Introduction

Let us assume an observation time T, hours or days, divided into ν equal observation time intervals ΔT such that $T = \nu \Delta T$. For each interval i, $1 \le i \le \nu$, a measuring instrument records voltage and current harmonic phasors $V_{hi}/\_\alpha_{hi}$ and $I_{hi}/\_\beta_{hi}$, ($h = 1,\ 2,\ 3,\ \ldots$ where h is the harmonic order), The rms voltage and current, V_i and I_i, as well as symmetrical components may also be computed and recorded. The intervals ΔT are small enough to consider these quantities as cyclic within each time interval or to assume the interval as a period of a recurrent set of signals. This approach enables, for each interval i, the computation and recording of apparent S_i, active P_i and different nonactive powers N_i, Q_i, D_{Ii}, D_{Vi} (for the definitions of these powers see reference [1]). Evidently each electrical quantity will be characterized by a certain probability distribution for the observation time T.

The main issue to be clarified in this approach is the formulation of correct definitions that characterize the flow of electric energy for the entire duration T. As will be explained, the mean value is the representative quantity for each of the total apparent components, regardless of the type of probability distribution. However, this is not true for the total apparent power. One significant fact is that if such probabilistic methods are used it is necessary to include a new type of nonactive power, the randomness power D_R.

Time-Varying Waveform Distortions in Power Systems   Edited by Paulo F. Ribeiro
© 2009 John Wiley & Sons, Ltd

4.2   Single-phase sinusoidal case

This is the simplest case and it will help prove the existence of the randomness power. For each interval i there are the active and reactive powers

$$P_i = V_i I_i \cos(\vartheta_i) \tag{4.1}$$

$$Q_i = V_i I_i \sin(\vartheta_i). \tag{4.2}$$

For the total observation time T the rms values of the voltage and current are

$$V = \sqrt{\frac{1}{T}\sum_{i=1}^{\nu} V_i^2 \Delta T} = \sqrt{\frac{1}{\nu}\sum_{i=1}^{\nu} V_i^2} \tag{4.3}$$

$$I = \sqrt{\frac{1}{T}\sum_{i=1}^{\nu} I_i^2 \Delta T} = \sqrt{\frac{1}{\nu}\sum_{i=1}^{\nu} I_i^2} \tag{4.4}$$

yielding a total apparent power

$$S = V I. \tag{4.5}$$

The total active or real power for the duration T is

$$P = \frac{1}{T}\sum_{i=1}^{\nu} P_i \Delta T = \frac{1}{\nu}\sum_{i=1}^{\nu} P_i = \bar{P}, \tag{4.6}$$

that is the mean of all the measured active powers P_i.

The reactive power can be defined starting from the total lost energy ΔW in the supply line. Assuming the resistance of the line that supplies the monitored load to be r, the total lost energy is

$$\Delta W = r \sum_{i=1}^{\nu} \frac{P_i^2 + Q_i^2}{V_i^2} \Delta T \tag{4.7}$$

and has two distinct terms: one proportional to the squared active power and the other proportional to the reactive powers squared. A power factor compensating capacitor meant to minimize ΔW will 'deliver' the reactive power Q_C. The optimum value of Q_C can be found by minimizing the function

$$F(Q_C) = \sum_{i=1}^{\nu} \frac{(Q_i - Q_C)^2}{V_i^2} \tag{4.8}$$

Table 4.1   Basic numerical example

i	1	2	3	4	5
V_i (V)	100.0	99.5	99.8	100.1	100.2
I_i (A)	100.0	180.0	160.0	80.0	60.0
P_i (W)	10 000.0	17 910.0	15 968.0	8008.0	6012

which leads to

$$Q_C = \frac{\sum_{i=1}^{\nu} \frac{Q_i}{V_i^2}}{\sum_{i=1}^{\nu} \frac{1}{V_i^2}} \approx \frac{1}{\nu} \sum_{i=1}^{\nu} Q_i = \bar{Q}. \tag{4.9}$$

This result helps define for the total reactive power when the powers are monitored as statistics and holds true for any type of probability distribution.

A simple numerical example will help shed light on this approach. A random load has its voltage, current and power recorded for five equal intervals (Table 4.1). This hypothetical load is assumed to be a resistance that varies randomly.

The total rms voltage (Equation (4.3)) and current (Equation(4.4)) are

$$V = 99.92 \text{ V} \quad \text{and} \quad I = 124.90 \text{ A}$$

giving an apparent power $S = 12\,480$ VA.

The active power (Equation (4.6)) is $P = 11\,579.6$ W leading to a power factor $PF = P/S = 0.928$. This result indicates that a certain type of nonactive power exists, otherwise the apparent and the active powers will be equal.

The nonactive power in question is the 'randomness power' and in this example it has the value

$$D_R = \sqrt{S^2 - P^2} = 4654.5 \text{ var}.$$

In the general case the substitution of Equations (4.1), (4.2), (4.3) and (4.4) in (4.5) gives

$$S^2 = V^2 I^2 = \frac{1}{\nu^2} \sum_{i=1}^{\nu} V_i^2 \sum_{i=1}^{\nu} I_i^2 = \frac{1}{\nu^2} \sum_{i=1}^{\nu} V_i^2 \sum_{i=1}^{\nu} [I_i \cos(\vartheta_i)]^2 + \frac{1}{\nu^2} \sum_{i=1}^{\nu} V_i^2 \sum_{i=1}^{\nu} [I_i \sin(\vartheta_i)]^2.$$

$$\tag{4.10}$$

By applying Lagrange's identity

$$\sum_{i=1}^{\nu} a_i^2 \sum_{i=1}^{\nu} b_i^2 = \left(\sum_{i=1}^{\nu} a_i b_i \right)^2 + \sum_{1 \le n < m \le \nu} (a_m b_n - a_n b_m)^2$$

to the squared apparent power in Equation (4.10), one finds

$$S^2 = \frac{1}{\nu^2}\left(\sum_{i=1}^{\nu}V_iI_i\cos(\vartheta_i)\right)^2 + \sum_{1\leq n<m\leq\nu}^{\nu}[V_mI_n\cos(\vartheta_n)-V_nI_m\cos(\vartheta_m)]^2$$

$$+\frac{1}{\nu^2}\left(\sum_{i=1}^{\nu}V_iI_i\sin(\vartheta_i)\right)^2 + \sum_{1\leq n<m\leq\nu}^{\nu}[V_mI_n\sin(\vartheta_n)-V_nI_m\sin(\vartheta_m)]^2 \tag{4.11}$$

$$= (\bar{P})^2 + (\bar{Q})^2 + D_R^2$$

where

$$D_R = \sqrt{\sum_{1\leq n<m\leq\nu}^{\nu}(V_mI_n)^2 + (V_nI_m)^2 - 2V_mV_nI_mI_n\cos(\vartheta_m-\vartheta_n)} \tag{4.12}$$

is the expression of the randomness power (measured in var). In practical situations S is computed from Equations (4.3), (4.4) and (4.5), $\bar{P}$ and $\bar{Q}$ are computed from Equations (4.6) and (4.9) and D_R from Equation (4.11), that is from the basic formula

$$D_R = \sqrt{S^2 - (\bar{P})^2 - (\bar{Q})^2}.$$

This approach enables the incorporation in D_R of the randomness reactive power, a minute component, otherwise neglected by the approximation of Equation (4.9).

4.3   Three-phase, sinusoidal and unbalanced

The IEEE Standard 1459–2000 [1] recommends the use of effective apparent power

$$S_e = 3V_eI_e \tag{4.13}$$

where V_e and I_e are the effective voltage and current, respectively

$$V_e^2 = (V^+)^2 + V_u^2 \tag{4.14}$$

$$I_e^2 = (I^+)^2 + I_u^2 \tag{4.15}$$

with V^+, I^+ the positive-sequence voltage and current and the residual voltage and current V_u, I_u, caused by the load imbalance. According to IEEE Standard 1459 the imbalance components V_u, I_u are functions of the negative- and zero-sequence components, that is

$$V_u^2 = [(V^-)^2 + (V^0)^2]/2 \quad \text{and} \quad I_u^2 = (I^-)^2 + 4(I^0)^2.$$

From Equations (4.13), (4.14) and (4.15) it is found that the effective apparent power S_e (Equation (4.13)), can be separated in two components:

$$S_e^2 = 9V_e^2 I_e^2 = (S^+)^2 + S_u^2 \tag{4.16}$$

where S^+ is the positive-sequence apparent power, considered as the most important component. The positive-sequence active and reactive powers which are the useful and dominant terms,

$$S^+ = \sqrt{(P^+)^2 + (Q^+)^2}; \quad P^+ = 3V^+ I^+ \cos(\vartheta^+), \quad Q^+ = 3V^+ I^+ \sin(\vartheta^+).$$

The second term in (4.16) is the unbalanced power (VA)

$$S_u = \sqrt{(V^+ I_u)^2 + (V_u I^+)^2 + (V_u I_u)^2}. \tag{4.17}$$

The unbalanced power has three subterms; the third term $V_u I_u$ contains the active and nonactive powers associated with the negative- and the zero-sequence components that can be easily separated from S_u.

The case we are interested in is when the stored information recorded for the ν equal time intervals includes the voltages V_i^+, V_{ui}, the currents I_i^+, I_{ui} and the powers P_i and Q_i. The effective voltage and current for the total duration of time T are

$$V_e = \sqrt{\frac{1}{\nu} \sum_{i=1}^{\nu} [(V_i^+)^2 + V_{ui}^2]} \tag{4.18}$$

$$I_e = \sqrt{\frac{1}{\nu} \sum_{i=1}^{\nu} [(I_i^+)^2 + I_{ui}^2]}; \tag{4.19}$$

thus, the total effective apparent power squared has four terms

$$S_e^2 = 9V_e^2 I_e^2 = \frac{9}{\nu^2} \left[\begin{array}{c} \sum_{i=1}^{\nu}(V_i^+)^2 \sum_{i=1}^{\nu}(I_i^+)^2 + \sum_{i=1}^{\nu}(V_i^+)^2 \sum_{i=1}^{\nu} I_{ui}^2 + \\ \sum_{i=1}^{\nu}V_{ui}^2 \sum_{i=1}^{\nu}(I_i^+)^2 + \sum_{i=1}^{\nu}V_{ui}^2 \sum_{i=1}^{\nu} I_{ui}^2 \end{array} \right]. \tag{4.20}$$

The first term is due to the positive-sequence components and just like in the previous case leads to three components:

$$(S^+)^2 = \frac{9}{\nu^2} \sum_{i=1}^{\nu}(V_i^+)^2 \sum_{i=1}^{\nu}(I_i^+)^2 = \frac{9}{\nu^2} \left[\left(\sum V_i^+ I_i^+\right)^2 + \sum_{1 \le n < m \le \nu} (V_m^+ I_n^+ - V_n^+ I_m^+)^2 \right]$$
$$= (\overline{P^+})^2 + (\overline{Q^+})^2 + (\overline{D_R^+})^2 \tag{4.21}$$

where

$$\overline{P^+} = \frac{1}{\nu}\sum_{i=1}^{\nu} P_i^+; \quad P_i = 3V_i^+ I_i^+ \cos(\vartheta_i^+) \tag{4.22}$$

$$\overline{Q^+} = \frac{1}{\nu}\sum_{i=1}^{\nu} Q_i^+; \quad Q_i = 3V_i^+ I_i^+ \sin(\vartheta_i^+) \tag{4.23}$$

are the total positive-sequence active and reactive powers and

$$
\begin{aligned}
\overline{D_R^+} &= \sqrt{\frac{9}{\nu^2}\sum_{1\le n<m\le\nu}^{\nu} [V_m^+ I_n^+ - V_n^+ I_m^+]^2} \\
&= \sqrt{\frac{9}{\nu^2}\sum_{1\le n<m\le\nu}^{\nu} (V_m^+ I_n^+)^2 + (V_n^+ I_m^+)^2 - 2V_m^+ V_n^+ I_m^+ I_n^+ \cos(\vartheta_m^+ - \vartheta_n^+)}
\end{aligned}
\tag{4.24}
$$

is the positive-sequence randomness power

The remaining three terms in Equation (4.20), correspond to the three terms in Equation (4.17), leading to the following expressions:

$$\frac{9}{\nu^2}\sum_{i=1}^{\nu}(V_i^+)^2 \sum_{i=1}^{\nu}(I_{ui})^2 = (\overline{S'_u})^2 + (D'_{uR})^2$$

with

$$\overline{S'_u} = \frac{3}{\nu}\left(\sum_{i=1}^{\nu} V_i^+ I_{ui}\right) \quad \text{and} \quad D'_{uR} = \sqrt{\frac{9}{\nu^2}\sum_{1\le n<m\le\nu}^{\nu}(V_m^+ I_{un} - V_n^+ I_{um})^2}.$$

Next is

$$\frac{9}{\nu^2}\sum_{i=1}^{\nu}(V_{ui})^2 \sum_{i=1}^{\nu}(I_i^+)^2 = (S''_u)^2 + (D''_{uR})^2$$

with

$$\overline{S''_u} = \frac{3}{\nu}\left(\sum_{i=1}^{\nu} V_{ui}^+ I_i^+\right) \quad \text{and} \quad D''_{uR} = \sqrt{\frac{9}{\nu^2}\sum_{1\le n<m\le\nu}^{\nu}(V_{um}^+ I_n^+ - V_{un}^+ I_m^+)^2},$$

followed by

$$\frac{9}{\nu^2}\sum_{i=1}^{\nu}(V_{ui})^2 \sum_{i=1}^{\nu}(I_{ui})^2 = (\overline{S'''_u})^2 + (D'''_{uR})^2$$

with

$$\overline{S'''_u} = \frac{3}{\nu}\left(\sum_{i=1}^{\nu} V_{ui} I_{ui}\right) \quad \text{and} \quad D'''_{uR} = \sqrt{\frac{9}{\nu^2}\sum_{1\le n<m\le\nu}^{\nu}(V_{um} I_n - V_{un} I_m)^2}.$$

Thus the final expression of the total effective apparent power for the duration $T = \nu \Delta T$ is

$$S_e^2 = \left(\overline{P^+}\right)^2 + \left(\overline{Q^+}\right)^2 + \left(\overline{S_u}\right)^2 + D_R^2 \qquad (4.25)$$

where

$$\left(\overline{S_u}\right)^2 = \left(\overline{S'_u}\right)^2 + \left(\overline{S''_u}\right)^2 + \left(\overline{S'''_u}\right)^2$$

and

$$D_R^2 = \left(D_R^+\right)^2 + \left(D'_{uR}\right)^2 + \left(D''_R\right)^2 + \left(D'''_{uR}\right)^2.$$

4.4 Three-phase systems with nonsinusoidal and unbalanced conditions

This is the general case. For each time interval i, the effective voltage and current are recorded; moreover, the fundamental component is separated from the total harmonics component, that is

$$V_{ei}^2 = V_{e1i}^2 + V_{eHi}^2 \quad \text{and} \quad I_{ei}^2 = I_{e1i}^2 + I_{eHi}^2$$

where V_{e1} and I_{e1} are the effective fundamental voltage and current and V_{eHi} and I_{eHi} are given by the expressions

$$V_{eHi} = \sqrt{\sum_{h\neq1}^{\infty} V_{ehi}^2} \quad \text{and} \quad I_{eHi} = \sqrt{\sum_{h\neq1}^{\infty} I_{ehi}^2}.$$

The total effective apparent power squared is

$$S_e^2 = 9 V_e I_e = \frac{9}{\nu^2} \sum_{i=1}^{\nu} (V_{ei})^2 \sum_{i=1}^{\nu} (I_{ei})^2 = \frac{9}{\nu^2} \left[\begin{array}{c} \sum_{i=1}^{\nu}(V_{e1i}^+)^2 \sum_{i=1}^{\nu}(I_{e1i}^+)^2 + \sum_{i=1}^{\nu}(V_{e1i}^+)^2 \sum_{i=1}^{\nu}I_{eHi}^2 + \\ \sum_{i=1}^{\nu}V_{eHi}^2 \sum_{i=1}^{\nu}(I_{e1i}^+)^2 + \sum_{i=1}^{\nu}V_{eHi}^2 \sum_{i=1}^{\nu}I_{eHi}^2 \end{array} \right].$$

$$(4.26)$$

The first term is due to the contributions of the fundamental voltage and current and has exactly the same terms as Equation (4.25),

$$\frac{9}{\nu^2} \sum_{i=1}^{\nu} (V_{e1i}^+)^2 \sum_{i=1}^{\nu} (I_{e1i}^+)^2 = \left(\overline{P_1^+}\right)^2 + \left(\overline{Q_1^+}\right)^2 + \left(\overline{S_{1u}}\right)^2 + D_{1R}^2.$$

The second term is due to the interaction between the fundamental voltages and the harmonic currents

$$\frac{9}{\nu^2} \sum_{i=1}^{\nu} \left(V_{e1i}^+\right)^2 \sum_{i=1}^{\nu} I_{eHi}^2 = \left(\overline{D_I}\right)^2 + D_{IR}^2$$

with

$$\overline{D_I} = \frac{1}{\nu} \sum_{i=1}^{\nu} 3V_{e1i}I_{eHi} \quad \text{and} \quad D_{IR} = \sqrt{\frac{9}{\nu} \sum_{1 \leq n < m \leq \nu} \left(V_{e1m}I_{eHn} - V_{e1n}I_{eHm}\right)^2}$$

being the current distortion power and the randomness current distortion power, respectively. The third term is due to the interaction between the harmonic voltages and the fundamental currents

$$\frac{9}{\nu^2} \sum_{i=1}^{\nu} V_{eHi}^2 \sum_{i=1}^{\nu} I_{e1i}^2 = \left(\overline{D_V}\right)^2 + D_{VR}^2$$

with

$$\overline{D_V} = \frac{1}{\nu} \sum_{i=1}^{\nu} 3V_{eHi}I_{e1i} \quad \text{and} \quad D_{VR} = \sqrt{\frac{9}{\nu} \sum_{1 \leq n < m \leq \nu} \left(V_{eHm}I_{e1n} - V_{eHn}I_{e1m}\right)^2}$$

being the voltage distortion power and the randomness voltage distortion.

The fourth term is due to the interaction between the harmonic voltages and the harmonic currents

$$\frac{9}{\nu^2} \sum_{i=1}^{\nu} V_{eHi}^2 \sum_{i=1}^{\nu} I_{eHi}^2 = \left(\overline{S_H}\right)^2 + D_{HR}^2$$

with

$$\overline{S_H} = \frac{1}{\nu} \sum_{i=1}^{\nu} 3V_{eHi}I_{eHi} \quad \text{and} \quad D_{HR} = \sqrt{\frac{9}{\nu} \sum_{1 \leq n < m \leq \nu} \left(V_{eHm}I_{eHn} - V_{eHn}I_{eHm}\right)^2}.$$

The total harmonic apparent power $\overline{S_H}$ can be further separated into the harmonic active power $\overline{P_H}$ and the nonactive harmonic power $\overline{N_H}$.

Finally, for the general case, the apparent power squared can be resolved using the expression

$$S_e^2 = \left(\overline{S_1^+}\right)^2 + \left(\overline{S_{1u}}\right)^2 + \left(\overline{D_I}\right)^2 + \left(\overline{D_V}\right)^2 + \left(\overline{S_H}\right)^2 + D_R^2 \tag{4.27}$$

where

$$D_R = \sqrt{D_{1R}^2 + D_{IR}^2 + D_{VR}^2 + D_{HR}^2}$$

is the total randomness power.

4.5 Conclusions

It is common practice to observe electrical quantities by means of statistics. A large time interval is divided into a finite number of small observation times and for each observation time the electrical quantities of interest are measured and recorded thus leading to probability densities for each one of the monitored quantities.

The present study proves that for the active and nonactive powers the equivalent value for the large time interval is always the mean value. For the apparent power, however, this property does not apply. The apparent power is determined from the product of the equivalent or effective rms voltages and currents. Under these circumstances the apparent power resolution has to include a new type of nonactive power, named randomness power. The randomness power is to be considered only when the apparent power is an equivalent value for a load or a cluster of loads that are monitored using a statistical approach.

Whenever randomness is present subharmonics can be detected. If the subharmonics and their associated powers are included, that is if the approach to the measurement for the total duration T is deterministic, then the randomness power is nil.

The randomness power does not carry net energy to the loads, nevertheless it is causing additional losses in the supplying lines and transformers' windings as well as power factor reduction.

Reference

[1] IEEE Standard 1459 – 2000, *Definitions for the Measurement of Electric Power Quantities Under Sinusoidal, Nonsinusoidal Balanced or Unbalanced Conditions*, Trial-use June 2000, full-use August 2002.

5

Visualizing Joseph Fourier's imaginative discovery via FEA

P. J. Masson, P. M. Silveira, C. A. Duque and P. F. Ribeiro

5.1 Introduction

Joseph Fourier was, without doubt, one of the greatest minds whose work changed the face of engineering. His contributions to physics and engineering are numerous; however, he is mostly remembered for his work on heat transfer that leads to the widely used transform named after him. Through a simple heat transfer experiment, Fourier noticed that the shape of temperature distribution in a ring was varying with time to become eventually a sinusoidal distribution. He then had the idea of representing the periodic temperature distribution by a sum of sinusoids giving birth to a novel signal analysis method which nowadays is used in all disciplines of engineering.

Fourier analysis is one of the most utilized mathematical techniques across a wide range of disciplines that covers from physics and engineering to biology, economics, oceanography and other areas. However, the genius of Joseph Fourier in the process of his discovery is not well known and recognized. This chapter attempts to reproduce some of the original experiments which Joseph Fourier conducted via the use of finite element analysis (FEA). The authors expect that the results will bring more insight into the understanding of this revolutionary technique as well as illustrating the fundamental need for an integration of physical intuition and mathematics for the progress of scientific developments.

Time-Varying Waveform Distortions in Power Systems   Edited by Paulo F. Ribeiro
© 2009 John Wiley & Sons, Ltd

This chapter presents Fourier's heat transfer experiment through the use of finite element analysis. An iron ring is modeled and transient thermal analysis is performed to reproduce the data Fourier obtained experimentally.

5.2   The historical context

The life of Jean Joseph Fourier was an exciting and controversial one, to say the least. Among the several facets of his life one can list: teacher, secret policeman, political prisoner, governor of Egypt, mayor, friend of Napoleon and, finally, secretary of the Academy of Sciences in Paris.

Joseph Fourier was born in Auxerre, France, as the son of a tailor. He was recommended to the Bishop of Auxerre, and was educated by the Benvenistes of the Convent of St Mark. He accepted a military lectureship in mathematics, took a prominent part in promoting the French Revolution and was rewarded by an appointment in 1795 to the École Normale Supérieure and by a chair at the École Polytechnique. A few years later, he traveled to North Africa with Napoleon Bonaparte and became Governor of Lower Egypt. After he came back to France to become prefect of Isère, he resumed his work on heat propagation. In 1822, he became permanent secretary of the French Academy of Sciences. In this year, Fourier published his *Theorie Analytique de la Chaleur*; in this work he claimed that any function of a variable can be decomposed into a sum of cosines of a multiple of that variable. Even though this statement was not entirely correct, the breakthrough was there. Dirichlet later determined the restrictions to Fourier's statement. The bases left by Fourier were reworked by numerous mathematicians until 1829 when a final demonstration was proposed by Sturm.

Fourier's work was not limited to heat transfer and numerous discoveries have been attributed to his genius such as greenhouse effect gases and our planet's energy balance [1–5].

5.3   The experiment

Joseph Fourier was working on heat transfer/propagation when he started experimenting and observing the heat transfer and temperature distribution in metallic materials. Having at first used a thin bar, he later on used a polished iron ring with a diameter of approximately 30 cm held in place by wooden supports. Holes were drilled into the ring for thermometers. A time dependent experiment was carried out where the ring was placed halfway into a furnace and then removed to an insulating bath of sand. The initial distribution of temperature was of uniformly hot around one half and cold around the other as shown in Figure 5.2. Figure 5.1 was constructed using FEA and the characteristics of the ring used by Joseph Fourier. The temperature then started to change as the heat flowed from hot to cold and illustrated in Figures 5.2 to 5.7. Fourier observed the nature of the distribution of the variation and associated it with simple sinusoidal patterns. Despite the imprecision of the instruments and tools used, Joseph Fourier's 'genius' mind was able to think of the initial distribution of temperature, and its shape as time went on, as a superposition of simple sinusoids, that varied from peak to trough, to peak an integer number of times along the circumference of the ring. As a result, Fourier could propose his creative and revolutionary equation:

$$f(t) = \frac{1}{2}a_0 + \sum_{h=1}^{\infty}[a_h \cos(h\omega_0 t) + b_h \sin(h\omega_0 t)].$$

(a)

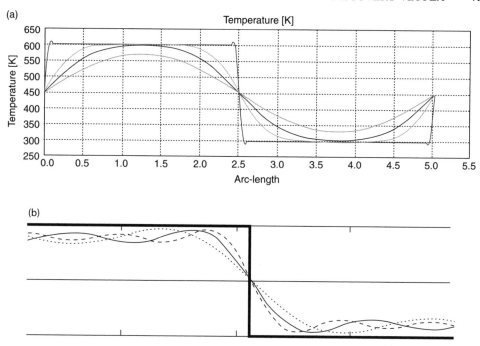

(b)

Figure 5.1   (a) Temperature distribution around the discontinuity obtained by FEA for different times (b) Temperature distribution around the discontinuity as a function of the number of Fourier coefficients

A summary of the idea that prompted the concept is depicted in Figures 5.1(a) and (b), where the behavior of the temperature distribution along the spatial discontinuity versus time (Figure 5.1(a)) and as a function of the number of Fourier series coefficients (Figure 5.1(b)) are respectively illustrated.

5.4   The finite element analysis modeling simulation

As in Fourier's experiment, an iron ring has been modeled in finite element analysis software COMSOL Multiphysics. The ring shown in Figure 5.2 is composed of iron with a thermal

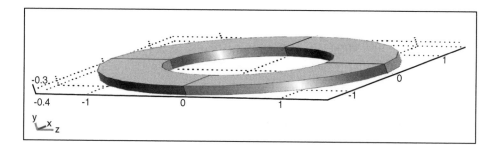

Figure 5.2   Geometry modeled in COMSOL Multiphysics

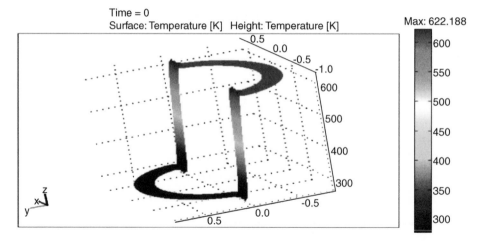

Figure 5.3   Initial temperature distribution

conductivity and heat capacity depending on temperature. The boundary is assumed to be adiabatic. Due to its large thermal diffusion time, iron can accommodate a pretty steep temperature gradient which is set to maximum at the initial time. The initial temperature conditions are 600 K in half of the ring and 300 K for the other half. At $t = 0$, heat transfer begins as represented in Figures 5.2 to 5.7. Of course, as no external heat transfer exists, the equilibrium temperature reached is the average 450 K. Figure 5.8 shows the evolution of temperature along the mean radius of the ring; as time passes, the temperature profile shows less harmonics content and ends up being a pure sinusoid.

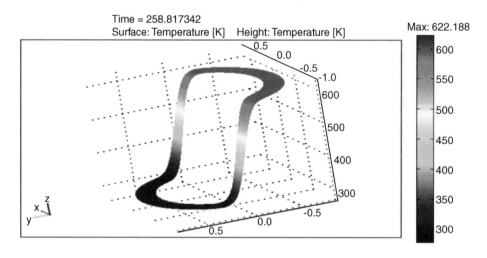

Figure 5.4   Time step after initial distribution – variation seems to take a simple sinusoidal pattern

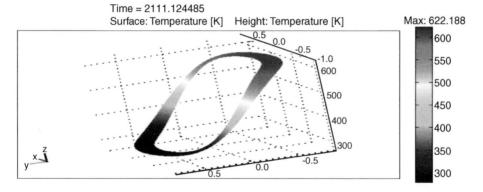

Figure 5.5   Additional time step

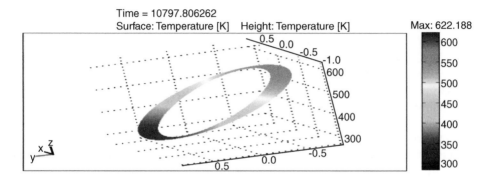

Figure 5.6   Additional time step

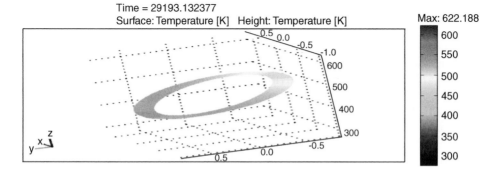

Figure 5.7   Additional time step

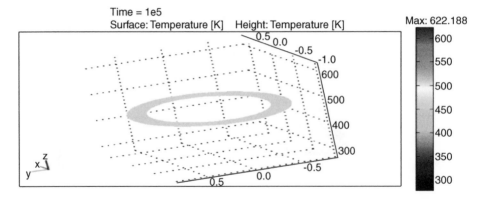

Figure 5.8   Final temperature distribution for the entire ring

5.5   Time–frequency analysis could have been discovered

Fourier reasoned that the higher frequency sinusoids would damp out rapidly. For example, a sinusoid with twice the frequency would imply that the distance between a hot peak and a cold trough was halved; on top of that, the temperature gradient would be doubled. As a result, a sinusoidal distribution with twice the frequency would dampen at four times the rate. Based on this observation, the Fourier series was developed for steady-state signals which assume that all the sinusoidal components would be present all the time. However, in reality Joseph Fourier's experiment was a time-dependent one and time-varying components could have been included.

In order to analyze the experiment with time-frequency resolution, the behavior of the temperature distribution can be approximated by a quadratic waveform, whose initial discontinuity changes and smooths over time as the temperature reaches stabilization, as illustrated in Figure 5.9.

The time–varying components could have been prompted and observed if Fourier had had more advanced mathematical tools (such as time–frequency analysis). Figure 5.10, for example, illustrates the solution of the time-dependent experiment using a methodology [6] in which the behavior time-varying harmonic components are observed as the temperature varies. This methodology, based on Fourier's analysis, can be accomplished by several approaches which use a sliding window to overcome the steady-state requirement of the Fourier series.

Considering that during the first cycles of the temperature, the behavior is like a square waveform, it is difficult to recompose the original signal using only sinusoidal components, even using a high order of components (N). In fact, since the square wave satisfies the Dirichlet conditions, the limit of $f(t)$ at the discontinuities, as $N \to \infty$, is the average value of the discontinuity. According to [7], when a signal has a discontinuity of unit height, the partial sum of the components results in a maximum value of 1.09, that is, there will be an overshoot of 9% of the height of the discontinuity, independently of how large N is. This behavior has been known as the Gibbs phenomenon, since it was the famous mathematical physicist, Josiah Gibbs, who investigated it and presented his explanation in 1899.

With his experience, Joseph Fourier had already inferred two important behaviors of the harmonic components: (i) a fast decay of the higher frequency components with time, as we

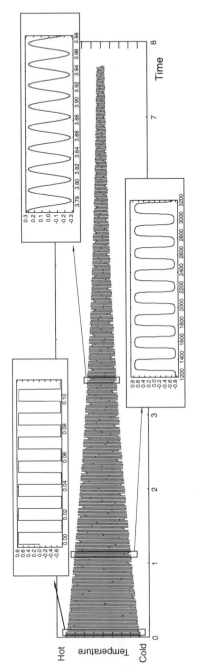

Figure 5.9   Representation of temperature variation versus time

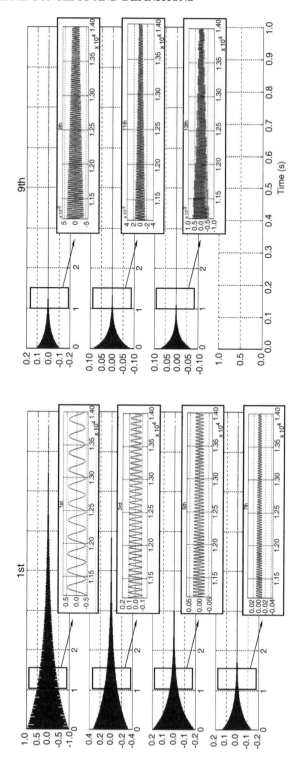

Figure 5.10   Decomposition of temperature variation by time-varying harmonics

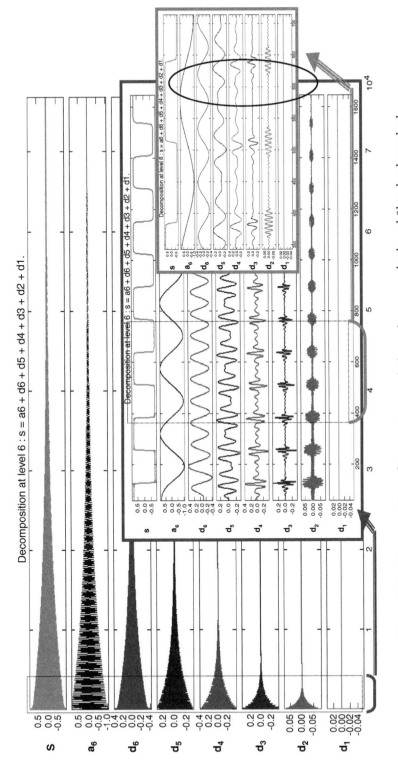

Figure 5.11   Decomposition of temperature variation using a wavelet-based filter bank method

can see in Figure 5.10 and (ii) a fast decay of the higher-order components at the discontinuities of the signal.

For this second behavior, although intuitive, it is possible to observe it easily by using another methodology of decomposition, instead of the Fourier technique. In this case, for example, the application of the wavelet transform provides different components, each one localized inside a different bandwidth. The higher-frequency components, which are present at the discontinuities of the signal, can be extracted and analyzed. Considering the signal shown in Figure 5.9, one can carry out its decomposition using a wavelet-based filter bank with six decomposition levels, whose results are shown in Figure 5.11. In this case the Meyer wavelet has been used, taking into account that it is a very suitable wavelet to extract signals with smoothed features. The Joseph Fourier's observations are clearly depicted in this figure. First, the higher the frequency component, the faster the decay with time, which can be seen by comparing the output signals (d1 to a6). Second, considering the zoomed details of Figure 5.11, one can observe the existence of high-frequency components in each discontinuity of the signal. These characteristics can be observed at the different detailed levels shown in Figure 5.11 where the decomposition of temperature variation using a wavelet-based filter bank method has been used. Figure 5.11 also shows that the temperature discontinuity can be adequately represented by only six components using a wavelet-based filter bank decomposition, whereas the traditional harmonic decomposition would require many times that number of components to achieve similar resolution since they are all of a steady-state nature. Direct physical interpretation of temperature variation from the higher detail coefficient levels would, however, be meaningless.

5.6   Conclusions

Joseph Fourier's many contributions to modern engineering science are so critically important and so pervasive that he is rightly regarded as the father of modern engineering. Great discoveries such as Fourier's transform can be found through basic physics experiments coupled with mathematics. Fourier's physical intuition lead to one of the most used analysis methods that changed the face of engineering. Finite element analysis allowed for a simple reproduction of Fourier's experiment of heat propagation/temperature variation through a metallic ring. Simulated data gave a clear view of how Fourier first thought of representing temperature distribution in a ring as a combination of sinusoidal functions and how this experiment gave information about how harmonics content is modified in time. The use of new signal processing methods, based on time–frequency decomposition, further illustrates Joseph Fourier's physical intuition to visualize the time-varying components long before the mathematical foundation was developed.

References

[1] D. A. Keston, 'Joseph Fourier – Politician and Scientist', *Today in Science* - http://www.todayinsci.com/F/Fourier_JBJ/FourierPoliticianScientistBio.htm.

[2] I. Grattam-Guiness, *Joseph Fourier (1768–1830): a survey of his life and work*, The MIT Press, 1972.

[3] J. Herivel, *Joseph Fourier: The Man and the Physicist*, Clarendon Press, Oxford, UK, 1975.

[4] J.-B.-J. Fourier, *Mémoires de l'Académie Royale des Sciences de l'Institut de France VII.* 1827, 570–604 (greenhouse effect essay).

[5] F. Arago, *The Project Gutenberg EBook of Biographies of Distinguished Scientific Men.* http://www.gutenberg.org/etext/16775.

[6] C. Duque, P. M. Silveira, T. Baldwin and P. F. Ribeiro, 'Novel Method for Tracking Time-Varying Power Harmonic Distortions without Frequency Spillover', IEEE PES General Meeting, Pittsburgh, July 2008, pages 1–6.

[7] A. V. Oppenheim and A. S. Willsky, *Signals and Systems*, Prentice-Hall, Englewood Cliffs, New York, USA, 1983.

Part II
CURRENT VARIATIONS

This part deals primarily with the harmonic current variations as opposed to voltage variations. Chapter 6 deals with the probabilistic modeling of harmonic currents. After some definitions about single current injection models, the vectorial summation of random harmonic currents characterized by distributions independent of time is considered. The basic methods presented in the literature for harmonic current summation modeling are described: the convolution method, the joint density method and Monte Carlo simulations. Applications of the methods to different case studies give an idea of the accuracy of the models considered. In Chapter 7 the probabilistic modeling of harmonic and interharmonic currents absorbed by single high-power loads is investigated with particular reference to the case of an AC/DC/AC power converter. The Monte Carlo method is utilized in a simplified scenario and numerical analysis gives an insight into harmonic and interharmonic distortion and its probabilistic modeling.

6

Summation of random harmonic currents

R. Langella and A. Testa

6.1 Introduction

This chapter deals with the probabilistic modeling of harmonic currents. After some definitions about single current injection models, the vectorial summation of random harmonic
currents characterized by distributions independent of time is considered. The basic methods
presented in literature for harmonic current summation modeling are discribed: the convolution method, the joint density method and Monte Carlo simulations. Applications of the
methods to different case studies give an idea of the accuracy of the models considered.

One aspect of assessing the harmonics tolerated by a power system is the estimation of
the statistic figures of harmonics arising from the various time-varying/probabilistic sources.
The assessment of harmonics is not exact or uniform, since there will be unpredictable
variations in either the nonlinear sources and/or parameters of the system which affect the
summation.

The combination of a number of harmonic time-varying sources will generally lead to less
than the arithmetic sum of the maximum values due to uncertainty of magnitude and phase
angle. Hence the resulting summation is extremely difficult to estimate accurately.

Before introducing the methods that can be adopted to solve the summation problem, it is
necessary to introduce and discuss some basic concepts: random harmonic and interharmonic
vectors, the harmonic summation principle and essential nomenclature.

Time-Varying Waveform Distortions in Power Systems   Edited by Paulo F. Ribeiro
© 2009 John Wiley & Sons, Ltd

6.2   Random harmonic and interharmonic vectors

The analysis of distortion at the interface busbar between utilities and consumers requires that the harmonic currents injected by disturbing loads are represented as random vectors. The random behavior of harmonic currents is related to the parameter of influence stochastic nature such as active and reactive powers, network configuration, nonlinear load operational conditions and so on.

The harmonic current of order h injected by a nonlinear load into the network can be represented as a vector $\bar{I}_h$, of amplitude I_h and phase φ_h and of Cartesian components X_h and Y_h. In any case, the statistical characterization of $\bar{I}_h$ requires the determination of the joint statistics of a pair of real random variables (I_h, φ_h) or (X_h, Y_h). With reference to (X_h, Y_h) and omitting the subscript h, the functions to be considered are:

- the joint distribution $F_{XY}(x,y)$, also called the joint cumulative probability function (jcpf), that is the probability of the event $\{X \leq x, Y \leq y\}$;

- the joint probability density function (jpdf), that is, by definition, the function:

$$f_{XY}(x, y) = (\partial^2 F_{XY}(x, y)/\partial x \partial y);$$

- the marginal distributions $F_X(x)$ and $F_Y(y)$, also called cumulative probability functions (cpfs);

- the marginal probability density functions (pdfs) $f_X(x)$ and $f_Y(y)$.

The graphs in Figure 6.1 show an example of the fifth harmonic time variability magnitude and of its pdf. Figure 6.1(a) shows the scatter plot of the vectors edges registered and Figure 6.1(b) the magnitude pdf.

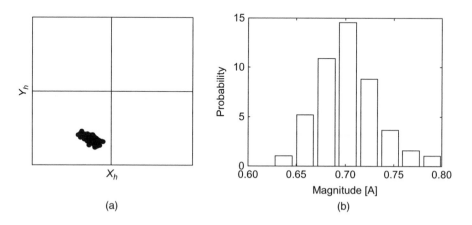

(a)   (b)

Figure 6.1   Fifth harmonic current from measurements: (a) scatter plot, (b) probability histogram of magnitude

6.3   Harmonic summation principle and essential nomenclature

The basis for harmonic combination is the superposition principle. To apply the superposition principle properly, a phasorial composition should be used. Here, the basic definitions and nomenclature are introduced together with some practical considerations about the computational problems.

6.3.1   Summation of random harmonic and interharmonic vectors

The problem of modeling the injection of the sum of numerous current vectors in the network needs technical analysis and statistical elaboration of measurements for each load and also knowledge and representation of the way in which random vectors combine during the time to give a resultant.

The sum of N random harmonic vectors gives:

$$\bar{I}_h = \sum_{k=1}^{N} \bar{I}_{h,k} = \sum_{k=1}^{N} X_{h,k} + j \sum_{k=1}^{N} Y_{h,k} = S_h + jW_h, \; I_h = \sqrt{S_h^2 + W_h^2}, \; \varphi_h = \tan^{-1}\left(\frac{W_h}{S_h}\right).$$

Obtaining the pdf of the harmonic current I_h is of great practical interest. The correct theoretical approach would be based on the use of the $2N$-dimensional joint probability density function

$$f_{Z_h}(z_h), \; \text{with} \; Z_h = [X_{h,1}, X_{h,2}, \ldots, X_{h,N}, Y_{h,1}, Y_{h,2}, \ldots, Y_{h,N}]. \tag{6.1}$$

The distribution of I_h is given by:

$$F_{I_h}(i_h) = \int_\Gamma f_{Z_h}(z_h) \prod_{i=1}^{2N} dz_{h,i}, \tag{6.2}$$

Γ being the $2N$-dimensional region of the hyperspace z where the constraint $I_h(z) \leq i_h$, is verified. Once the integral has been solved, it is trivial to obtain the pdf of I_h.

6.3.2   Basic considerations

The vectorial summation of N random harmonic and interharmonic components, in a defined scenario of space (loads, utility network characteristics, etc.) and of time (the year, the annual maximum load day, etc.), is in principle very simple and comprehensive if the $2N$-dimensional jpdf of the $2N$ real random variables representing the N vectors involved is available for the scenario at hand.

In practice, also having in mind only numerical approaches that discretize the hyperspace z, assuming M discrete values for each coordinate in its definition interval, the following dramatic computational problems arise:

- the matroid to be utilized for representing the jpdf f_Z given in Equation (6.1) assumes dimension $D = 2M^{2N}$ (f.i. if $M = N = 10$, that are very low values, then $D = 2 \times 10^{20}$);

- the solution of integral Equation (6.2) is very time consuming;

- the determination of the jpdf is difficult to obtain by both experimental analyses and simulation approaches.

If the dependence amongst the different random vectors is ignored (hypothesis A) or accounted for outside the summation stage, the problem reduces to consider N matroids, each of dimension $D_i = 2M^2$, to represent N bidimensional jpdfs ($M = N = 10$ gives $D = ND_i = 2000$).

Moreover, if the dependence amongst the pairs of real random variables utilized to represent each vector is ignored (hypothesis B) or accounted for outside the summation stage, the problem reduces to consider $2N$ matrices, each of dimension $D_i = 2M$, to represent $2N$ marginal pdfs ($M = N = 10$ gives $D = 2ND_i = 400$).

Different procedures have been used to approach the summation of stationary random vectors, in order to avoid the Monte Carlo simulation burden or the dramatic computational problems deriving from the theoretical formulation reported in nomenclature; reviews are available in [4–6, 8] and [11]. Among these procedures the first part of this section summarizes those used most widely.

The various methods proposed have originated from the appropriate application of probability theory [18]. Most of them are based on analytical models founded on fully developed convolutions or on the central limit theorem; differences consist of the more or less restrictive hypotheses required by each of them (mainly the statistical dependence or independence amongst the random variables representing the Cartesian coordinates of the vectors and the number of components). All of the methods presented in the following section are founded on two general hypotheses:

- the random vectors are statistically independent;

- the distributions of harmonic vectors are independent of time.

6.4   Basic methods

The basic methods proposed in the literature are the convolution method, the joint density method and, finally, Monte Carlo simulations.

6.4.1   Convolution method

The first method [4] assumes that the resultant resolved components are statistically independent. The method goes:

$$f_{S_h}(s_h) = f_{X_{h,1}}(x_{h,1}) * f_{X_{h,2}}(x_{h,2}) * \ldots * f_{X_{h,N}}(x_{h,N}),$$

$$f_{W_h}(w_h) = f_{Y_{h,1}}(y_{h,1}) * f_{Y_{h,2}}(y_{h,2}) * \ldots * f_{Y_{h,N}}(y_{h,N}),$$

$$f_{S_h^2}(s_h^2) = \frac{1}{2\sqrt{s_h^2}} \left[f_{S_h}(\sqrt{s_h^2}) + f_{S_h}(-\sqrt{s_h^2}) \right],$$

$$f_{W_h^2}(w_h^2) = \frac{1}{2\sqrt{w_h^2}} \left[f_{W_h}(\sqrt{w_h^2}) + f_{W_h}(-\sqrt{w_h^2}) \right],$$

$$f_{S_h^2 + W_h^2}(s_h^2 + w_h^2) = f_{S_h^2}(s_h^2) * f_{W_h^2}(w_h^2),$$

$$f_{I_h}(i_h) = 2i_h f_{S_h^2 + W_h^2}(i_h^2), \tag{6.3}$$

where * denotes convolution.

In general the method requires the knowledge of each vector resolved component pdfs. A simplification is possible when, for a large number N of harmonic vectors, the central limit theorem is applicable to the summation of the $X_{h,k}$ and of the $Y_{h,k}$. In this case S_h and W_h pdfs approximate to two Gaussian distributions of known means and variances. However, Equation (6.3) requires the statistical independence between S_h^2 and W_h^2.

6.4.2   Joint density method

Another method [5] does not need the independence between S_h and W_h but it requires N to be so high to make the central limit theorem applicable. Therefore, S_h and W_h are jointly normal with their joint density given by:

$$f_{S_h W_h}(s_h, w_h) = \frac{e^{-\frac{\eta}{2(1-r^2)}}}{2\pi\sigma_S\sigma_W\sqrt{1-r^2}},$$

where r is the correlation coefficient and

$$\eta = \frac{(s-\mu_S)^2}{\sigma_S^2} - \frac{2r(s-\mu_S)(w-\mu_W)}{\sigma_S\sigma_W} + \frac{(w-\mu_W)^2}{\sigma_W^2}. \tag{6.4}$$

The density function of I_h is directly derived by the following relation:

$$f_{I_h}(i_h) = \int_0^{2\pi} f_{S_h W_h}(i_h \cos\phi_h, i_h \sin\phi_h) i_h d\phi_h. \tag{6.5}$$

When r equals 0, S_h and W_h are independent because they are jointly normal.

In the same hypotheses, the density function of the phase Φ is derivable solving the following integral:

$$f_\Phi(\phi) = \int_0^\infty f_{SW}(i \cos\phi, i \sin\phi) i \, di. \tag{6.6}$$

6.4.3   Monte Carlo method

Usually, Monte Carlo methods [17, 19] are used to simulate a prescribed random behavior of the network loads. That is, random number generators are used to assign specific probability distributions to certain parameters of the loads, thus reflecting the random variations in the loads' operating conditions. In this way, deterministic models of the load can then be used to generate the random harmonic current injection. Subsequently, the statistics of the resulting harmonic voltages are numerically determined, usually from the linear propagation of these harmonic currents through the system impedances. The advantage of this approach is in the possibility of simulating a wide variety of random load characteristics until the resulting statistics agree with the available field measurements. The disadvantage is that this method is computationally intensive and time consuming since it is difficult to determine how to adjust the load random models in order to produce desired results. That is, although the method is flexible, the direct relation between the load probability models and the resulting harmonics is not apparent. This means that if a given set of load random characteristics yields unsuccessful results (simulated harmonics do not match the available measurements), then a limited insight is gained from this simulation trial in terms of altering load models for more accurate results in the next trial. Unless equipped with accurate information on the nature of the random loads, it is extremely difficult to develop simulation models which generate adequate results.

6.5   Magnitude and phase distribution obtainable starting from the joint density method [16]

In this section we recall some brief remarks on analytical distributions, firstly for the magnitude and then for the phase. The integral forms Equations (6.5) and (6.6) are not easy to solve, due to the complicated structure of f_{SW}. Nevertheless, it is worth noting that if a priori it is possible to assume that r is equal to zero, then S and W are independent, because they are jointly normal, and the integral forms in Equations (6.5) and (6.6) become more straightforward to handle. In this case the integration of Equations (6.5) and (6.6) can also be performed by opportune convolutions, as fully shown in [1–3], so obtaining a formal but not effective simplification because in the most general case difficulties arise due to numerical problems and to the sensitivity to the real axes partition choice.

The rotation aim is to determine an opportune rotation angle θ of the original Cartesian reference Π_{SW} that gives a new Cartesian reference $\Pi_{S'W'}$ in which $r_{S'W'} = 0$. The angle θ can be obtained as a function of variances and covariances of the original resolved components [3]:

$$\theta = \frac{1}{2}\tan^{-1} 2\frac{\sigma_{SW}^2}{\sigma_S^2 - \sigma_W^2}. \tag{6.7}$$

S and W result in two normal and correlated random variables changed by the rotation Equation (6.7) into S' and W', normal and uncorrelated, that is to say, also independent, so simplifying the integral forms Equations (6.5) and (6.6) and also solving the theoretical problem of the original distribution concerning the assumption of normal jpdf. The outcomes of the rotation are fully shown in [3], also referring to summation methods not utilizing the bivariate normal distribution.

6.5.1 Magnitude

The aim is to have a comprehensive insight into the different fully analytical or empirical solutions, in order to compare the regions of applicability, the performances and, mainly, the usefulness in practical engineering problems in which sometimes only some statistic parameters are available.

6.5.1.1 Closed form solutions holding under particular hypotheses

In the hypotheses of $r_{SW}=0$ and $\mu_W=0$, the integration of Equation (6.5) provides a complicated solution [6], based on a series of products of I_k, the modified Bessel function of integral order k:

$$f_I(i) = \frac{i}{\sigma_S\sigma_W}\exp\left[-\frac{i^2}{4}\left(\frac{1}{\sigma_S^2}+\frac{1}{\sigma_W^2}\right)\right]\cdot\exp\left[-\frac{1}{2}\left(\frac{\mu_S}{\sigma_S}\right)^2\right]\cdot\xi, \tag{6.8}$$

where

$$\xi = I_0\left[\frac{i^2}{4}\left(\frac{1}{\sigma_S^2}-\frac{1}{\sigma_W^2}\right)\right]I_0\left(i\frac{\mu_S}{\sigma_S^2}\right) + \dots + 2\sum_{k=1}^{\infty}(-1)^k I_k\left[\frac{i^2}{4}\left(\frac{1}{\sigma_S^2}-\frac{1}{\sigma_W^2}\right)\right]I_{2k}\left(i\frac{\mu_S}{\sigma_S^2}\right). \tag{6.9}$$

In some special cases, the resultant magnitude density given by Equation (6.8) turns into a simple form. First of all, if $\sigma_S=\sigma_W=\sigma$ then Equation (6.8) becomes:

$$f_I(i) = \frac{i}{\sigma^2}\exp\left[-\frac{\mu_S^2+i^2}{2\sigma^2}\right]\cdot I_0\left(\frac{\mu_S\cdot i}{\sigma^2}\right); \tag{6.10}$$

Equation (6.10) is called the Rician probability density function.

Moreover, when $\mu_S\cdot i\gg\sigma^2$, that is the argument of $I_0(.)$ is large, the following approximation of Equation (6.9) arises:

$$f_I(i) = \frac{1}{\sqrt{2\pi\sigma^2}}\exp\left[-\frac{1}{2}\left(\frac{i-\mu_S}{\sigma}\right)^2\right]\cdot\sqrt{\frac{i}{\mu_S}}, \tag{6.11}$$

which is a Gaussian law except for the factor $(i/\mu_S)^{1/2}$, and for this reason it is called 'almost Gaussian'. It is interesting to note that the previous relations can also be utilized when $\mu_W\neq0$ by the substitution of μ_S with $(\mu_S^2+\mu_W^2)^{1/2}$.

Finally, when $\mu_S=\mu_W=0$ then Equation (6.9) becomes the Rayleigh distribution:

$$f_I(i) = \frac{i}{\sigma^2}\exp\left[-\frac{i^2}{2\sigma^2}\right]. \tag{6.12}$$

As is well known, this is the case of random vectors with phases distributed in the whole interval $(0, 2\pi)$.

Table 6.1   Methods for obtaining $f_I(i)$ from the bivariate normal distribution $f_{SW}(s,w)$: closed-form solution holding under particular hypotheses

Name	Hypotheses	Expression
Bessel	$r_{SW}=0,$ $\mu_W=0$	$f_I(i) = \dfrac{i}{\sigma_S \sigma_W} \exp\left[-\dfrac{i^2}{4}\left(\dfrac{1}{\sigma_S^2} + \dfrac{1}{\sigma_W^2}\right)\right] \cdot \exp\left[-\dfrac{1}{2}\left(\dfrac{\mu_S}{\sigma_S}\right)^2\right] \cdot \xi$
Rician	$r_{SW}=0,$ $\mu_W=0,$ $\sigma_S=\sigma_W=\sigma$	$f_I(i) = \dfrac{i}{\sigma^2} \exp\left[-\dfrac{\mu_S^2 + i^2}{2\sigma^2}\right] \cdot I_0\left(\dfrac{\mu_S \cdot i}{\sigma^2}\right)$
'Almost Gaussian'	$r_{SW}=0,$ $\mu_W=0,$ $\sigma_S=\sigma_W=\sigma,$ $\mu_S \cdot i \gg \sigma^2$	$f_I(i) = \dfrac{1}{\sqrt{2\pi\sigma^2}} \exp\left[-\dfrac{1}{2}\left(\dfrac{i-\mu_S}{\sigma}\right)^2\right] \cdot \sqrt{\dfrac{i}{\mu_S}}$
Rayleigh	$r_{SW}=0,$ $\mu_W=\mu_S=0,$ $\sigma_S=\sigma_W=\sigma$	$f_I(i) = \dfrac{i}{\sigma^2} \exp\left[-\dfrac{i^2}{2\sigma^2}\right]$

Table 6.1 summarizes the methods for obtaining $f_I(i)$ from the bivariate normal distribution $f_{SW}(s,w)$ in closed-form solution holding under particular hypotheses.

6.5.1.2   Empirical solutions

Other methods seem to follow a not fully analytical demonstration for obtaining a closed form for the solution of Equation (6.5). Basically, they impose the equating of some moments of the actual f_I with the corresponding moments of an approximate distribution. In spite of all that, they obtain very powerful and accurate results.

In the hypothesis of $r_{SW}=0$, in [9] a χ^2 distribution is introduced which gives for the magnitude density:

$$f_I(i) = \frac{2^{(1-v/2)} i^{(v-1)}}{\eta^{v/2}\Gamma(v/2)} \exp\left[-\frac{i^2}{2\eta}\right], \tag{6.13}$$

with the parameters expressed by the following relations:

$$v = \frac{(\mu_S^2 + \mu_W^2 + \sigma_S^2 + \sigma_W^2)^2}{2\mu_S^2\sigma_S^2 + 2\mu_W^2\sigma_W^2 + \sigma_S^4 + \sigma_W^4},$$

$$\eta = \frac{2\mu_S^2\sigma_S^2 + 2\mu_W^2\sigma_W^2 + \sigma_S^4 + \sigma_W^4}{(\mu_S^2 + \mu_W^2 + \sigma_S^2 + \sigma_W^2)}.$$

Instead, also in the hypothesis of $r_{SW}\neq0$, in [15] a method for assimilating the resultant magnitude distribution with a generalized gamma distribution (ggd) is proposed:

$$f_I(i) = \frac{2\alpha^\alpha i^{(2\alpha-1)}}{\Gamma(\alpha)\beta^{2\alpha}} \exp\left[-\alpha(i/\beta)^2\right], \tag{6.14}$$

where Γ is the Gamma function and the parameters α and β can be expressed in terms of the bivariate normal distribution model five parameters:

$$\beta = \sqrt{\mu_S^2 + \mu_W^2 + \sigma_S^2 + \sigma_W^2},$$

$$\alpha = \frac{\beta^4}{4\mu_S^2\sigma_S^2 + 4\mu_W^2\sigma_W^2 + 2\sigma_S^4 + 2\sigma_W^4 + C},$$

$$C = 4r_{SW}\sigma_S^2\sigma_W^2(2\mu_S\mu_W + r_{SW}\sigma_S^2\sigma_W^2).$$

It is easy to demonstrate that the expression in Equation (6.12) arises from the simplification of the more general expression in Equations (6.13) and (6.14). Parameters ν, η, α and β can be derived from a moment fitting procedure as follows. The moment generating function of two jointly normal random variables X and Y is:

$$M(t_1, t_2) = \exp[\mu_X t_1 + \mu_Y t_2 + (\sigma_X^2 t_1^2 + 2r\sigma_X\sigma_Y + \sigma_Y^2 t_2^2)/2],$$

with their joint moment of order $j + k$ given by:

$$m_{jk} = E[X^j Y^k] = \frac{\partial^j \partial^k}{\partial t_1^j \partial t_2^k} M(t_1, t_2) \Big|_{\substack{t_1 = 0 \\ t_2 = 0}}.$$

On the other hand, the kth order moment of the ggd can be obtained by means of the expression:

$$m_k = \frac{\Gamma(r + k/2)}{\Gamma(r)} \left(\frac{s}{\sqrt{r}}\right)^k.$$

Equating the same order moments gives the ggd model parameters.

Table 6.2 summarizes the methods for obtaining $f_I(i)$ from the bivariate normal distribution $f_{SW}(s,w)$ giving 'empirical' solutions.

6.5.2 Phase

Attention is also paid to the resultant phase distribution which until now has received little or no importance in the literature but which can considerably help in understanding the role played by clusters of harmonic injections at different buses of the network.

Table 6.2   Methods for obtaining $f_I(i)$ from the bivariate normal distribution $f_{SW}(S, W)$: empirical solution

Name	Hypotheses	Expression
χ^2 procedure	$r_{SW} = 0$	$f_I(i) = \dfrac{2^{(1-\nu/2)} i^{(\nu-1)}}{\eta^{\nu/2}\Gamma(\nu/2)} \exp\left[-\dfrac{i^2}{2\eta}\right]$
χ^2 procedure + rotation		$f_{I'}(i') = f_I(i) = \dfrac{2^{(1-\nu'/2)} i'^{(\nu'-1)}}{\eta'^{\nu'/2}\Gamma(\nu'/2)} \exp\left[-\dfrac{i'^2}{2\eta'}\right]$
Generalized gamma distribution		$f_I(i) = \dfrac{2\alpha^\alpha i^{(2\alpha-1)}}{\Gamma(\alpha)\beta^{2\alpha}} \exp\left[-\alpha(i/\beta)^2\right]$

The density function of the phase Φ is derivable solving the integral form:

$$f_\Phi(\varphi) = \int_0^\infty f_{SW}(i\cos\varphi, i\sin\varphi)i\,di. \tag{6.15}$$

The solutions require $r_{SW} = 0$. Also in this case it is possible to take advantage of an axis rotation. Nevertheless, it is necessary to consider that an angle θ rotation does change the phase density:

$$f_\Phi(\varphi) = f_{\Phi'}(\varphi - \vartheta).$$

6.5.2.1   Closed form solution

In the same hypotheses already assumed to represent the distribution for the resultant magnitude with a Rayleigh distribution:

$$f_\Phi(\varphi) = \frac{1}{2\pi}, \tag{6.16}$$

that is the resultant is phase uniform in $(0, 2\pi)$.

6.5.2.2   Almost analytical solution

In the hypothesis of $r_{SW} = 0$, the solution of the integral form Equation (6.15) is obtained in the Appendix of [10] and it is given by:

$$f_\Phi(\varphi) = K\frac{2 + \sqrt{\pi}\zeta\exp(\zeta^2/4)(1 + \mathrm{erf}(\zeta/2))}{4A}, \tag{6.17}$$

where the auxiliary variable ζ, which is a function in φ, is defined by means of the relation:

$$\zeta = 0.5B/\sqrt{A} \tag{6.18}$$

and the quantities K, A and B are valuable as shown in the Appendix of [16].

It is easy to demonstrate that the expression in Equation (6.16) arises from the simplification of the more general expression in Equation (6.17).

6.6   Case studies

6.6.1   Numerical case studies

All the procedures described above were tested in different case studies, also performing Monte Carlo simulations to obtain a reference.

Case studies of Rayleigh and Rician magnitude distributions applicability, not reported here for the sake of brevity, were performed obtaining very good results also applying χ^2 and ggd procedures. Instead, the cases fully reported in the following refer:

- case 1A to completely verify 'almost Gaussian' applicability conditions, while case 1B and 1C move toward 'almost Gaussian' non-applicability conditions starting from 1A;

Table 6.3   Case studies parameters

Case study	Cartesian reference	θ	Bivariate normal distribution					ggd		χ^2 Distribution	
			μ_S	σ_S	μ_W	σ_W	r_{SW}	α	β	ν	η
1A	Π_{SW}	-	20.0	1.7	0.0	1.7	0.0	34.7	20.2	69.4	2.4
1B	Π_{SW}	-	10.0	1.7	0.0	1.7	0.0	9.3	10.3	18.5	2.4
1C	Π_{SW}	-	2.5	1.7	0.0	1.7	0.0	1.4	3.5	2.8	2.1
2A	Π_{SW}	-	3.0	1.0	5.2	1.3	0.6	4.6	6.2	-	-
2B	$\Pi_{S'W'}$	$\pi/3$	6.0	1.5	0.0	0.7	0.0	4.6	6.2	9.2	2.1

- case 2 to a scenario where the bivariate distribution is characterized by an elliptic symmetry; since the principle axes of the ellipse are not parallel to the Cartesian axes, the correlation coefficient is different from zero (case 2A) but it becomes equal to zero (case 2B) after a $\pi/3$ angle rotation.

In Table 6.3 the parameter values utilized are reported for the original bivariate distribution, for the ggd and the χ^2 distribution; for the Monte Carlo simulation a minimum of 15 000 samples have been considered in order to solve integral forms of Equations (6.5) and (6.15) directly.

Figures 6.2 and 6.3 report the scatter plot, the magnitude and phase pdfs for case studies 1A, 1B and 1C while Figures 6.4 and 6.5 report the same qualities for case 2B.

Figure 6.2 shows that Gaussian and 'almost Gaussian' give results very similar in 1A, similar in 1B and appreciably wide apart in 1C. Concerning the 'almost Gaussian' results, it is worth emphasizing that the right tail approximates to the solution better, that is to say the higher percentiles are better estimated; this is due to the better verification of the condition $\mu_S \cdot i \gg \sigma^2$, especially when this condition is not verified for all the possible values.

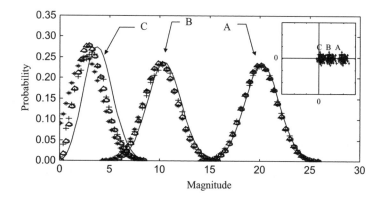

Figure 6.2   Case studies 1A,1B and 1C, current magnitude pdf, obtained by Monte Carlo simulation (+), χ^2 procedure ($\bigcirc$), ggd model (Δ), 'almost Gaussian' model (——) and Gaussian assumption (*)

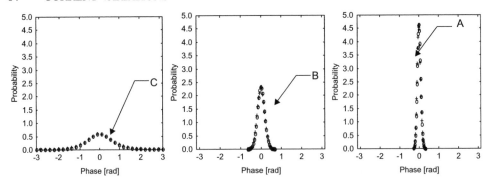

Figure 6.3   Case studies 1A, 1B and 1C, current phase pdf. obtained by Monte Carlo simulation (+), and by Equation (6.21) (○)

It is worth noting that the χ^2 and ggd procedure results are very close to the actual (Monte Carlo) results in all of the cases considered, performing as well as the closed forms, when applicable. Moreover, χ^2 distribution and ggd procedures have been tested in more general scenarios, such as those reported in [7] and [12–14], not necessarily characterized by scatter plot symmetry but which give good results.

6.6.2   Real cases

In order to have an insight into the applicability of the proposed methods to harmonics measured in medium voltage distribution networks, measurement results were processed for

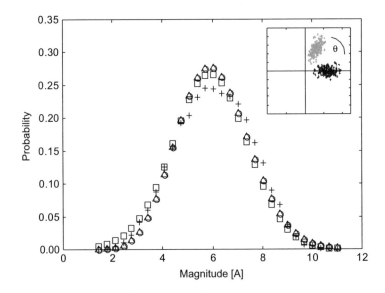

Figure 6.4   Case study 2B, current magnitude pdf, obtained by Monte Carlo simulation (+), ggd model (Δ), χ^2 procedure (○) and by Equation (6.13) (□) after a π/3 angle rotation

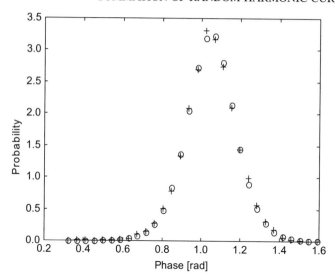

Figure 6.5   Case study 2B current phase pdf. obtained by Monte Carlo simulation (+), and by Equation (6.21) (○), after a π/3 angle rotation

comparing estimated probability density functions with sample histograms of measured harmonic currents/voltages.

The measurements consisted of 90 readings collected during a working day morning, in fall 1997, over three medium voltage lines: line #1 supplying a large plant, line #2 a cluster of small industrial laboratories and line #3 predominantly residential loads (Figure 6.6). Measurements were collected simultaneously, between 11.00 a.m. and 11.40 a.m., storing a spectrum approximately each half minute. The observation interval was relatively short, but not too short for practical purposes, for example, estimating the network behavior during a time interval deemed as the most critical in the working cycle. Even if the 37 min are not a long time, data can exhibit nonstationary behavior due to the particular time of the day when some consistent modification are likely to occur in the production processes and in the way residential customers utilize electric energy.

Here emphasis is given only to lowest-order harmonic, that is, the fifth. The graphs in Figures 6.7 and 6.8 show the fifth harmonic magnitude histograms along the pdf estimated by

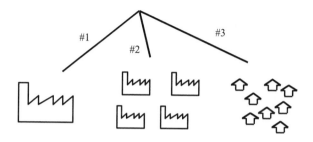

Figure 6.6   Line #1 supplying a large plant, line #2 a cluster of small industrial laboratories and line #3 predominantly residential loads

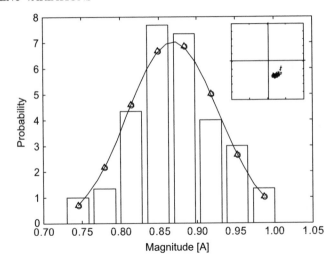

Figure 6.7   Probability of fifth harmonic current magnitude estimated from sample and Gaussian assumption (continuous line); data are relative to line #2, feeding a cluster of industrial customers

Equations (6.13) and (6.14), starting from measured μ_S, μ_W, σ_S^2, σ_W^2 and r_{SW}, and utilizing axis rotation when necessary. Figure 6.9 also reports the time behavior of the line #3 current. Figure 6.10 shows the resultant of the line currents, evaluated by adding up the contributions from each line. Harmonic busbar voltage (Figure 6.11) is considered too. Then, the 'almost Gaussian' method has been applied to the case of the voltage, by utilizing the substitution of μ_S

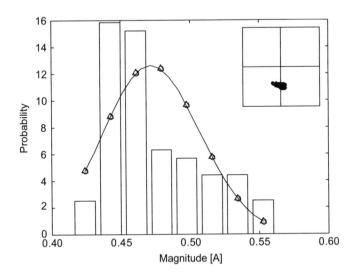

Figure 6.8   Probability of fifth harmonic current magnitude estimated from sample histogram and Gaussian assumption (continuous line); data are relative to line #3, feeding a cluster of residential customers

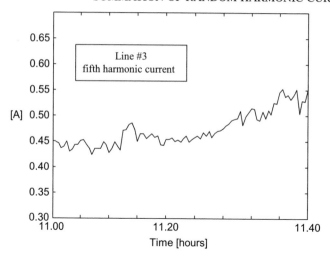

Figure 6.9   Fifth harmonic current in line #3, feeding a cluster of residential customers

with $(\mu_S^2 + \mu_W^2)^{1/2}$ and the assumption $\sigma = (\sigma_S^2 + \sigma_W^2)^{1/2}$. Finally, the phase of line #2 current is analyzed in Figure 6.12 the histograms along the pdf estimated by Equation (6.15). The scatter plots of the recorded vectors in the complex plane are included in all the graphs.

A critical observation reveals that:

- both line #1 and line #2 are described fairly well resorting to Gaussian modeling;

- line #3 results show, even in the limited observation period, a multimodal behavior;

- all of the scatter plots show elliptic shapes, suggesting that the hypothesis of Gaussian distribution is verified for the resolved components; in fact, an elliptic scatter plot will

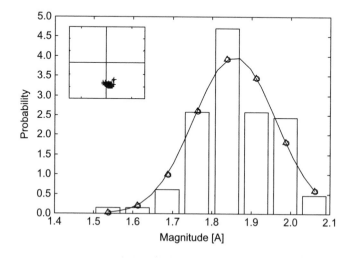

Figure 6.10   Probability of fifth harmonic current magnitude estimated from sample histogram and Gaussian assumption (continuous line); data are relative to the sum of all lines

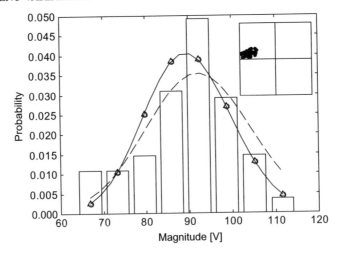

Figure 6.11   Probability of fifth harmonic voltage magnitude estimated from sample (+),
χ² procedure (◯), ggd model (Δ) and 'almost Gaussian' model (——)

still be displayed by vectors whose resolved components are distributed according to
$f(Q)$, f being some positive function with unitary integral and Q the quadratic form in
Equation (6.4); the nonstationary line #3 can be observed either in the scatter plot (two
groups of data can be detected) or in the time domain; an increasing trend in harmonic
injection can be commonly detected and associated to residential activities and it is not a
surprise, therefore, that the discussed algorithms do not fit accurately the considered
process;

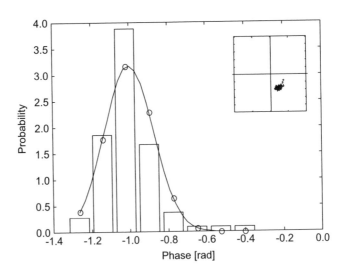

Figure 6.12   Probability of fifth harmonic current phase estimated from sample histogram
and Gaussian assumption (continuous line); data are relative to line #2, feeding a cluster of
industrial customers

- increasing trends can be detected, even if in a less significant way, in all of the fifth harmonic currents measured in the three lines; it is interesting to observe that the assumption of independence between loads would support the belief that, adding all the harmonic currents, the resultant resolved components jpdf would be closer to the normal jpdf than that of the contributions. In reality, trends seem to add up so that the estimates of the current flowing into the supply transformer are less accurately fitted by the analytical equations described here. For this purpose, Figures 6.11 and 6.12 report the magnitude pdf for both the current resultant and the voltage measured at the supply busbar for both quantities, estimates are not accurate as it would be expected owing to theoretical considerations;

- the 'almost Gaussian' method performs very well when it can take advantage of the high mean and low variance values;

- the line current phase estimation by the almost analytical solution gives an excellent performance.

For higher-order harmonics, a tendency to conform more easily to the Gaussian model has been detected. These harmonics show, as anticipated, the tendency to display a single mode and are, therefore, more likely to be described by the probability density functions discussed in the chapter.

6.7   Conclusions

This chapter has shown opportune rotations of the Cartesian axes in which the random vectors to be summed are represented given favorable consequences, reducing to zero the correlation between the resolved components, that in actual cases may be relevant mainly to low-order harmonics. Moreover, the rotation application has also been used for the summation of random vectors in the presence of low dependence among them.

Afterwards, Gaussian modeling of harmonic vectors in power systems have been covered with the aim of validating and/or proposing methods for resultant vectors magnitude and phase distribution evaluation. Among the tested methods it is possible to distinguish those which are analytical from those which are empirical. Analytical approaches must satisfy certain assumptions and therefore are not suited for every possible field situation. Empirical solutions do not pass through any rigorous analytical validation, but they perform rather well in a broad band of simulated situations. Even if prevalent attention has been dedicated to magnitude distribution, as it is normally done in the literature, phase distributions have also been derived by an analytical procedure.

Besides presenting the application of all these methods to simulated situations, the chapter has also focused on a practical use of these distributions. For this purpose, harmonic measurements performed on a node of the Italian medium voltage network have been processed. The results have shown that most of the recorded quantities can be successfully described by the empirical procedures for the magnitude distribution estimation. Moreover, the proposed relationship for the phase distribution evaluation has shown to perform satisfactorily. Furthermore, when some assumptions are met, analytical solutions can be successfully employed.

The results obtained lend hope that when a Gaussian modeling of harmonic vectors is considered not far from reality, the possibility of performing and processing measurements in a simple way is given. In fact, provided that in the considered observation interval the signals are stationary, the useful information about a single harmonic vector can be derived by the collection of only five real numbers: the means of the resolved components and the (symmetric) covariance matrix elements. This can be very useful being necessary for the application of the standards to cumulate the efforts of numerous different stationary intervals to obtain statistics extended to the whole reference time to be considered.

References

[1] IEEE Standard 519-1992 *IEEE Recommended Practices and Requirements for Harmonic Control in Electrical Power Systems*, 1993.

[2] M. Lemoine, 'Quelques aspects de la pollution des reseaux par les distortion harmoniques de la clientele', *RGE*, **85**, 1976, 247–255.

[3] R.E. Morrison and A.D. Clark, 'Probabilistic representation of harmonic currents in AC traction systems', *IEE Proc. B*, 1984, **131**, 181–189.

[4] Y. Baghzouz and O.T. Tan, 'Probabilistic modeling of power system harmonics', *IEEE Trans. on I.A.*, **23**, 1987, 173–180.

[5] E. Kazibwe, T.H. Ortmeyer and M. S. A. A. Hamman, 'Summation of probabilistic harmonic vectors', *IEEE Trans. on P.D.*, **4**, 1989, 621–628.

[6] L. Pierrat, 'A unified statistical approach to vectorial summation of random harmonic components', 4th European Conference on Power Electronics and Applications, Florence, Italy, pp. III. 100–III.105, 1991.

[7] Y.J. Wang, L. Pierrat and L. Wang, 'Summation of harmonic currents produced by ac/dc static Power Converters with randomly fluctuating loads', IEEE/PES Summer Meeting, Vancouver, 93 SM 413-5 PWRD, July 1993.

[8] R. Carbone, G. Carpinelli, M. Fracchia, L. Pierrat, R. E. Morrison, A. Testa and P. Verde, 'A Review of Probabilistic Methods for the Analysis of Low Frequency Power System Harmonic Distortions', IEE Conference on Electromagnetic Compatibility,Manchester, UK, September 1994.

[9] A. Cavallini and G. C. Montanari, 'A simplified solution for bidimensional random-walks and its application to power quality related problems', IEEE IAS Annual Meeting, Orlando, USA, 1995.

[10] A. Cavallini, 'Stochastic approach to the problem of harmonics in power system' (in Italian), Ph.D. Thesis, Bologna, Italy, 1995.

[11] P. Marino, F. Ruggiero and A. Testa, 'On the vectorial summation of independent random harmonic components', 7th ICHQP, Las Vegas, USA, 1996.

[12] F. Ruggiero, 'On the summation of random harmonic vectors in power systems', Ph.D. Thesis, Napoli, Italy, 2000.

[13] R. Langella, 'Probabilistic Modeling of Harmonic and Interharmonic Distortion in Electrical Power Systems' Ph.D. Thesis, Aversa (CE), Italy, 2001.

[14] R. Langella, P. Marino, F. Ruggiero and A. Testa, 'Summation of random harmonic vectors in presence of statistic dependences', Proceedings of the Fifth PMAPS, Vancouver, Canada, 1997.

[15] L. Pierrat and Y. J. Wang, 'Summation of randomly varying harmonics - towards a univariate distribution function using generalized gamma distribution', Proceedings of the Fifth PMAPS, Vancouver, Canada, 1997.

[16] A. Cavallini, R. Langella, F. Ruggiero and A. Testa, 'Gaussian Modeling of Harmonic Vectors in Power Systems', Proceedings of the Eighth IEEE International Conference on Harmonics and Quality of Power, Athens, Greece, Vol. 2, pp. 1010–1017, October 1998.

[17] Y. Rubinstein, *Simulation and the Monte Carlo method*, John Wiley and Sons, Inc., New York, USA, 1981.

[18] A. Papoulis, *Probability, Random Variables and Stochastic Processes*, Third edition, McGraw Hill, New York, USA, 1991.

[19] S. R. Kaprielian, A. E. Emanuel, R. V. Dwyer and H. Mehta, 'Predicting Voltage Distortion in a System with Multiple Random Harmonic Sources', *IEEE Transactions on Power Distribution*, **9**, 1994, 1632–1638.

7

Probabilistic modeling of single high-power loads

R. Langella and A. Testa

7.1 Introduction

In this chapter, the probabilistic modeling of harmonic and interharmonic currents absorbed by single high-power loads is investigated with particular reference to the case of AC/DC/AC power converters. The Monte Carlo method is utilized in a simplified scenario and numerical analysis gives an insight into harmonic and interharmonic distortion and its probabilistic modeling.

Among all the disturbing loads, power converters for the control of electrical energy are used more and more, in particular in industrial systems. AC/DC/AC power converters, such as modern multistage adjustable speed drives, are well known to be generators of interharmonic current components, in addition to harmonic current components typical for single stage converters [1–4] causing a reduction in the quality of the electrical energy. Nowadays, particular attention is devoted to interharmonics which are very difficult to eliminate by means of conventional filters because of the variability of their frequencies versus the output frequency of the conversion system.

An important aspect correlated to the presence of interharmonics is that the signal periodicity varies as the output frequency of the conversion system varies. In fact, when interharmonics affect the signal waveforms the signal periodicity results are different from the supplying system's fundamental period. This aspect creates serious problems in terms of signal numerical analysis. These problems can become dramatic in a probabilistic scenario in which, in principle, all of the working conditions of the conversion system under study have to be considered.

The latest IEC Standard 61000-4-30, contains methods of measurement and interpretation of results for harmonics and interharmonics. The 'interharmonic grouping technique' has been introduced to contain the overwhelming computational burden in managing interharmonics,

---

Time-Varying Waveform Distortions in Power Systems   Edited by Paulo F. Ribeiro
© 2009 John Wiley & Sons, Ltd

whose frequencies may assume all values, allowing an easier probabilistic modeling of interharmonic current pollution. This is a consequence of the elimination of the random variable constituted by the frequency of a single interharmonic component.

Starting from the IEC proposal and referring to modern adjustable speed drives, a probabilistic approach for the preventive evaluation of harmonic and interharmonic distortion, able to take into account the variability of the conversion system working conditions, is presented and discussed in this chapter. Its usefulness is for verifying harmonic and inter-harmonic current pollution limits, for filter sizing and so on. The main characteristics of the proposed approach are:

- converter deterministic modeling by means of accurate electromagnetic transient program (EMTP) solutions that include the control;

- proper selection operating points for the system solution, according to a proper output frequency partitioning related to the interharmonic grouping;

- results interpolation to extend the obtained solutions to the whole range of the possible working points;

- high efficiency Monte Carlo simulation, utilizing interpolated solution results.

Some numerical analyses are presented and their results are reported and discussed to show the usefulness of the proposed approach and to give an insight into the probabilistic interharmonic behavior.

7.2   Deterministic modeling

The conversion systems considered here are AC/DC/AC converters whose typical scheme is shown in Figure 7.1: the supply AC system, with frequency f_1, and the AC load, with frequency f_2, are tied together by a DC link. In the case of line commutated inverter (LCI) drives, the rectifier and the inverter are both based on fully controlled thyristor switches and the DC link has an inductive filter.

Various models are available to obtain the AC absorbed currents [4]. Each of them is characterized by a different degree of complexity in representing the AC supply system, the converters, the DC link and the motor. In the following sections, three different model types are discussed.

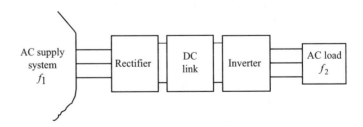

Figure 7.1   Scheme of a typical AC/DC/AC conversion system

7.3   Experimental analog models

This representation consists of a number of scale-model converters and model power system components such as high Q factor chokes, low magnetizing current transformers and capacitors. The rectifier and inverter models consist of small thyristor six-pulse bridges. The supply is replaced by a motor/generator set synchronized to the mains frequency.

In principle, any actual system nonideal condition can be reproduced by means of an appropriate choice of single components. However, in practice this is difficult to achieve. The model utilization requires laboratory measurements and their analysis. The component parameter values are chosen to obtain correspondence between the model and the real system. The analogue model is simple to set up but there are complications in obtaining results. However, the waveforms do appear in real time.

7.4   Frequency domain analytical models

Rectifier and inverter bridge models in the frequency domain, with different degrees of complexity, have been proposed; some of the most recent ones utilize the switching function approach.

With specific reference to a conventional AC/DC rectifier, the model calculates the DC voltage, $v_{DC}(t)$, starting from the AC supplying side-line voltages, $v_a(t)$, $v_b(t)$, $v_c(t)$:

$$v_{DC}(t) = S_{va}(t)\, v_a(t) + S_{vb}(t)\, v_b(t) + S_{vc}(t)\, v_c(t), \tag{7.1}$$

where $S_{va}(t)$, $S_{vb}(t)$ and $S_{vc}(t)$ are proper voltage modulation functions. Once calculated the DC-side voltage and, starting from the DC load modeling, the DC side current, $i_{DC}(t)$, can also be calculated.

The AC line currents (i.e. the phase a current, $i_a(t)$) can be obtained from the DC current by multiplying it for proper current modulation functions, $S_i(t)$:

$$i_a(t) = S_{ia}(t)\, i_{DC}(t). \tag{7.2}$$

This approach has been extended [1, 5] to include the inverter to account for inter-harmonics. Comprehensive frequency domain models are also able to take into account nonideal supply conditions.

Frequency domain models do not require significant computational effort and are very quick in execution. Then, it seems desirable to overcome some inaccuracies related to the approximate estimation of the converter switching functions and to include the control modeling to manage the load variability.

7.5   Time domain models

The time domain models can refer in principle to very general schemes. Figure 7.2 shows that for each part of the physical system depicted in Figure 7.1, it is possible to refer to very complete models, taking into account any kind of nonideal conditions such as background distortion, unbalances, magnetic material saturation, firing asymmetries and so on.

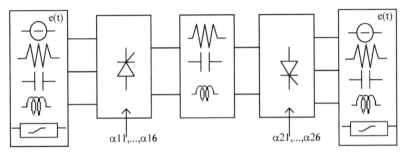

Figure 7.2   Time domain model general scheme

No simplifying assumptions are strictly needed, although in practice limitations in detail are forced by the need to contain the computational effort.

With reference to the well-known EMTP package, it is necessary to conduct the following sequence:

- to set up differential equations for each part of system;

- to use interconnection methods to connect the parts of the models;

- to solve the differential equations using numerical integration until steady-state conditions are reached;

- to subject the waveforms of interest to Fourier's analysis, for appropriate time intervals.

The model couples all the well-known advantages of detailed models that are high accuracy and well-developed software availability. Disadvantages are of excessive data requirements, difficulties in the use of the model and the long calculation time.

Bearing in mind the utilization of a deterministic model in a Monte Carlo simulation approach, the time domain models offer the advantage of accuracy, software availability and flexibility together with the possibility of including the converter control. So, if some expedients are introduced to limit the computational burden limiting the number of fully developed simulations, this model seems to be the most attractive.

7.6   Probabilistic modeling

The harmonic and interharmonic currents injected in the supplying system by the conversion systems depend on the load demand in terms of power, speed, and so on. As the load demand often has a random nature then current frequencies, amplitudes and phases are random variables.

Also, equipment uncertainty of the drive components, such as filters and/or DC link and/or motor and control systems, should be included in order to obtain a more comprehensive probabilistic current pollution model. Moreover, in [6], referring to both currents and voltages, it was shown that it was mainly the supply system parameter uncertainty which can have a great relevance. Here only the load demand uncertainty is taken into account for its prevalent interest and for the sake of simplicity.

7.6.1 Monte Carlo simulation

The complexity and the nonlinearity of the analytical relations among the conversion system parameters suggest the use of the well-known Monte Carlo method [7, 8]. Further, nonlinearities are introduced by the grouping of harmonics and interharmonics via the relations reported in the Appendix to this chapter (Section 7.9). The main advantage of such a solution is the possibility of simulating a wide variety of random load characteristics, avoiding simplifying hypotheses assumptions.

Starting from the determinations of the load resistant torque and of the motor speed, the deterministic model of the conversion system can be utilized to obtain the corresponding determinations of the harmonic and interharmonic components of the injected currents. The determinations of the resistant torque and the motor speed have to be generated according to proper probability density functions and also accounting for statistic dependences. The required pdf can be obtained by starting from the expected working conditions of the specific mechanical load considered, or from the experience of similar applications.

A simulation procedure could be reassumed by means of the flow diagram of Figure 7.3. If accurate EMTP converter modeling, including the control system, is utilized, the high number of solutions, requested to ensure the Monte Carlo simulation convergence, determines a very considerable computational burden.

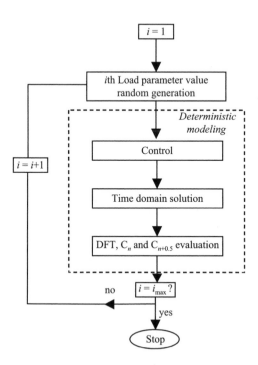

Figure 7.3   Flow diagram of the simulation procedure

7.6.2 Simplified procedure

As an alternative, a new and computationally more efficient procedure can be followed; it is designed to reduce significantly the computational burden by reducing the number of deterministic solutions without any negative effect on convergence. It takes advantage of the conversion system behavior knowledge. In practice, the simplified procedure consists of the following fundamental steps:

- starting from the load demand (i.e. the resistant torque and speed demand probability distributions), based on the conversion system control characteristics (i.e. $V/f=$ constant for AC motor drives), the range of the possible output frequencies, f_2, is evaluated;

- referring to the grouping technique, the maximum order of harmonic and interharmonic groups is fixed;

- the output frequency range is partitioned in an opportune number of frequency subintervals according to Equation (7.3) (see Section 7.7.2); each subinterval produces interharmonic components each remaining in a single group without going from one group to the next when the frequency changes (see Figure 7.4);

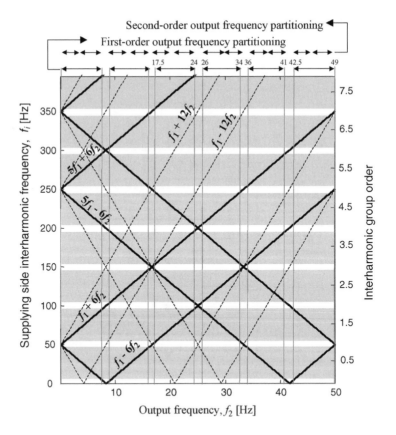

Figure 7.4   LCI drive interharmonic frequencies and groups versus the output frequencies: - first-order interharmonics, — second-order interharmonics

- for each sub-interval of f_2, two frequencies are selected to solve the conversion system: they are the frequencies closest to the edges of the interval, that result in an integer multiple of 2.5 Hz for signal processing needs [9–12];

- for the intermediate frequencies in a given f_2 sub-interval, interpolation between the edge results is utilized to obtain a sufficiently high number of determinations of corresponding interharmonic amplitudes;

- finally, the statistic of the results is performed for each group according to the specific load demand probability density function by means of a specific Monte Carlo simulation.

The partitioning of the conversion system output frequencies can be operated by taking into account the maximum order of interharmonics to be computed: the higher the inter-harmonic order is, the higher the number of output frequency intervals results. It is generally enough to consider the first and the second interharmonic orders, and by using this simplified approach the number of simulations, and the computational burden, is drastically reduced, without affecting convergence.

7.7 Numerical analyses

Some numerical analyses on a test system have been carried out to test the effectiveness of the proposed approach and to give an insight into the probabilistic behavior of the interharmonic distortion.

7.7.1 Test system

The considered system is a high-power LCI asynchronous motor drive like that used in electrical power stations for the motion of the auxiliary service pumps and fans [13]. Its scheme is shown in Figure 7.5 and the system main parameters are reported in Table 7.1. The diverter is utilized only at the motor start-up and at very low speeds and, consequently, it has not been considered in the analysis.

The inverter operation frequencies (output frequencies, f_2) have been considered to be variable between 17.5 and 50 Hz, according to the two motor speed scenarios of Figure 7.6. The scenarios are both represented in terms of output frequency probability density functions; the first refers to a uniform distribution while the second refers to more realistic conditions in which two main speed subintervals occur.

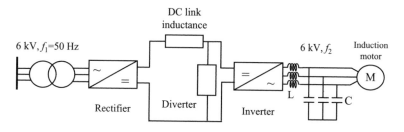

Figure 7.5   Electrical scheme of the test system

Table 7.1   Test system parameters

Supply transformer		DC Link	
Rated power [MVA]	2.5	Six-step load commutated	
Ratio V_1/V_2	1	LC output filter	
Resistance [Ω]	0.0500	L = 12.7 mH	C = 237 μF
Inductance [mH]	2.30	**Motor**	
Rectifier, Inverter		Rated power [MVA]	2.5
Resistance [Ω]	0.400	Resistance [Ω]	0.418
Inductance [mH]	40.2	Inductance [mH]	6.87

The first is a simplified scenario useful for investigating the conversion system behavior. The second scenario specifically refers to the fans control strategy and to the power generation in thermo-electrical Italian power plants. The output power and voltage have been considered to be proportional to the output frequency; they are controlled by means of the rectifier delay angle, α_r.

7.7.2   Deterministic analysis

Based on time domain simulation executed on the test system under study, a brief description of an attempt to classify interharmonics and of their amplitude and frequency variability is reported.

When a six-pulse rectifier is utilized to drive a six-pulse line commutated inverter of the type commonly used for large synchronous and induction motor drives, as the frequency of the inverter is varied, to change the speed of the motor, the converter at each end will generate

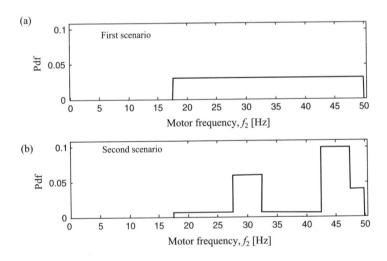

Figure 7.6   Drive output frequency scenarios: (a) uniform distributed pdf and (b) realistic distributed pdf

interharmonics which will change in amplitude and frequency as the output frequency is changed.

Referring to the frequencies of the injected current interharmonic components, f_i, they result in a function of the conversion system output frequencies, f_2. For AC/DC/AC high-power converters this gives:

$$f_i = |\{[(n_1 q_1 \pm 1)f_1] \pm [n_2 q_2 f_2]\}|, \ n_1 = 0, 1, 2, 3, \ldots n_2 = 1, 2, 3, \quad (7.3)$$

where f_1 is the supplying system fundamental frequency and q_1 and q_2 are the number of pulses of the converters. It is worth noting that for $n_2 = 0$ Equation (7.3) gives only the harmonic frequencies.

In order to recognize the origin of a given interharmonic component, it seems useful to introduce the symbol $I^u_{|n1q1\pm1|,n2}$ for the component originated from the difference into $\{\cdot\}$ of Equation (7.3) and the symbol $I^d_{|n1q1\pm1|,n2}$ for the component originated from the addition. So doing, the first subscript gives the order of the rectifier harmonics while the second gives the order of the inverter DC-ripple harmonics of the reference frequency $q_2 f_2$. Since n_2 assumes values 1, 2, 3 ... it could to be useful to consider it as the order of a specific interharmonic subset. In practice, the interharmonic components that are caused by the modulation of the first harmonic component of the inverter DC-side current with all of the harmonic components of the rectifier switching function are defined as 'first-order interharmonics'. The interharmonic components caused by the modulation of the second harmonic component of the inverter DC-side current with all of the harmonic components of the rectifier switching function are 'second-order interharmonics', and so on for higher orders.

The higher-order interharmonic components have an even smaller amplitude and they tend to manifest themselves in systems where there is resonance, especially with low dumping or high Q factor.

By characterizing Equation (7.3) to the specific case of a six-pulse rectifier/inverter drive ($q_1 = 6$, $q_2 = 6$, $f_1 = 50$ Hz, $f_2 \in [0,50]$ Hz) it is possible to calculate from the graphical representation of Equation (7.3) the interharmonic frequencies produced by all the possible drive configurations.

Figure 7.7 reports the first- and second-order interharmonic components ($n_1 = 1$ and $n_2 = 1$ and 2, that is to say $f_i = |(6 \pm 1) 50 \pm n_2 6 f_2|$) produced by the changing of the drive output frequencies. The plot is limited to the components generated by only the fundamental, the fifth and the seventh harmonic components of the rectifier switching function.

From this graph it is easy to deduct the interharmonic components produced, for example, when the drive is running with an output frequency of 30 Hz, by drawing a vertical line at 30 Hz and intersecting the lines plotted (Figure 7.7).

As a confirmation of the prediction already done, in Figure 7.8 the absorbed current interharmonic amplitudes versus frequency are reported for an output frequency equal to 30 Hz. The lower part of the same figure can easily be obtained by rotating by 90° the right side of Figure 7.7. In this way it is possible to recognize, for each interharmonic component, its order and the rectifier and inverter modulating frequencies.

It is useful to note how the two main interharmonics ($I^d_{1,1}$, $I^u_{1,1}$) are more than four times greater than all of the other components. These two components originate from the modulation of the rectifier switching function fundamental component with the first-order

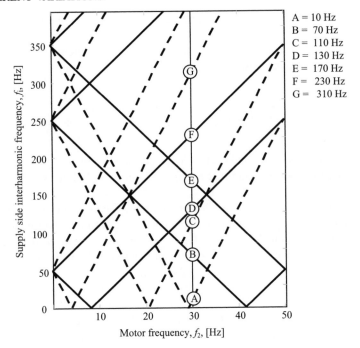

Figure 7.7   LCI drive interharmonic frequencies and groups versus the output frequencies: - first-order interharmonics, — second-order interharmonics

DC-side inverter harmonic component; other interharmonics originate from the rectifier harmonics.

The interharmonic, in practice, is of low amplitude and especially in variable speed drives can drift around the harmonic spectrum, changing in amplitude and frequency with varying motor operating conditions. Figure 7.9 reports the interharmonic spectra of the test system for eight different operating conditions.

Figure 7.10 shows the superimposition of all of the spectra of Figure 7.9. This shows the richness of the current spectrum if a statistic record is made. Of course, this implies great difficulties in storage, analyses and presentation of the results. The time domain window length to be used to detect all of the possible interharmonic components should be infinite if it is assumed that the output frequency of the inverter changes with continuity.

Finally, Figure 7.11 gives a quantitative idea of the interharmonic amplitude (and frequency) variability in terms of $I_{1,1}^u$, $I_{1,1}^d$, $I_{5,1}^u$, $I_{5,1}^d$, for motor frequency varying from 17.5 to 47.5 Hz. The amplitude variations are very high (about four times from the minimum to the maximum value).

7.7.3   Probabilistic analysis

According to the scenario of Figure 7.6(a), a first probabilistic analysis has been carried out starting from the results of five time domain simulations. Each of the first four simulations

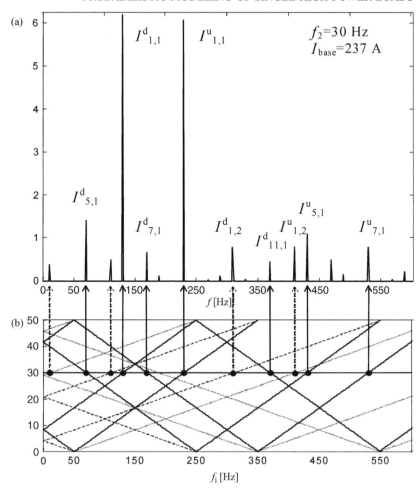

Figure 7.8   Absorbed current interharmonic for $f_2 = 30\,Hz$: (a) amplitudes versus frequency and (b) frequencies (abscissa) versus motor frequency, f_2 (ordinate)

is representative of one motor frequency in the first-order partitioning intervals (Figure 7.4); the fifth simulation is representative of the rated working condition ($f_2 = 50\,Hz$).

Figure 7.12 reports the results in terms of frequency histograms for the amplitude of the interharmonic group of order 4.5, $C_{4.5}$. It is worth noting that the choice of utilizing only five simulations leads to poor resolution results which do not represent the probabilistic behavior well.

Further histograms have been obtained by applying the Monte Carlo method as described in Section 7.6.2. For the sake of brevity only some groups have been selected and the relative results reported. The choice for the 4.5, 5 and 5.5 groups involves the main harmonic and its nearest interharmonics, one of them reaching very high values (Figure 7.11).

In particular, Figures 7.13 and 7.14 refer to relative and cumulative frequency histograms for the first and the second load scenarios, respectively.

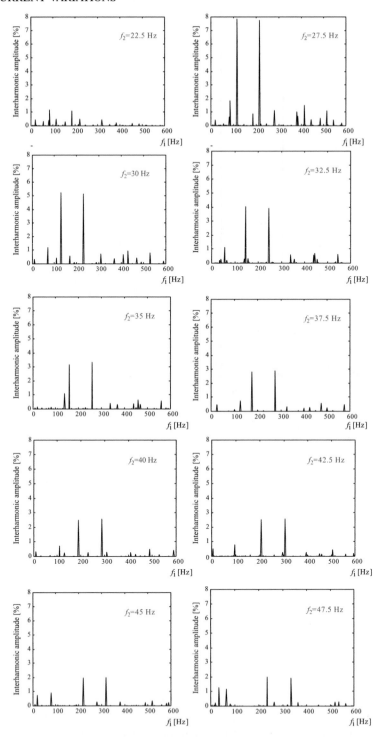

Figure 7.9   Interharmonic spectra for eight operating conditions of the test system

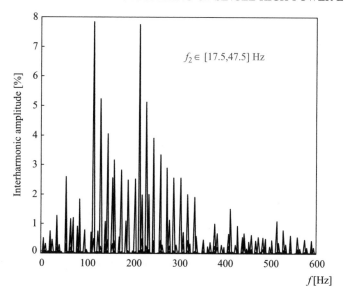

Figure 7.10   Superimposition of the spectra of Figure 7.9

It can be observed that:

- interharmonic group (IG) amplitudes start from zero and cover only some value subinterval, three $(0 \div 7.5\%, 25 \div 35\%, 50 \div 100\%)$ for $C_{4.5}$ and two $(0 \div 35\%, 77.5 \div 100\%)$ for $C_{5.5}$;

- the harmonic group (HG) amplitude starts from 40% and covers all of the values of a unique subinterval $(40 \div 100\%)$;

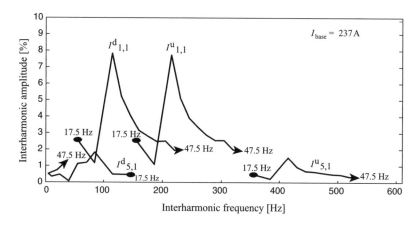

Figure 7.11 $I_{1,1}^{u}$, $I_{1,1}^{d}$, $I_{5,1}^{u}$, $I_{5,1}^{d}$, interharmonic amplitudes versus frequency, for motor frequency varying from 17.5 to 47.5 Hz

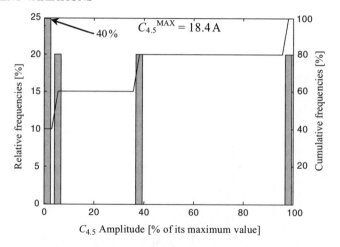

Figure 7.12   C_{4.5} relative and cumulative frequency histograms for the first load scenario

- the IG highest values are always characterized by low probabilities, while the contrary happens for HG;

- the IG zero class $(0 \div 2.5\%)$ always assumes significant probability values reaching its maximum (more than 20%);

- the cumulative frequency functions behave differently for IGs and HGs giving sharp shapes for HG highest amplitudes and smooth shapes for IG higher values.

To the knowledge of the authors of [14] no 'total interharmonic distortion' factors have yet been introduced. Once the need to separate interharmonics from harmonics is assumed, the authors consider a total interharmonic distortion (TID) factor defined as:

$$TID_\% = \frac{\sqrt{\sum_{n=0}^{N} C_{n+0.5-200-ms}^2}}{C_{1-200-ms}} \cdot 100. \tag{7.4}$$

Figure 7.15 represents the relative and cumulative frequency histograms of the TID respectively for the first and the second load scenarios.

It is worth noting that:

- TID reaches very relevant values (about 12%) even if, in both cases, they are characterized by low probability values;

- the cumulative frequency functions have the same smooth shapes for highest amplitudes of those of the IGs.

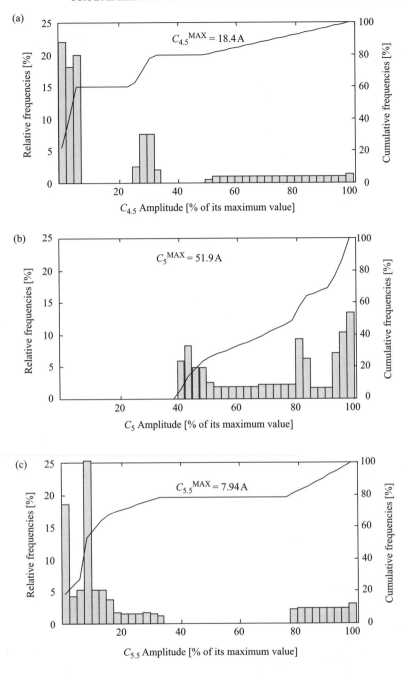

Figure 7.13 Relative and cumulative frequency histograms for the first load scenario:
(a) $C_{4.5}$, (b) C_5 and (c) $C_{5.5}$

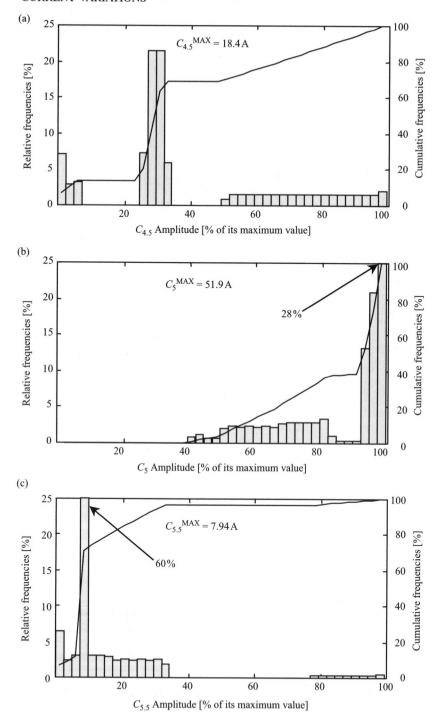

Figure 7.14   Relative and cumulative frequency histograms for the second load scenario: (a) $C_{4.5}$, (b) C_5 and (c) $C_{5.5}$

(a)

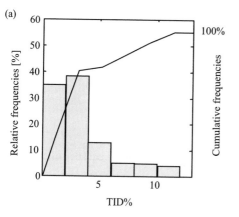

(b)

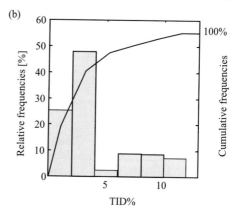

Figure 7.15   TID % relative and cumulative frequency histograms: (a) first scenario and (b) second scenario

Table 7.2   Definitions

Harmonic frequency, f_n	A frequency which is an integer multiple of the power supply frequency, f_1.
Harmonic order, n	The ratio of f_n to f_1.
Harmonic component, C_{10n}	The rms value of any of the components having a harmonic frequency as the result of a discrete Fourier transform (DFT).
Harmonic group (HG) of amplitude $C_{n\text{-}200\text{-}ms}$	The output 'bins' C each 5 Hz of the DFT grouped, as shown in Figure 7.16, according to: $$C^2_{n-200-ms} = \sum_{k=-1}^{1} C^2_{10n+k}.$$
Interharmonic frequency	A frequency which is not an integer multiple of the fundamental frequency.
Interharmonic component	The rms value of a spectral line with a frequency between the harmonic frequencies.
Interharmonic group, IG, of amplitude $C_{n+0.5\text{-}200\text{-}ms}$	The output bins of the DFT grouped as follows: $$C^2_{n+0.5-200-ms} = \sum_{k=2}^{8} C^2_{10n+k}.$$
IG frequency	The center frequency of the harmonic frequencies between which the group is situated.

Table 7.3   Recommended signal processing

Discrete Fourier transform (DFT)
Window width, T_W: exactly 10 periods of the fundamental, corresponding approximately to 200 ms
Sampling frequency, f_s: synchronized to the fundamental, sufficiently high to make possible the analysis of frequency components up to 2 kHz
Rectangular window (RW): no window weighting

Table 7.4   Measurement evaluation[1]

Very short time rms harmonic measurement[2]
The rms value $C_{n\text{-}3\text{-s}}$ over each 3 s interval is calculated with the following equation using the 15 instantaneous values assessed during the interval:

$$C^2_{n-3-s} = 1/15 \cdot \sum_{k=1}^{15} C^2_{n,i-200-\text{ms}}.$$

Short time rms harmonic measurement[3]
The rms value $C_{n\text{-}10\text{-min}}$ over each 10 min interval is calculated with the following equation using the 200 instantaneous values assessed during the interval:

$$C^2_{n-10-\text{min}} = 1/200 \cdot \sum_{k=1}^{200} C^2_{n,i-200-\text{ms}}$$

Long time rms harmonic measurement[4]
The rms value $C_{n\text{-}2\text{-h}}$ over each 2 h interval is calculated with the following equation using the 12 instantaneous values assessed during the interval

$$C^2_{n-2-h} = 1/12 \cdot \sum_{k=1}^{12} C^2_{n,i-10-\text{min}}.$$

---

[1] A basic rms harmonic measurement $C_{n\text{-}200\text{-ms}}$ is marked 'nonconforming' during either a voltage dip, a swell, or an interruption.

[2] For Class A reference performances, if eight or more of the 15 200 ms values are either 'nonconforming' or unavailable due to lack of measurement data, then the 3 s value(s) shall be marked as 'nonconforming'.

[3] For Class A reference performances: each 10 min interval shall commence at a 10 min boundary on the real-time clock. If 100 or more of the 200 3 second values are either 'nonconforming' or unavailable due to lack of measurement data, then the 10 min value(s) shall be marked as 'nonconforming'.

[4] The 2 h rms harmonic value will be obtained by taking the rms of the 10 min rms values during each 10 min interval. Each 2 h interval shall be contiguous and nonoverlapping. For Class A reference performances: each 2 h interval shall commence at a 2 h boundary on the real-time clock. If six or more of the 12 10 min values are either 'nonconforming' unavailable due to lack of measurement data, then the 2 h value(s) shall be marked as 'non- conforming'.

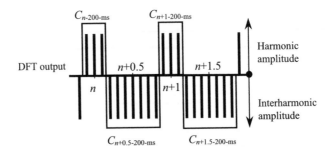

Figure 7.16   IEC grouping of 'bins' for harmonics ↑and for interharmonics ↓

7.8   Conclusions

The probabilistic modeling of harmonic and interharmonic distortion caused by single high-power loads has been investigated.

With reference to AC/DC/AC drives, the Monte Carlo method has been utilized in a simplified scenario by taking advantage of a reduced number of proper accurate time domain solutions of the conversion system. In order to perform the statistical analysis of the results in terms of current interharmonic amplitudes, the requested high number of determinations has been obtained by a simple interpolation procedure.

The numerical analyses performed on a high-power conversion test system have demonstrated how, in the case considered:

- the amplitudes of the current interharmonic groups start from zero and cover only a limited number of subinterval values;

- the current total interharmonic distortion (TID) factor can reach very relevant values;

- the highest values of both single current interharmonic groups and current total interharmonic distortion factor are characterized by very low probability values.

7.9   Appendix: IEC Standard 61000-4-30 – general guide on harmonics and interharmonics measurements and instrumentation for power supply systems and equipment connected thereto

The IEC Standard, IEC 61000-4-30, contains methods of measurement and interpretation of results for voltage harmonics and interharmonics. In Tables 7.2, 7.3 and 7.4 they are synthesized with reference to 50 Hz systems.

References

[1] L. Hu and R. Yacamini, 'Calculation of harmonics and interharmonics in HVDC schemes with low DC side impedance', *IEE Proc.C*, **140**, 1993, 469–475;E. V. Person, 'Calculation of Transfer Functions in Grid-Controlled Convertor Systems', *IEE Proc.*, **117**, 1970, 989–997.

[2] L. Hu and R. Yacamini, 'Harmonic Transfer through Converters and HVDC Links', *IEEE Transactions on Power Electronics*, **7**, 1992, 514–525.

[3] R. Carbone, D. Menniti, R. E. Morrison, E. Delaney and A. Testa, 'Harmonic and Interharmonic Distortion in Current Source Type Inverter Drives'. *IEEE Transactions on Power Delivery*, **10**, 1995, 1576–1583.

[4] R. Carbone, D. Menniti, R. E. Morrison and A. Testa, 'Harmonic and Interharmonic Distortion Modeling in Multiconverter System'. *IEEE Transactions on Power Delivery*, **10**, 1995, 1685–1692.

[5] L. Hu and R. Morrison, 'The Use of the Modulation Theory to Calculate the Harmonic Distortion in HVDC Systems Operating on an Unbalanced Supply', *IEEE Transactions on Power Systems*, **12**, 1997, 973–980.

[6] R. Carbone, D. Castaldo, R. Langella, P. Marino and A. Testa, 'Network Impedance Uncertainty in Harmonic and Interharmonic Distortion Studies', Proceedings of IEEE International Conference Power Technology, Budapest, Hungary, 1999.

[7] R. Y. Rubinstein, *Simulation and the Monte Carlo method*, John Wiley and Sons, Inc., New York, USA, 1981.

[8] R. N. Allan and R. Billinton, *Reliability Evaluation of Engineering systems*, Second edition, Plenum Press, New York, USA, 1992.

[9] D. Gallo, R. Langella and A. Testa, 'On The Processing Of Harmonics And Interharmonics In Electrical Power Systems', Proceedings of the IEEE Power Engineering Society Winter Meeting, Singapore, January 2000.

[10] D. Gallo, R. Langella and A. Testa, 'Double Stage Harmonic and Interharmonic Processing Technique', Proceedings of the IEEE Power Engineering Society Summer Meeting 2000, Seattle, USA, July 2000.

[11] D. Gallo, R. Langella and A. Testa, 'Comparison Among Techniques for Distorted Waveforms Analysis in Power Systems', Proceedings of the Ninth IEEE International Conference on Harmonics and Quality of Power, Orlando, Florida, USA, October 2000.

[12] D. Gallo, R. Langella and A. Testa, 'Self tuning Harmonic and Interharmonic Processing Technique', Proceedings of the Fifth Workshop on Power Definitions and Measurements under non Sinusoidal Conditions, Milan, Italy, October 2000.

[13] P. Caramia, R. Carbone and A. Russo, 'Attenuation of Harmonic Pollution due to Adjustable Speed Drives in the Electrical Circuits of the Power Plant Auxiliary Services', Proceedings of the IEEE Power Engineering Society Winter Meeting, Singapore, January 2000.

[14] R. Carbone, R. Langella and A. Testa, 'Simplified Probabilistic Modeling of AC_DC_AC Power Converter Interharmonic Distortion', Sixth PMAPS, Madeira (PG), September 2000.

Part III
VOLTAGE VARIATIONS

This part investigates the behavior of variations in voltages. In Chapter 8 some models for the probabilistic network analysis of transmission and distribution power systems in the presence of harmonics are dealt with. The models considered derive from the classical models for deterministic analysis: direct method, harmonic power flow and iterative harmonic analysis. For each method the theoretical aspects are presented and followed by numerical examples on transmission and distribution test power systems follow to illustrate the methods. All models described assume deterministic values for network impedances.

While Chapter 8 models for probabilistic network analysis have been considered assuming, for simplicity's sake, the network component impedances to be constant, in Chapter 9 the impedance variability is considered. The probabilistic aspects of harmonic impedances are introduced with reference to a case study.

8

Probabilistic modeling
for network analysis

P. Caramia, P. Verde, P. Varilone and G. Carpinelli

In this chapter, some models for the probabilistic network analysis of transmission and distribution power systems in the presence of harmonics are dealt with. The models considered derive from the classical models for deterministic analysis: direct method, harmonic power flow and iterative harmonic analysis. The chapter is organized so that for each method the theoretical aspects are presented first, then numerical examples on transmission and distribution test power systems follow to illustrate the methods. All models described assume deterministic values for network impedances.

8.1 Introduction

The probabilistic harmonic modeling presented in the previous chapters is concerned with the characterization of injected current harmonics at a given busbar by one or several nonlinear loads.

In recent International Standards, however, the interest is devoted not only to current but also to voltage harmonic probability density functions, and in particular to the percentiles and maximum value evaluation [1, 2]: the IEC 1000-3-6 in the assessment procedure refers to 95% probability daily value and to maximum weekly value of voltage and current harmonics while in the European Standard EN 50160 the 95% probability weekly value should not exceed the specified limit. Moreover, probabilistic aspects are also included in the recent draft revision of the IEEE 519 [3, 4]. On the other hand, knowledge of the voltage probability density functions is mandatory when the problem of the effects of distortions on electrical components is dealt with [5, 6]. The probabilistic network analysis

Time-Varying Waveform Distortions in Power Systems   Edited by Paulo F. Ribeiro
© 2009 John Wiley & Sons, Ltd

of transmission and distribution power systems in the presence of harmonics can be effected by:

- direct methods, in which the probabilistic evaluation of the voltage harmonics at system busbars is effected neglecting the interactions between the supply voltage distortion and the nonlinear load current harmonics;

- integrated methods, in which the probabilistic evaluation of the voltage and current harmonics are calculated together, properly taking into account the above mentioned interactions. Only two deterministic integrated methods have been extended in the probabilistic field: the harmonic power flow and the iterative harmonic analysis.

8.2 Probabilistic direct method for network analysis

In the deterministic field, the direct method for network analysis consists of two steps [7]. The method starts with the evaluation of the converter current waveform assuming that the voltage at the converter terminal is an ideal pure sinusoid; the waveforms are obtained by time domain simulation or with closed form relations. Then, the Fourier analysis is carried out and the current harmonics are obtained. This process is repeated for all converters present in the system.

The second step consists of the evaluation of the voltage harmonics at system busbars; the following network harmonic relations are applied:

$$\bar{V}_1^h = \sum_{j=1}^{N} \dot{Z}_{1j}^h \bar{I}_j^h$$
$$\cdots\cdots\cdots$$
$$\cdots\cdots\cdots \qquad , \qquad (8.1)$$
$$\bar{V}_N^h = \sum_{j=1}^{N} \dot{Z}_{Nj}^h \bar{I}_j^h$$

where $\dot{Z}_{ij}^h$ is the (i,j)-term of the harmonic impedance matrix and $\bar{I}_j^h, \bar{V}_j^h$ are the current harmonic of order h injected into the network at bus j and the voltage harmonic of the same order at the same bus, respectively. The current harmonic $\bar{I}_j^h$ can be due to the presence of either one or more converters at the same bus j.

The Equation (8.1) are N sum of N phasors and they correspond to the following $2N$ sum of $2N$ scalar quantities:

$$V_{iR}^h = \sum_{j=1}^{N} I_j^h [R_{ij}^h \cos(\varphi_j^h) - X_{ij}^h \sin(\varphi_j^h)]$$
$$\qquad\qquad (8.2)$$
$$V_{iI}^h = \sum_{j=1}^{N} I_j^h [R_{ij}^h \sin(\varphi_j^h) + X_{ij}^h \cos(\varphi_j^h)], \qquad i = 1, \ldots, N$$

being $\bar{V}_j^h = V_{jR}^h + jV_{jI}^h$, $\dot{Z}_{ij}^h = R_{ij}^h + jX_{ij}^h$ and $\bar{I}_j^h = I_j^h \angle \varphi_j^h$.

In the probabilistic field, the input data and output variables of direct methods are random variables and therefore they must be characterized by probability functions, in particular probability density functions (pdfs), either joint or marginal.

The probabilistic direct methods require two steps to be performed: the pdfs of the current harmonics injected into the network busbars are evaluated first, for fixed pdfs of the input data and employing proper harmonic current models (Figure 8.1); the voltage harmonic pdfs at

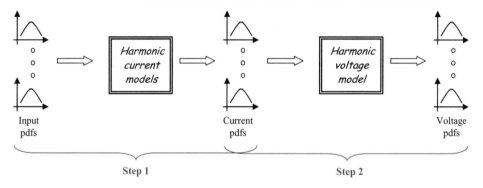

Input
pdfs

Current
pdfs

Voltage
pdfs

Step 1

Step 2

Figure 8.1   Probabilistic direct method

system buses are then evaluated by assuming the harmonic current pdfs of Step 1 as input data
and using the relations of Equation (8.1) as a harmonic voltage model.

8.2.1   Theoretical aspects

The harmonic current models and the probabilistic techniques to perform the evaluation of the
current harmonics pdfs (Step 1, Figure 8.1) are shown in details in the previous chapters and
in [8], which deal with the representation of a random harmonic phasor (only one converter is
present at the same bus) or of a sum of random harmonic phasors (more converters are present
at the same bus). Generally, the rectangular components or x–y projections are chosen for
phasor representation, because of the convenience they offer when adding phasors; in such
cases, once the pdfs of the sum of x–y projections are known, the pdf of the magnitude of the
sum is obtained. To obtain pdfs, analytical expressions, approximate solutions, the convolution
approach, the Monte Carlo simulation or the central limit theorem are applied.

With reference to the evaluation of the voltage harmonic pdfs (Step 2, Figure 8.1), the
analysis of Equation (8.1) clearly shows that each voltage harmonic phasor of order h at system
busbars can be obtained as a sum of random phasors, each one being the product of a harmonic
current phasor times a harmonic impedance matrix term. In practice, to obtain the voltage
harmonic pdfs, the evaluation of the pdfs of a sum of phasors must be carried out once again, as
in the case of current harmonic evaluation. Hence, the probabilistic techniques of the first step
(analytical expressions, approximate solutions, the Monte Carlo simulation, the central limit
theorem and so on) can be utilized to derive the voltage harmonic pdfs too.

8.2.2   Numerical applications

For illustration purposes, there follows the probabilistic direct method applied to the 18-bus
distribution test system of Figure 8.2 [9]. The values of electrical parameters of system
components are reported in [10].

The aim is the evaluation of the fifth harmonic voltage probability density functions caused
by the presence of one or more static converters at different busbars.

The following cases will be presented:

- one static converter at bus 14;

- eight static converters of the same power at different busbars;

- eight static converters at different busbars with one of them of dominant power.

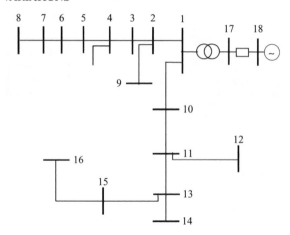

Figure 8.2   18-bus distribution test system [9]

In all cases, for the evaluation of the pdfs of the fifth harmonic current amplitude and argument we used the following very simple AC/DC six-pulse static converter model [11]:

$$I^5 = \frac{18\sqrt{2}V \cos\alpha}{5\pi^2 R}$$
$$\varphi^5 = -5\alpha + \pi, \tag{8.3}$$

V being the rms of the converter voltage, α the converter firing angle and R the DC converter resistance.

Assuming the firing angle as the only input random variable [11] with a uniform pdf, the fifth harmonic amplitude pdf is shown in Figure 8.3(a);[1] the pdf of the harmonic current argument is a uniform pdf, as clearly indicated by Equation (8.3).

With reference to Case (i), in Figure 8.3(b) the pdfs of the fifth harmonic voltage amplitude at buses $j = 1$, $j = 11$ and $j = 14$ are shown.

From the analysis of Figure 8.3(b) it can be deduced that the mean values and variances are increasing when the converter bus 14 is approached. This result could be forecasted analytically by recalling that the harmonic impedance matrix term modulus increases as the converter bus is approached with obvious consequences on mean values and variances.

With reference to Case (ii), in Figure 8.4 the pdfs of the real component, imaginary component and amplitude of the fifth harmonic voltage at busbars $j = 1$ and $j = 14$ are shown, assuming that eight converters are placed at busbars 3, 4, 7, 10, 13, 14, 15 and 16.

From the analysis of Figure 8.4 it clearly appears that no contribution to real or imaginary components of the harmonic voltage is dominant so that these components tend to approach a normal distribution, as suggested by the central limit theorem application. This is confirmed by the inspection of all the pdfs (not shown in the figures) of terms of the sum on the right-hand side of Equation (8.2); for example, the pdfs of the eight terms on the right-hand side of the real

---

[1] All the pdfs of Figures 8.3, 8.4 and 8.5 have been estimated with a Monte Carlo simulation procedure.

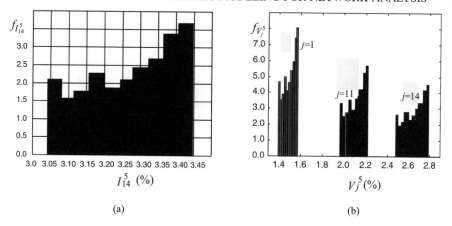

Figure 8.3   *Fifth harmonic (a) current and (b) voltage amplitude probability density functions*

or imaginary component of the voltage phasor at busbar 14 have similar minimum and maximum values.

With reference to Case (iii), in Figure 8.5 the pdfs of the real component, imaginary component and amplitude of the fifth harmonic voltage at busbars $j = 1$ and $j = 14$ are shown, assuming that one power prevailing MW converter is placed at bus 14 and seven lower power converters are placed at busbars 3, 4, 7, 10, 13, 15 and 16.

From the analysis of Figure 8.5 it clearly appears that one contribution to real or imaginary components of the harmonic voltage is dominant and, then, the real and imaginary components of the voltage harmonic are far from the normal distribution. This is confirmed by the inspection of all the pdfs of terms of the sum on the right-hand side of Equation (8.2): for example, the pdfs of seven terms on the right-hand side of the real or imaginary component of the voltage amplitude at busbar 14 have similar minimum and maximum values, all included in the interval $[-0.15, 0.1]$, while the eighth term has minimum and maximum values included in the interval $[-1.5, 1]$.

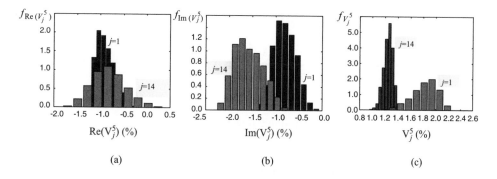

Figure 8.4   *Fifth harmonic voltage probability density functions without dominant converters: (a)real component, (b) imaginary component and (c) amplitude*

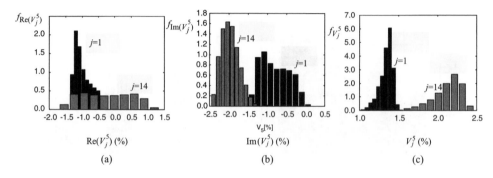

Figure 8.5   Fifth harmonic voltage probability density functions with a dominant converter: (a) real component, (b) imaginary component and (c) amplitude

8.3   Probabilistic harmonic power flow

The probabilistic harmonic power flow (PHPF) represents the natural extension to the probabilistic field of the harmonic power flow firstly proposed by Heydth *et al.* [12].

8.3.1   Theoretical aspects

The harmonic power flow model considered here is expressed by the following equations:

$$
\begin{aligned}
(\mathbf{P}_1^1)^{sp} &= \mathbf{P}_1^1(\mathbf{U}^1, \mathbf{\Phi}^1) \\
(\mathbf{Q}_1^1)^{sp} &= \mathbf{Q}_1^1(\mathbf{U}^1, \mathbf{\Phi}^1)
\end{aligned}
\tag{8.4}
$$

$$
\begin{aligned}
(\mathbf{P}_{nl})^{sp} &= \mathbf{P}_{nl}(\mathbf{U}, \mathbf{\Phi}) \\
(\mathbf{S}_{nl})^{sp} &= \mathbf{S}_{nl}(\mathbf{U}, \mathbf{\Phi})
\end{aligned}
\tag{8.5}
$$

$$
\begin{aligned}
(\mathbf{P}_{gen}^1)^{sp} &= \mathbf{P}_{gen}^1(\mathbf{U}^1, \mathbf{\Phi}^1) \\
(\mathbf{U}_{gen}^1)^{sp} &= \mathbf{U}_{gen}^1
\end{aligned}
\tag{8.6}
$$

$$
\begin{aligned}
\mathbf{0} &= \mathbf{I}_r(\mathbf{U}, \mathbf{\Phi}) \\
\mathbf{0} &= \mathbf{I}_x(\mathbf{U}, \mathbf{\Phi})
\end{aligned}
\tag{8.7}
$$

$$
\begin{aligned}
\mathbf{0} &= \mathbf{g}_r(\mathbf{U}, \mathbf{\Phi}, \mathbf{X}) \\
\mathbf{0} &= \mathbf{g}_x(\mathbf{U}, \mathbf{\Phi}, \mathbf{X})
\end{aligned}
\tag{8.8}
$$

$$
\mathbf{0} = \mathbf{R}(\mathbf{U}, \mathbf{\Phi}, \mathbf{X}),
\tag{8.9}
$$

where in the case of probabilistic harmonic power flow:

$(\mathbf{P}_1^1)^{sp}$, $(\mathbf{Q}_1^1)^{sp}$ are input random vectors of active and reactive powers specified at fundamental for each linear load busbar,

$(\mathbf{P}_{nl})^{sp}$, $(\mathbf{S}_{nl})^{sp}$ are input random vectors of total active and apparent powers specified for each nonlinear load busbar,

$(\mathbf{P}^1_{gen})^{sp}$ is an input random vector of active power specified at fundamental for each generator busbar without the slack,

$(\mathbf{U}^1_{gen})^{sp}$ is an input random vector of voltage magnitude at fundamental specified for each generator busbar,

$\mathbf{U}, \mathbf{\Phi}, \mathbf{X}$ are input random vectors of voltage magnitude and argument at all harmonics and at fundamental, of angles of commutation and of the auxiliary parameters α_i and β_i,

$\mathbf{U}^1, \mathbf{\Phi}^1$ are input random vectors of voltage magnitude and argument at the fundamental.

Equations (8.4) and (8.5) represent the power balance equations (active, reactive and apparent) at linear and nonlinear load busbars, Equation (8.6) represent the active power and voltage regulation balance equations at generator busbars, Equation (8.7) represent the harmonic current balance equations at linear load busbars, Equation (8.8) represent the harmonic and fundamental current balance equations at nonlinear load busbars and, finally, Equation (8.9) represents the commutation angle equations at nonlinear load busbars.

From Equations (8.4) to (8.9) it was assumed that in the converter busbars the firing angle α_i and the DC load auxiliary parameter β_i (i.e. the resistance R in the case of a DC passive load and the voltage E in the case of an active load) are unknown and that total active and apparent powers are specified.[2]

Equations (8.4) to (8.9) can be expressed in a compact form as:

$$g_s(\mathbf{N}) = \mathbf{T_b} \qquad (8.10)$$

where $\mathbf{T_b}$ is the input random vector including active, reactive and apparent powers, total or at fundamental, and $\mathbf{N}$ is the state random vector including amplitude and arguments of voltages, at fundamental and at harmonic orders.

Several probabilistic techniques have been applied to the nonlinear system of Equation (8.10) in order to obtain the pdfs of the state random vector starting from the knowledge of the pdfs of the input random vector [13–16]. They are:

- nonlinear Monte Carlo simulation;
- linear Monte Carlo simulation;
- convolution process approach;
- approximate distribution approach.

8.3.1.1 Nonlinear Monte Carlo simulation

The Monte Carlo (MC) procedure requires the knowledge of the probability density functions (pdfs) of the input variables; for each random input datum, a value is generated according to its proper probability density function.

---

[2] As is well known, the specifications for the converter busbars are of various types depending on the nature of the DC side that the six-pulse converter feeds and on the data available. When the firing angle α_i and the DC load auxiliary parameter β_i are not specified in advance, equations which involve active and reactive powers at fundamental (P^1, Q^1) or total active and apparent powers (P, S), are usually considered.

According to the generated input values, the operating steady-state conditions of the harmonic power flow system are evaluated solving the nonlinear system of Equation (8.10) by means of an iterative numerical method. Once convergence is achieved, the state of the multiconvertor power system is completely known and the values of all the variables of interest are stored.

The preceding procedure is repeated a sufficient number of times, L, to obtain a good estimate of the probability density functions of the output variables according to stated accuracy [17–20]. A useful stopping criterion to be adopted can be based on the use of a coefficient of variation tolerance, as proposed in [21].

8.3.1.2   Linear Monte Carlo simulation

If the vector $\mu(\mathbf{T_b})$ is the expected values of $\mathbf{T_b}$ and a deterministic harmonic power flow is calculated using $\mu(\mathbf{T_b})$ as input data, the solution of Equation (8.10) will give the state vector $\mathbf{N_0}$ such that:

$$g_S(\mathbf{N_0}) = \mu(\mathbf{T_b}). \tag{8.11}$$

Linearizing Equation (8.10) around the point $\mathbf{N_0}$ and recalling Equation (8.11), gives the result:

$$\mathbf{N} \cong \mathbf{N_0} + \mathbf{A}\Delta\mathbf{T_b} = \mathbf{N'_0} + \mathbf{A}\mathbf{T_b} \tag{8.12}$$

where:

$$\Delta\mathbf{T_b} = \mathbf{T_b} - \mu(\mathbf{T_b})$$
$$\mathbf{N'_0} = \mathbf{N_0} - \mathbf{A}\mu(\mathbf{T_b}),$$

the matrix $\mathbf{A}$ being the inverse of the Jacobian matrix evaluated in $\mathbf{N_0}$.

The linear Equation (8.12) can be included in a Monte Carlo simulation to obtain approximate probability density functions of the output $\mathbf{N}$ vector components starting from the pdfs of the input $\mathbf{T_b}$ vector components.

It should be noted that the use of the linear MC simulation allows a significant reduction in computational efforts compared with nonlinear MC. However, it should also be noted that since the harmonic power flow equations are linearized around an expected value region, any movement away from this region produces an error. The errors can grow with the variance of the input random vector components, with values linked to the nonlinear behavior of the equation system.

8.3.1.3   Convolution process approach

The convolution process can be applied to Equation (8.12) in the form:

$$f(N_i) = N_{0i} + [f(y_{i1})*f(y_{i2})*\ldots\ldots*f(y_{in})] \tag{8.13}$$

where:

f represents the probability density function;
$*$ represents the convolution;
N_{0i} represents the ith term of the vector $\mathbf{N_0}$;
y_{ij} represents the (i-j) term $A_{ij}[T_{bj} - \mu(T_{bj})]$;
A_{ij} represents the (i-j) terms of the matrix $\mathbf{A}$;
T_{bj} represents the jth term of the vector $\mathbf{T_b}$.

Equation (8.13) can be applied using numerical methods based on Laplace transforms or, more efficiently in term of execution time and precision, transforming the equations into the frequency-domain using fast Fourier transform.

The computational efforts depend on the number of input data normally distributed; in fact, as is well known, all normally distributed functions can be easily grouped in one unique normal equivalent since only the expected value and covariance matrix are required to define this function. Equation (8.13) contains discrete or other pdfs and this normal equivalent, with obvious computational advantages. The accuracy of the method is strongly linked to the degree of dependence among the nonnormal random variables to be summed. As the dependence increases, the errors can increase accordingly.

8.3.1.4   Approximate distribution approach

The approximate distribution approach consists of approximating the true pdfs of the voltage and current harmonics with pdfs whose analytical expressions are determined once a finite number of moments of the true pdfs are known. In this way, the problem of defining the whole pdf is confined to the evaluation of these moments. It is obvious that this approach makes sense if the moments of the true pdfs are available or can be estimated with accuracy and less computational effort.

Several classical approaches [20, 22] have been tested to find the best in the field of current and voltage harmonics. They are:

- Gram–Charlier's approach;

- Edgenworth's approach;

- Pearson's approach;

- Johnson's approach.

They all require significantly less computational effort.

Summarizing the approaches, Edgenworth's and Gram–Charlier's approximate the true pdfs with a series in the derivative of the normal pdf. The biggest problem in these approaches is linked to the choice of the most appropriate number of the series terms to be employed.

The Pearson approach is based on a particular family of pdfs (called a system of distributions) which represents a large number of observed distributions which are used to approximate to the true pdfs. The Johnson's approach is based on the choice of a proper transformation to convert a given pdf into another known form. This approach requires knowing only the first four moments of the pdf to be approximated.

All the above approaches have been tested with several current and voltage pdfs, considering all voltage levels and unimodal and multimodal pdfs [23]. They were compared on the basis of the 95, 99 and maximum values; the approach that showed the best behavior was the Pearson approach. Therefore, the Pearson approach has been applied as a probabilistic technique for probabilistic harmonic power flow (PHPF).

8.3.2   Numerical applications

The probabilistic techniques analyzed above to perform PHPF have been applied to evaluate the voltage and current harmonics of the multiconvertor transmission test system in Figure 8.6. The busbar 3 is a linear load busbar and the busbars 4 and 5 are nonlinear load busbars.

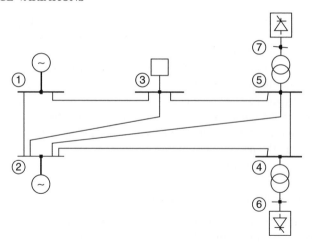

Figure 8.6   Five-bus transmission test system

To provide an acid test, the nonlinear loads are six-pulse bridge converters of high power, therefore producing very high distortion levels. The values of electrical parameters of generators, transformers and lines are reported in [16].

The following cases are discussed below:

- Case A: reference case;

- Case B: the case in which the standard deviation of active and reactive powers is increased by 200%.

In the reference case the standard deviations of nonlinear load powers range between 15 and 20% while the standard deviations of linear load powers are about 10%.

To compare all the probabilistic techniques mentioned above, besides the probability density functions of voltage and current harmonics, the following statistical measures have been calculated:

- expected value;

- 95% percentile;

- 99.9% percentile, assumed as maximum value.

With reference to Case A, Tables 8.1 and 8.2 show the mean errors on the expected value, 95% and 99.9% percentiles of voltage and current harmonics calculated with all the considered methods and for the harmonics of order 5, 7, 11 and 13.

The mean errors are defined as:

$$\varepsilon(SM) = \frac{\displaystyle\sum_{i=1}^{N} \frac{|SM_{\mathrm{NLMC},i} - SM_{\mathrm{L},i}|}{|SM_{\mathrm{NLMC},i}|}}{N} \tag{8.14}$$

Table 8.1   Case A: errors on voltage harmonics of order 5, 7, 11 and 13th

| H | Voltage harmonic mean errors | | | | | | | | |
| | Linear Monte Carlo | | | Convolution process | | | Pearson approach | | |
	$\varepsilon\%\ (\mu)$	$\varepsilon\%$ (pc95)	$\varepsilon\%$ (pc99.9)	$\varepsilon\%\ (\mu)$	$\varepsilon\%$ (pc95)	$\varepsilon\%$ (pc99.9)	$\varepsilon\%\ (\mu)$	$\varepsilon\%$ (pc95)	$\varepsilon\%$ (pc99.9)
5	1.201	0.878	0.824	1.201	2.476	4.762	1.201	1.380	3.910
7	1.550	5.115	7.027	1.550	8.220	13.18	1.550	4.620	9.140
11	0.416	1.326	1.475	0.417	38.35	58.40	0.416	1.540	3.440
13	0.436	0.994	4.335	0.436	6.755	11.10	0.436	1.650	2.380

where $SM_{NLMC,i}$, $SM_{L,i}$ are the value of the generic statistical measure (SM = expected value, percentile, etc.) of each voltage and current harmonic calculated via nonlinear Monte Carlo simulation (assumed as exact value) and via one of the other methods, respectively; N is the number of system busbars.

From the analysis of Tables 8.1 and 8.2 it follows that:

- both voltage and current harmonics calculated by the linear Monte Carlo and by the Pearson approaches are affected by some errors;

- voltage harmonics calculated by the convolution process are affected by nonnegligible errors;

- the errors generally increase with percentile value;

- there is no correlation between errors and harmonic order.

The nonnegligible errors on the voltage harmonics calculated by the convolution process are due to the presence of significant values of correlation factors between the powers of nonlinear loads.

With reference to the execution time, assuming the nonlinear Monte Carlo simulation time is equal to 1 p.u., using a PC Pentium III 800 MHz, the time taken for the different approaches is:

Table 8.2   Case A: errors on current harmonics of order 5, 7, 11 and 13

| H | CURRENT HARMONIC MEAN ERRORS | | | | | | | | |
| | Linear Monte Carlo | | | Convolution process | | | Pearson approach | | |
	$\varepsilon\%\ (\mu)$	$\varepsilon\%$ (pc95)	$\varepsilon\%$ (pc99.9)	$\varepsilon\%\ (\mu)$	$\varepsilon\%$ (pc95)	$\varepsilon\%$ (pc99.9)	$\varepsilon\%\ (\mu)$	$\varepsilon\%$ (pc95)	$\varepsilon\%$ (pc99.9)
5	0.855	0.987	12.28	0.885	10.520	14.559	1.029	0.370	0.690
7	5.765	7.095	5.821	1.814	9.530	12.654	1.770	4.410	8.800
11	2.976	3.060	3.509	0.676	8.155	11.649	0.675	1.850	4.080
13	5.541	5.757	5.421	0.976	2.305	3.1093	0.957	2.310	6.030

Table 8.3   Cases A and B: errors on all the voltage harmonics

CASE	Voltage harmonic mean errors								
	Linear Monte Carlo			Convolution process			Pearson approach		
	$\varepsilon\%$ (μ)	$\varepsilon\%$ (pc95)	$\varepsilon\%$ (pc99.9)	$\varepsilon\%$ (μ)	$\varepsilon\%$ (pc95)	$\varepsilon\%$ (pc99.9)	$\varepsilon\%$ (μ)	$\varepsilon\%$ (pc95)	$\varepsilon\%$ (pc99.9)
A	0.901	2.078	3.415	0.901	13.950	21.860	0.901	2.298	4.718
B	2.132	8.270	14.690	2.132	19.350	31.950	2.132	8.358	17.393

- for the linear Monte Carlo was 0.0091 p.u.;
- for the convolution process approach was 0.004 p.u.;
- for the Pearson approach was 0.0003 p.u.

It should be stressed that the above times must be considered with care because the algorithms and the software have not been optimized for computational speed.

In Tables 8.3 and 8.4 the errors on mean values and percentiles of both current and voltage harmonics are reported for Case B. To simplify the comparison, the errors reported in these tables are obtained extending Equation (8.14) to all the harmonics (the sum includes all the system busbars and all the considered harmonics); moreover, as reference, Case A data are also reported.

From the analysis of Tables 8.3 and 8.4, it arises that there is a no negligible influence of variance, mainly on voltage harmonics. It should be noted that the performance of linear models is acceptable within a range of uncertainty of random variation of the input data lower than the range which characterizes the behavior of the probabilistic load flow at fundamental frequency. This is presumably due to the significant nonlinear behavior of the harmonic power flow equations.

Finally, the case of multimodal distribution for voltage and current harmonics was also considered. The multimodal distribution was obtained forcing the nonlinear loads powers to be discrete distributions. As an example, Figure 8.7 shows the probability density functions of the fifth harmonic voltage at busbar 4.

Table 8.4   Cases A and B: errors on all the current harmonics

CASE	CURRENT HARMONIC MEAN ERRORS								
	Linear Monte Carlo			Convolution process			Pearson approach		
	$\varepsilon\%$ (μ)	$\varepsilon\%$ (pc95)	$\varepsilon\%$ (pc99.9)	$\varepsilon\%$ (μ)	$\varepsilon\%$ (pc95)	$\varepsilon\%$ (pc99.9)	$\varepsilon\%$ (μ)	$\varepsilon\%$ (pc95)	$\varepsilon\%$ (pc99.9)
A	3.784	4.225	6.759	1.108	7.627	10.490	1.108	2.235	4.900
B	4.290	6.014	8.904	1.477	5.499	6.513	1.502	3.925	8.083

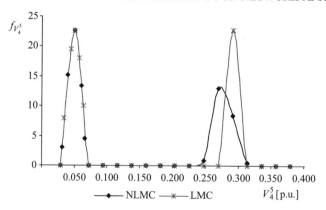

Figure 8.7   Probability density functions of the fifth harmonic voltage at busbar 4 in the case of multimodal distribution

8.4   Probabilistic iterative harmonic analysis

Probabilistic iterative harmonic analysis (PIHA) can be considered as an extension of the well known iterative harmonic analysis proposed by Yacamini, Arrillaga *et al.* [24, 25].

In the deterministic field, the iterative method is based on a procedure in which the converter equations, as well as the network equations, are solved in a successive manner. In solving the converter equations, the current harmonics injected into the network are estimated by assuming the voltage supply at each converter busbar to be fixed. When the network model is solved, the busbar voltage harmonics are evaluated by assuming that the current harmonics injected by converters are fixed. The evaluation of the current harmonics is effected by making use of one of the available converter models; the voltage harmonics at the network busbars can be calculated by means of Equation (8.1).

The iterative procedure requires the preliminary knowledge of the steady-state system operation at the fundamental frequency and this is obtained by performing a single-phase or three-phase AC/DC power flow.

8.4.1   Theoretical aspects

Since an iterative procedure is included in the solution, expressions to describe the harmonic parameters are not available and it is necessary to resort to the Monte Carlo technique to achieve a probabilistic method [26, 27]. As the flow chart in Figure 8.8 shows the PIHA requires, as first step, an AC/DC load flow. In this step the determination of the input random variables is obtained by means of suitable estimation codes which take into account the pdfs of the input data.

From a knowledge of the operating conditions of the system at the fundamental frequency, the method proceeds with the evaluation of the current harmonics of each converter and of the voltage harmonics at the busbars of the network. Such an estimation is brought about by solving the same models used in the deterministic method; it is repeated until the difference between the voltage vectors obtained on successive cycles is below a predetermined value.

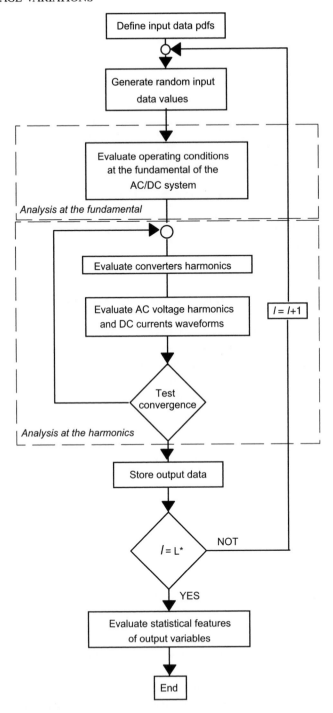

Figure 8.8   Monte Carlo simulation procedure for probabilistic iterative harmonic analysis

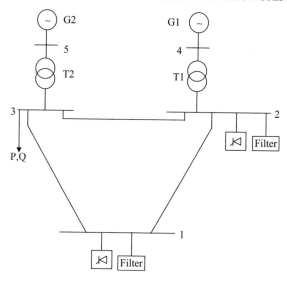

Figure 8.9   Three-bus transmission test system

The Monte Carlo simulation allows us to consider the system configuration among the input random variables. It is a discrete random variable and the sample space of all the possible network configurations is generally constituted by a few numbers of elements.

8.4.2   Numerical applications

The probabilistic IHA has been applied to evaluate the voltage harmonics of the multiconvertor transmission test system in Figure 8.9. The busbar 3 is a linear load busbar and the busbars 1 and 2 are nonlinear load busbars. At each bus with a converter, a filtering system with branches tuned at the fifth, seventh, 11th and 13th harmonics is provided. The values of electrical parameters of generators, transformers, lines and filtering systems and the values of the characteristic parameters of the input data pdfs are reported in [27].
    The following cases are discussed:

- reference case (Case A), where the active and reactive powers of linear loads were assumed to be uncorrelated, Gaussian and where the direct currents of converters were considered beta distributed;

- the case of outage of the 11th harmonic filter at bus 1 (Case B);

- the case in which the network configuration is considered among the input random variables (Case C), with the following configurations considered: Configuration 1: as in Figure 8.9; Configuration 2: as in Figure 8.9 without line 1–2; Configuration 3: as in Figure 8.9 without line 1–3; Configuration 4: as in Figure 8.9 without line 2–3.

In Figure 8.10 the estimated 11th and 17th voltage harmonic pdfs at busbar 1 are shown for all the considered cases. It should be noted that:

- the outage of a filter branch causes a considerable increase in the corresponding harmonic voltage;

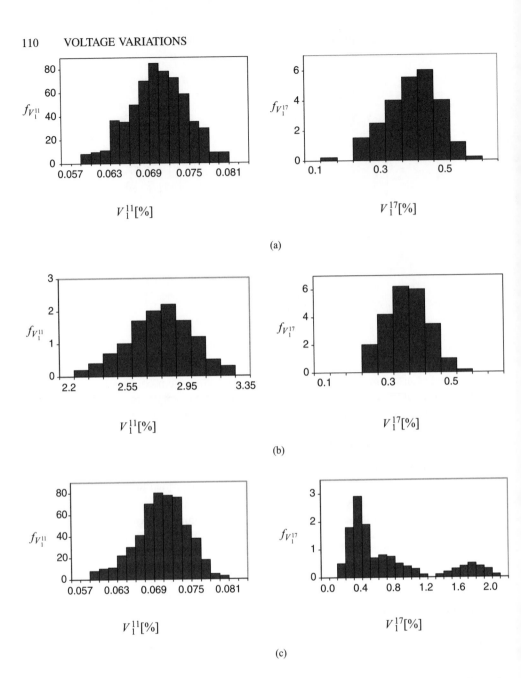

Figure 8.10   11th and 17th harmonic voltage probability density functions: (a) Case A, (b) Case B and (c) Case C

- the 17th harmonic voltage in Case C is characterized by a bimodal pdf so that even if its mean value is only 0.74%, the maximum value is about 2% as a consequence of a high value of standard deviation;

- the bimodal behavior of pdfs in Case C is due to the high values of the 17th harmonic voltages in configuration 2 being all other configurations characterized by pdfs whose maximum values are always lower than 1%.

8.5   Conclusions on methods for network analysis

The following main conclusions can be drawn. The direct methods require considerably less computational effort, mainly when harmonic currents are present at only one bus of the power system in the study, but they guarantee high accuracy results only in the presence of low voltage distortion levels. Moreover, when this method is applied, attention should be paid in applying the central limit theorem.

The probabilistic harmonic power flow in its present form can be applied to balanced power systems. The computational efforts and result accuracy of this method depend on the probabilistic technique involved. In particular:

- the Monte Carlo simulation procedure applied to the nonlinear equation system (nonlinear Monte Carlo simulation) guarantees high result accuracy but requires very high computational effort;

- the simplified probabilistic techniques based on linearized models (linear Monte Carlo simulation, convolution approach, Pearson's approach) guarantee a significant reduction of computational effort but not an acceptable result accuracy in the case of both significant values of input data pdfs variance and multimodal distributions.

The probabilistic iterative harmonic analysis can be applied to both balanced and unbalanced power systems. It is characterized by high result accuracy but requires high computational effort, mainly in the case of unbalanced power systems.

It should be noted that nowadays massive computation is becoming accessible with new machines and configurations (parallel/distributed processing environment) so that the computational efforts required by Monte Carlo simulations do not appear to be such insuperable handicaps as in the past. Moreover, variance reduction techniques could be applied to reduce the computational effort.

References

[1] IEC 1000-3-6 *Assessment of Emission Limits for Distorting Loads in MV and HV Power Systems*, 1996.

[2] EN 50160, *Voltage Characteristics of Electricity Supplied by Public Distribution Systems – European Standard, CLC, BTTF 68-6*, November 1994.

[3] IEEE Standard 519 Draft, *Recommended Practices and Requirements for Harmonic Control in Electric Power Systems*, 20 February 2005.

[4] G. Carpinelli and P. Ribeiro, *IEEE Standard 519 Revision: The Need for Probabilistic Limits of Harmonics* – IEEE PES Summer Meeting, Vancouver, Canada, July 2001.

[5] P. Caramia, G. Carpinelli, A. Russo, P. Varilone and P. Verde, 'An Integrated Probabilistic Harmonic Index', IEEE PES Winter Meeting, New York, USA, January 2002.

[6] P. Caramia, G. Carpinelli, A. Cavallini, G. Mazzanti, G.C. Montanari and P. Verde, 'An Approach to Life Estimation of Electrical Plant Components in the Presence of Harmonic Distortion', IEEE PES International Conference ICHQP, Orlando, USA, October 2000, pp. 887–891.

[7] IEEE Task Force on Harmonics Modeling and Simulation: Modeling and Simulation of the Propagation of Harmonics in Electric Power Networks – Part I: Concepts, Models and Simulation Techniques, *IEEE Transactions on Power Delivery*, **11**, 1996, 452–465.

[8] Probabilistic Aspects Task Force of Harmonics Working Group: Time-Varying Harmonics: Part II Harmonic Summation and Propagation, *IEEE Transactions on Power Delivery* **17**, 2002, 279–285.

[9] W. K. Chang and W. M. Grady, 'Minimizing Harmonic Voltage Distortion with Multiple Current-Constrained Active Power Line Conditioners', *IEEE Transactions on Power Delivery*, **12**, 1997, 837–843.

[10] G. Carpinelli, A. Russo, M. Russo and P. Verde, 'On the Inherent Structure Theory of Network in Presence of Harmonics', *IEE Proc. Gen., Transm. and Distr.*, **145**, 1998, 123–132.

[11] Y. J. Wang, L. Pierrat and L. Wang, 'Summation of Harmonic Produced by AC/DC Static Converters with Randomly Fluctuating Loads', *IEEE Transactions on Power Delivery*, **9**, 1994, 1129–1135.

[12] D. Xia and G.T. Heydt, 'Harmonic Power Flow Studies: Part I and II', *IEEE Trans. Power Apparatus and Systems*, **101**, 1982, 1257–1280.

[13] G. Carpinelli, T. Esposito, P. Varilone and P. Verde, 'First-order Probabilistic Harmonic Power Flow', *IEE Proc. Gen., Transm. and Distr.*, **148**, 2001, 541–548.

[14] G. Carpinelli, T. Esposito, P. Varilone and P. Verde, 'Probabilistic Harmonic Power Flow for Percentile Evaluation', IEEE PES 2001 Canadian Conference on Electrical and Computer Engineering, Toronto, Canada, May 2001, pp. 831–838.

[15] T. Esposito, A. Russo and P. Varilone, 'Probabilistic Modeling of Converters for Power Evaluation in Nonsinusoidal Conditions', IEEE PES 2001 Canadian Conference on Electrical and Computer Engineering, Toronto, Canada, May 2001, pp. 1059–1066.

[16] P. Caramia, G. Carpinelli, T. Esposito and P. Varilone, 'Evaluation Methods and Accuracy in Probabilistic Harmonic Power Flow', International Conference on PMAPS, Naples, Italy, September 2002.

[17] R. Y. Rubinstein, *Simulation and the Monte Carlo Method*, John Wiley & Sons, Inc., New York, USA, 1981.

[18] G. J. Anders, *Probability Concepts in Electric Power Systems*, John Wiley & Sons, Inc., New York, USA, 1990.

[19] M. V. F. Pereira and N. J. Balu, 'Composite Generation/Transmission Reliability Evaluation', *Proc. of IEEE*, **80**, 1992, 470–491.

[20] P. A. Stuart and K. J. Ord, *Kendall's Advanced Theory of Statistics*, Sixth edition. Vol. I, Distribution Theory, E. Arnold (ed.), London, 1994.

[21] R. N. Allan, A. M. Leite da Silva and R. C. Burchett, 'Evaluation Methods and Accuracy in Probabilistic Load Flow Solutions', *IEE Transactions on Power Apparatus and Systems*, **100**, 1981, 2539–2546.

[22] G. J. Hahn and S. S. Shapiro, *Statistical Models in Engineering*. John Wiley & Sons, Inc., New York, USA, 1967.

[23] T. Esposito and P. Varilone, 'Some Approaches to Approximate the Probability Density Function of Harmonics', IEEE Proceedings of 10th ICHQP 2002, Rio De Janeiro, Brazil, October 2002.

[24] R. Yacamini and J. C. de Oliveira, 'Harmonics in Multiple Convertor Systems: a Generalized Approach', *IEE Proc. Part B*, **127**, 1980, 96–106.

[25] J. Arrillaga, J. F. Eggleston and N. R. Watson, 'Analysis of the AC Voltage Distortions Produced by Converter-fed DC Drives', *IEEE Trans. on IA*, **21**, 1985, 1409–1417.

[26] G. Carpinelli, F. Gagliardi and P. Verde, 'Probabilistic Modelings for Harmonic Penetration Studies', Fifth International Conference on Harmonics in Power Systems, Atlanta, USA, September 1992, pp. 35–40.

[27] P. Caramia, G. Carpinelli, F. Rossi and P. Verde, 'Probabilistic Iterative Harmonic Analysis', *IEE Proc. Gen., Transm. and Distr.*, **141**, 1994, 329–338.

9

Probabilistic modeling of harmonic impedances

R. Langella and A. Testa

In the previous chapter models for probabilistic network analysis were considered assuming, for simplicity sake, the network component impedances remain constant. This assumption can be justified by the complexity of taking into account the time variance of the network configuration and the existing correlations present; the assumption is near to be verified in short-term scenarios. In this chapter, the single node analysis is considered in long-term scenarios so that the dimensions of the problem are reduced and impedance variability cannot be ignored. The probabilistic aspects of harmonic impedances are introduced with reference to a case study.

9.1 Introduction

Considerable attention has been devoted to probabilistic approaches for the prediction of the harmonic currents injected by loads and of the corresponding nodal harmonic voltages [1–3]. In this framework, minor importance has been devoted to the variability of the system impedances [4–6].

The network component parameter uncertainty and/or the network, supply and load configuration changing may cause significant variations in the amplitude and phase of the nodal equivalent impedances. Parameter uncertainty is a consequence of their variability in time and with temperature or of the inherent difficulty in forecasting actual values in the designing stage for a new plant. Change is a consequence of obvious operational needs; in particular, the effect of supply or load changing is relevant mainly when resonances are present. Under these conditions, a voltage distortion evaluation [7–8], based on the distorted load current injection, even if probabilistically modeled, but referring to a single reference condition for impedances, may produce a result that is incorrect and sometimes unacceptable.

Time-Varying Waveform Distortions in Power Systems   Edited by Paulo F. Ribeiro
© 2009 John Wiley & Sons, Ltd

In the following sections the effects of correlations among currents, impedances and voltages actually existing in a system are discussed. Then, the system impedance probabilistic model developed in [6], and taking into account parameter uncertainty as well as load and supply configuration variability, is recalled. Finally, the effects of modeling accuracy are discussed with reference to three methods for evaluating voltage distortion: the first does not take into account impedance variability; the second neglects the actual correlation effects between currents and impedances; the third takes into account the actual behavior of currents and impedances.

9.2   General considerations

International standards suggest the study of frequency dependent behavior of the supply system impedances in opportune conditions. The most critical conditions are related to the presence of strong parallel resonances causing very high values of the impedances around given frequencies. Resonance condition characteristics are determined in distribution networks by different factors among which the power factor correction capacitors, together with the load demand, play a very relevant role.

Thus, assuming invariant reference conditions for the study of the behavior of the supply impedance, means ignoring the variability of the relevant quantities with time. This assumption reflects on both the resonance frequency positions and the impedance amplitudes at resonant frequencies. On the other hand, it is well known that load variability with time requires tuning of the capacitors for the power factor correction.

Then, assuming impedance variability uncorrelated with current and voltage variability means not taking into account dependences which couple the behavior of all the involved quantities with time. Having again in mind the power factor correction, the higher the load demand is the higher the capacitance value; moreover, the lower the resonant frequencies, the higher are the dumping effects of the loads acting as ohmic-inductive components in parallel with the supply impedances. Furthermore, the higher the whole load demand at a given node is, the higher the probability of having higher injected distorted currents and higher voltage background distortion.

The main links existing between the values assumed by the final quantity of interest (the harmonic voltage V^h) and the values of the quantities of direct influence and by quantities of indirect influence are reported in Figure 9.1. The main direct influence quantities are: the

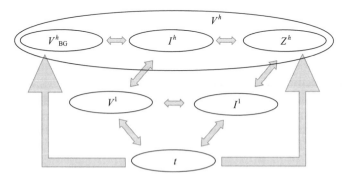

Figure 9.1   Main links existing among the harmonic voltage V^h, and quantities of direct influence (harmonic current I^h, supply impedance Z^h, background voltage V^h_{BG}) and of indirect influence (V^1, I^1, ... and time t)

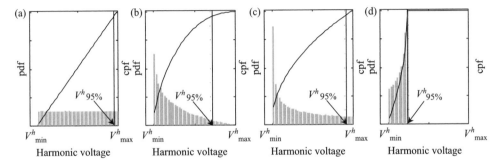

Figure 9.2   Pdf and cpf of harmonic voltage for harmonic current and impedance uniformly distributed assuming: (a) nonvariability of the impedance, (b) Z^h and I^h independence, (c) Z^h and I^h direct correlation and (d) Z^h and I^h cross correlation

harmonic current I^h, the supply harmonic impedance Z^h and background voltage V^h_{BG}. The main indirect influence quantities are: the total active and reactive load current I^1 at fundamental frequency, the supply fundamental voltage V^1 ... and the time t. Of course, in real cases further links do exist and their importance depends on the specific situation being analyzed.

To have an idea also of the quantitative influence on the statistic parameters of interest, different correlation conditions can be analyzed with reference to a simple example. Both impedance and current are assumed to be uniformly distributed in a range of values between a minimum and a maximum value, $Z^h_{min} = 0$, $Z^h_{max} = 1$ and $I^h_{min} = 0$ and $I^h_{max} = 1$ p.u., respectively; the absence of background voltage distortion is also hypothesized. The harmonic voltage behavior can be obtained simply by applying Ohm's law and utilizing the Z^h and I^h determinations obtained by means of a Monte Carlo simulation routine.

Figure 9.2 reports the distributions of the harmonic voltage obtained in four different reference conditions:

- constant value for the impedance, which means for example:

$$Z^h_{max} = Z^h = 1 \text{ p.u.};$$

- independence, which means that:

$$P(Z^h|I^h) = P(Z^h) \quad \text{and} \quad P(I^h|Z^h) = P(I^h),$$

for each determination of Z^h and I^h;

- 'direct' dependence given by:

$$P(Z^h|I^h) = 1 \quad \text{if} \quad \frac{Z^h - Z^h_{min}}{Z^h_{max} - Z^h_{min}} = \frac{I^h - I^h_{min}}{I^h_{max} - I^h_{min}}, \quad \text{otherwise} = 0,$$

that means the determination of Z^h has a distance from its minimum value, which is proportional to the distance of the corresponding determination of I^h from its minimum value;

- 'cross' dependence given by:

$$P(Z^h|I^h) = 1 \quad \text{if} \quad \frac{Z^h - Z^h_{min}}{Z^h_{max} - Z^h_{min}} = \frac{I^h_{max} - I^h}{I^h_{max} - I^h_{min}}, \quad \text{otherwise} = 0,$$

that means the determination of Z^h has a distance from its minimum value, which is proportional to the distance of the corresponding determination of I^h from its maximum value.

The observations that can be made are trivial and the differences among the shapes are very evident; 95 % percentile values are also very different (the distances from the minimum values respectively: 0.95, 0.70, 0.90 and 0.25 times p.u.).

Finally the importance, in principle, of taking into account not only impedance variability but also the actual correlation between impedances and currents results, is evident. Of course, the higher the variations of the impedance are, the higher their effects, hence the importance of taking them into account. An idea of impedance variability in a realistic situation is given below.

9.3 Impedance variability

In [9], with reference to the IEEE industrial test system proposed by the IEEE Task Force on harmonic modeling and simulation in [10], the impedance variability has been analyzed. The causes of impedance variability are: the system component parameter uncertainty, the load time variability and the supply time variability.

The industrial system considered consists of 13 buses and is representative of a medium-sized industrial plant. Due to the balanced nature of the system, only the unifilar scheme is presented (Figure 9.3); the positive sequence rated data are provided in [6] and are referred to 60 Hz fundamental frequency and to the 13.8 kV voltage.

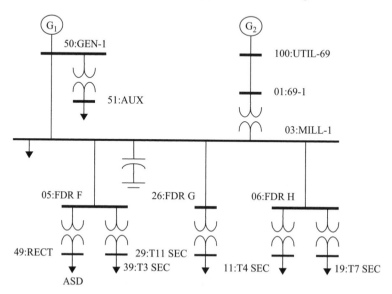

Figure 9.3   The IEEE industrial test system

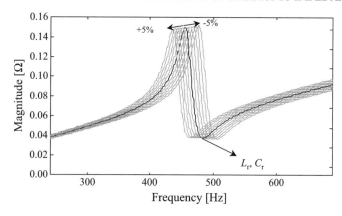

Figure 9.4 System impedance magnitude versus frequency (zoom from 240 to 680 Hz), at node 49:RECT, for different values of L and C

The plant is fed from a utility supply at 69 kV. Moreover, a local generator is directly connected to the local plant distribution system operating at 13.8 kV. The automatic tuning of the power factor correction capacitor is assumed.

Explicit reference is made to the self-impedance of the bus 49:RECT supplying the unique nonlinear load. Here and in the following sections, the impedance values are all referred to the 480 V level. The behavior of this impedance, obviously, can give an idea of what can happen to all the network buses except for the generation buses. Adequate equivalents to represent the system components for harmonic and interharmonic penetration analyses are needed. The authors refer to classical models adopted in literature [9] and fully described in [6] for supply, lines and cables, transformers and loads.

9.3.1   Parameter uncertainty

The system component parameters L and C value uncertainty has been modeled in a Gaussian scenario. In particular, the majority of determinations (99.57%) of L and C fall in the intervals $(L_r-5\%L_r, L_r + 5\%L_r)$ and $(C_r-5\%C_r, C_r + 5\%C_r)$, respectively, L_r and C_r being the rated values. The effects on the bus 49:RECT impedance magnitude are reported in Figure 9.4, for different values of L and C parameters versus frequency (zoom from 240 to 680 Hz).

9.3.2   Load time variability

In order to take into account the effects of the load variations, all the loads present in the system have been modeled by means of a unique time-varying equivalent circuit as described in [6]. The parameter values of the equivalent load are set according to the values of the power consumption. The capacitor varies its capacitance exactly compensating for the reactive power requested by the equivalent load.

For each time interval of a typical working day, Table 9.1 reports the active power consumption, the power factor and their corresponding frequencies of occurrence, which can be considered as the probability of occurrence in a generic instant of a working day.

Table 9.1   Active power consumption, power factor and probability for different time intervals

	T_1	T_2	T_3	T_4	T_5	T_6
P [p.u.]	0.2	0.5	0.8	0.7	1.05	0.6
$\cos\varphi$	0.83	0.82	0.77	0.83	0.75	0.80
$P(T_i)$	0.459	0.125	0.125	0.083	0.125	0.083

Figure 9.5 shows the effects of the different power consumption on the system equivalent impedance as seen from the bus 49:RECT. It can be observed that:

- the resonant frequency varies over a wide range from a minimum value (450 Hz) to a maximum value (1120 Hz); such a variability depends basically on the adjustments of the capacitance combining with the equivalent inductance at the node 03:MILL-1 practically fixed by the generator equivalent inductances;

- the impedance amplitude peak grows, as expected, as the load power consumption decreases;

- the load variability strongly influences the system impedance at a given frequency, in a wide range of frequencies (e.g. from 500 to 800 Hz).

9.3.3   Supply time variability

Supply characteristics and configuration may vary in time, for exampleaccording to duty cycles depending on different economic and technical reasons. For the sake of simplicity outages are here considered as the only cause of variability. Table 9.2 contains the local generator and the utility supply reliability parameters [11], in terms of failure rate, λ, unavailability, U, and of probability of outage in a generic time instant, $P(G_{iOFF})$.

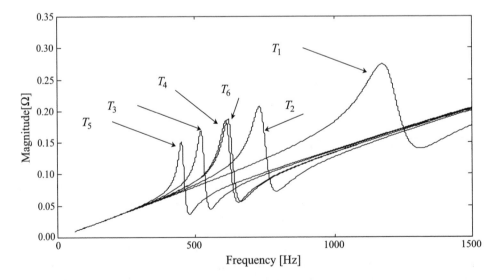

Figure 9.5   System impedance magnitude versus frequency, at node 49:RECT, for different power consumption values

Table 9.2   Supply reliability parameters

	λ [failures/year]	U [h/year]	$P(G_{iOFF})$
G_1	1.64	2.4	$4 \cdot 10^{-3}$
G_2	4.5	32	$0.3 \cdot 10^{-3}$

Figure 9.6 shows the effects of the different supply configurations on the system equivalent impedance as seen from the bus 49:RECT. It can be seen how the supply configuration variability influences the equivalent system impedance much more than the other phenomena. Anyway, it must be considered that the probability of occurrence of the events $G_{1OFF}G_{2ON}$ and $G_{1ON}G_{2OFF}$ are very small so, in practice, their mean influence is not relevant.

9.3.4   Model

In order to obtain the probabilistic model of the system impedance, Z, all the impedance variability causes, analyzed in Sections 9.3.1 to 9.3.3, must be taken into account. A possibility is to take advantage of the Bayes formula [12] as described in [6].

In [6] probabilistic modeling was made for different time scenarios: (i) the entire week (week), (ii) the working day (day) and (iii) the working-hours (w-hours) of the working day. For the weekend, the power consumption was assumed equal to that of time interval T_1 of a working day.

In the following pdf diagrams, the abscissa axis is divided into a finite number of equal classes from 0 to 100% of the maximum value, for each impedance. Figure 9.7 shows the probability density function (pdf) and cumulative probability function (cpf) of the system impedance magnitude for different frequencies referred to only working day scenarios. In particular Figures 9.7(a) and (b) refer to harmonic frequencies (fifth and 11th) while Figure 9.7(c) refers to the interharmonic frequency of 636 Hz. It is possible to distinguish the contribution of the different time intervals, T_i, of the day: the color characterizing each time interval is chosen so that the darker the color is the lower the power load consumption. It is evident how the fifth harmonic impedance has monomode behavior, with a very low variance,

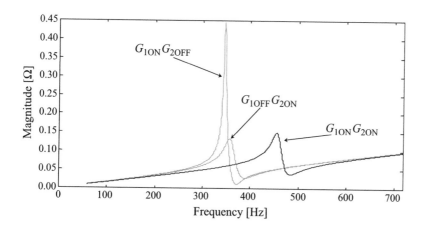

Figure 9.6   *System impedance magnitude versus frequency, at node 49:RECT, for different supply configurations*

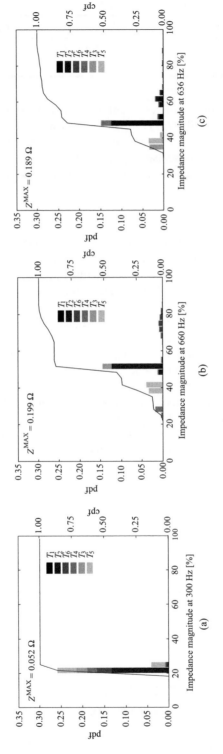

Figure 9.7   *Pdf and cpf of the system impedance magnitude, at bus 49:RECT, [%] of its maximum value during a working day: (a) fifth harmonic, (b) 11th harmonic and (c) interharmonic at 636 Hz*

Table 9.3   49:RECT bus impedance (whole week)

	Mean	Max	$Z_{95\%}$	Rated
Z_{300Hz} [Ω]	0.048	0.052	0.049	0.049
Z_{660Hz} [Ω]	0.106	0.199	0.152	0.087
Z_{636Hz} [Ω]	0.102	0.189	0.131	0.083

due to its distance (left side) from the shifting resonance peak: also a deterministic modeling could give equivalent information.

The 11th harmonic impedance presents a quadrimode behavior characterized - obviously – by a great dispersion – and reaches the greatest values for the medium load condition (T_4); this can be explained (see Figure 9.5) considering that the equivalent impedance resonance peak approaches and crosses the frequency of interest. The interharmonic impedance behavior is similar to that outlined for the 11th harmonic due to the short distance (24 Hz) between the two frequencies.

Table 9.3 reports some statistical parameters for the same (aforementioned) frequencies but with reference to a whole week. In the last column of the table the rated value is also reported, defined as that characterizing the IEEE test system in the condition considered in [10]. The table confirms the considerations developed for Figure 9.7.

9.4   Effects on distortion modeling results

The results of the previous section together with the data characterizing the ASD, which is the harmonic source connected to the bus 49:RECT, are utilized to perform the analysis of the plant harmonic voltage distortion with methods assuming different representations of time variability and correlations among currents and impedances.

It has been assumed that the ASD changes its working point according to the same time intervals utilized for the plant power consumption reported in Table 9.1. The corresponding asynchronous motor frequencies and active power consumption vary as described in [6]. This shows evident strong correlation between current and impedance.

9.4.1   Distortion models

Three different methods were utilized to evaluate the harmonic and interharmonic voltages at bus 49:RECT:

- the first applies Ohm's law to all the determinations of Z and I and utilizes Bayes relations to obtain the voltage probability, taking into account the actual behavior of the involved quantities;

- the second is the simplified procedure which takes into account impedance variability but applies the product $V = ZI$ to the corresponding statistical parameters (mean, max, etc.), so ignoring the actual correlation effects;

- the third takes into account only current variability, assuming the rated value for the impedances.

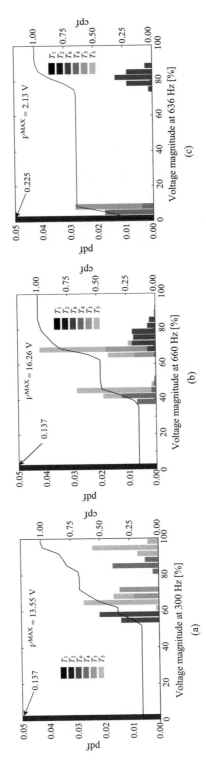

Figure 9.8   Pdf and cpf of the voltage magnitude, at bus 49:RECT, [%] of its maximum value during a working day by means of the first procedure: (a) fifth harmonic, (b) 11th harmonic and (c) interharmonic at 636 Hz

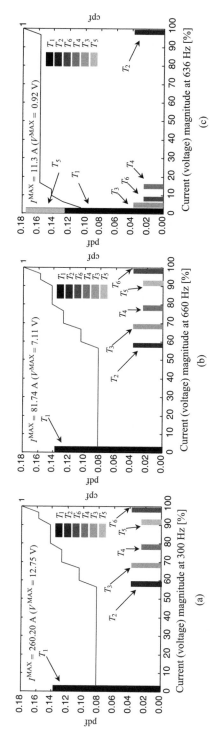

Figure 9.9   Pdf and cpf of the current magnitude (and of voltage magnitude for constant impedances of rated values reported in Table 9.3), at bus 49:RECT, [%] of its maximum value during a working day: (a) fifth harmonic, (b) 11th harmonic and (c) interharmonic at 636 Hz

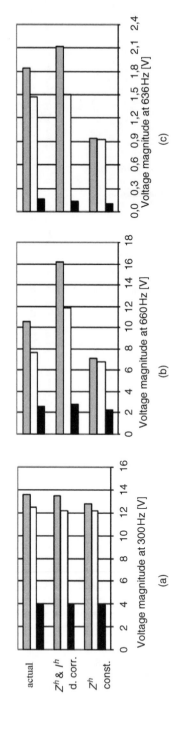

Figure 9.10   Comparison of the statistic parameters of the voltage at bus 49:RECT during a whole week (■ max □ 95% ■ mean), obtained 'Z^h constant' by the first method, 'Z^h and I^h directly correlated' by the second method and 'actual' by the third method: (a) fifth harmonic, (b) 11th harmonic and (c) interharmonic at 636 Hz

9.4.2   Results

For the sake of brevity, the resulting voltages reported in this section once again refer only to 49:RECT bus. Figure 9.8 shows the pdf and cpf of the harmonic and interharmonic voltage magnitudes, as a percentage of their maximum values, in a working day, obtained by means of the first procedure that gives the actual values due to the absence of simplifying assumptions. The pdf and cpf cannot be obtained by the second method. Figure 9.9 shows the pdf and the cpf of the 300, 660 and 636 Hz currents obtained referring to a working day. This figure can be still used for the results of the third method, considering that the voltages are obtained from the currents simply applying a constant scale factor.

The differences between the histograms of Figure 9.8 and Figure 9.9 are evident. It is clear that the pdfs of Figure 9.8 are characterized by a greater dispersion of the results; the values of the main statistical parameters are heavily influenced.

A comparison between the values of the statistic parameters, with reference to a whole week, obtained by means of the three methods, can be done with reference to Figure 9.10. It suggests the following considerations:

- mean values are close enough for the harmonics (no more than 20 % of difference), very different for the interharmonic (reaching 60 % of differences);

- the maximum value for the 11th harmonic (and for the interharmonic at 636 Hz) is overestimated by the second and underestimated by the third procedure;

- 95% percentile values behave similarly to the maximum values.

9.5   Conclusions

The main outcomes of this chapter are:

- it is necessary to take into account system impedance variability;

- procedures taking into account impedance time variability but neglecting the actual correlation between impedances and currents, that are particularly relevant in an industrial system, give unreliable results;

- assuming direct correlations between impedances and currents, in particular the same behavior with time – as done with the second method – gives precautionary (sometimes too pessimistic) results;

- interharmonics require the most accurate analysis also due to their ability in producing immediately perceptible damages (light flicker) once a certain level has been reached.

References

[1] P. Marino, F. Ruggiero and A. Testa, 'On the vectorial summation of independent random harmonic components', Seventh ICHQP, Las Vegas, USA, 1996.

[2] R. Langella, P. Marino, F. Ruggiero and A. Testa, 'Summation of random harmonic vectors in presence of statistic dependences', Fifth PMAPS, Vancouver, Canada, 1997.

[3] A. Cavallini, R. Langella, F. Ruggiero and A. Testa, 'Gaussian Modeling of Harmonic Vectors in Power Systems', Eighth ICHQP, Athens, Greece, 1998.

[4] P. Caramia, G. Carpinelli, F. Rossi and P. Verde. 'Probabilistic iterative harmonic analysis of power systems', *IEE Proc. on General Transmission Distribution*, **141**, 1994, 329–338.

[5] R. Carbone, D. Castaldo, R. Langella, P. Marino and A. Testa, 'Network Impedance Uncertainty in Harmonic and Interharmonic Distortion Studies', IEEE Budapest Power Tech '99, Budapest, Hungary, 1999.

[6] R. Carbone, D. Castaldo, R. Langella and A. Testa, 'Probabilistic Modeling of Industrial Systems for Voltage Distortion Analysis', Ninth ICHQP, Orlando, Florida, USA, 2000.

[7] IEC 1000-3-6, *Electromagnetic compatibility, assessment of emission limits for distorting loads in MV and HV power systems*.

[8] IEEE Standard 519-1992, *IEEE Recommended Practices and Requirements for Harmonic Control in Electrical Power Systems*, 1993.

[9] D. Castaldo, R. Langella and A. Testa, 'Probabilistic Aspects of Harmonic Impedances', IEEE Winter Power Meeting 2002, New York, USA, January 2002.

[10] Task Force on Harmonics Modelling and Simulations' 'Test Systems for Harmonics Modelling and Simulations', *IEEE Transactions on Power Delivery*, **142**, 1999 579–587.

[11] G. Carpinelli, V. Mangoni, S. Oliva, M. Russo and A. Testa 'Scelta dello schema elettrico e funzionamento di un sistema elettrico industriale con cogenerazione', Autoproduzione e Cogenerazione nell'Industria, Cassino, Italy, February 1995

[12] A. Papoulis, *Probability, Random Variables and Stochastic Processes*, third edition, McGraw Hill, New York, USA, 1991.

Part IV

STANDARDS AND MEASUREMENT ISSUES

In this part standards and measurement issues are considered. In Chapter 10 time-varying and probabilistic considerations are taken into account when setting harmonic limits for harmonic standards such as IEEE 519 and IEC 61000-3-6. These limits are usually expressed in the form of different indices which are covered in Chapter 11. The indices can refer either to a single customer point-of-common coupling or to a segment of the distribution system or, more generally, to the utility's entire distribution system. Chapter 11 deals with both single site and system probabilistic harmonic indices.

Chapter 12, on the other hand, presents a brief description of measurement techniques used to obtain the harmonic content of current or voltage signals. First, the front-end signal chain is outlined, the output of which is assumed to be a digital data stream. Then three methods are discussed to produce the harmonic signal content based on this digital, sampled data stream.

The first method is the well-known fast Fourier transform (FFT). The second one is a model-based technique. These two methods are essentially finite impulse response (FIR) and infinite impulse response (IIR) filters, an important property to consider when applying them to nonstationary signals (i.e. signals in which the harmonic content varies with time). The third method uses wavelets.

10

Time-varying and probabilistic considerations: setting limits

T.H. Ortmeyer, W. Xu and Y. Baghzouz

10.1 Introduction

Harmonic limits have been established and widely applied over the last 20 years. The primary harmonics standards are IEEE 519 and IEC 61000-3-6 [1, 2]. IEEE 519 was first published in 1981, was substantially revised in 1992, and is currently undergoing another substantial revision. IEC 61000-3-6 has undergone a similar cycle of development. The two standards are currently in widespread use throughout the world, and the steady-state limits they impose are not particularly controversial.

Harmonics are, in theory, steady-state quantities. Power systems, however, are continually changing, and harmonic levels vary regularly. Figure 10.1 shows measured total harmonic distortion (THD) measured on a 138 kV bus. These levels are generally within the harmonic limits, except for a 5.3 min excursion and a later 3.2 min excursion above 2%.

At this time, it is unclear how to deal with these short-term excursions above the steady-state harmonic limits. While it is widely recognized that some level of short-term excursions above the limits should be allowable, the area remains under active research, and there is at this point no consensus on short-term harmonic limits.

This chapter provides a perspective on limiting factors on short term harmonic levels.

10.2 Harmonic definition

Clearly, rapid variations in distortion levels fall into the category of transients rather than harmonics. As the speed of the variations decrease, at some point the distortion is more clearly described as being a set of harmonics of varying magnitude rather than a transient phenomena.

Time-Varying Waveform Distortions in Power Systems   Edited by Paulo F. Ribeiro
© 2009 John Wiley & Sons, Ltd

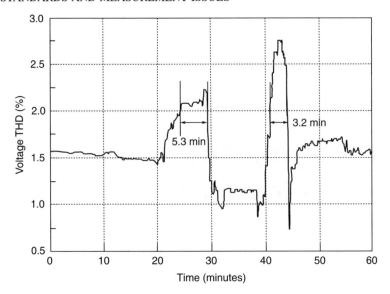

Figure 10.1   138kV bus voltage THD as a function of time

IEC 61000-4-3 sets a window of 3 s as being the shortest window over which harmonics should be measured, and IEC 61000-3-2 allows a 50% increase in distortion level for this period, for certain types of equipment. IEEE 519-1992 does not address this measurement issue, but does recognize that short-term excursions above the limits are to be expected.

10.3   Harmonic effects

The fundamental principles guiding harmonic standards are to:

- avoid system and load damage and disruption due to high harmonic levels,
- limit harmonic losses to an acceptable level,
- when mitigation is necessary, find an economical and equitable solution.

In developing short-term harmonic limits, the primary consideration will be the first item, avoiding damage or disruption of system and load equipment.

Harmonic effects are tied to either voltage magnitudes or current flows. One primary impact of voltage magnitude on system equipment is the insulation issue – and electrical overstress, subsequent corona and breakdown are generally high-speed phenomena. Load disruption is another primary impact of voltage distortion. Load disruption effects are also short-term in nature, with nuisance tripping having been reported to occur regularly as distortion levels approach 10%. The impact of overcurrent on system equipment, on the other hand, is most directly tied to I^2R heating, and resulting temperature increases, which is a slow phenomena. Load performance is much more complex – distorted voltage applied to a load causes distorted current draw. Heating effects on induction and synchronous machines are a primary concern, although torque ripple and increased noise also can be a limiting factor. In electronic devices, high-frequency ripple and resulting effects on input filters can raise both heating and insulation overstress concerns. On the other hand, certain waveforms can cause power supplies to prematurely fail to meet their specifications at low voltage, for example.

Thermal time constants on power equipment are relatively large, and are an obvious choice for appropriate time limitations for short-term harmonics. Voltage overstress and load disruption, being fast phenomena, can theoretically provide upper limits on acceptable levels of short-term harmonics. The problem in applying these concepts is that the power system is an amalgamation of equipment with widely varying thermal time constants and insulation levels.

10.4   Proposed limit methodologies

Figure 10.3 shows a histogram and cumulative distribution function for a typical harmonic voltage measurement, in this case, for a week-long measurement. While these data yield information on the time duration at a given distortion level, they do not contain information on the duration of individual events at that distortion level. Therefore, data presented in this form do not give full information for determining acceptability for short-term bursts of harmonics. An alternate method for displaying harmonic data is presented in Figure 10.2, where curves are presented for both the maximum lengths observed for individual bursts, and the total cumulative length of bursts during a measurement period. This curve also includes a conceptual limit for short-term harmonic limits.

It seems clear that there is a need for limits on harmonic bursts for individual events. A statement that harmonic limits should be met 95% of the time is insufficient in that it does not

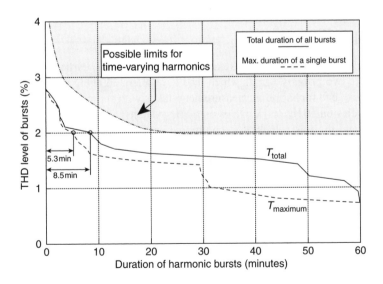

Figure 10.2   Drawing of cumulative distribution (T_{total}) and the maximum duration of individual burst ($T_{maximum}$) for a harmonic measurement. The curve also includes a conceptual limit for short-term harmonic levels

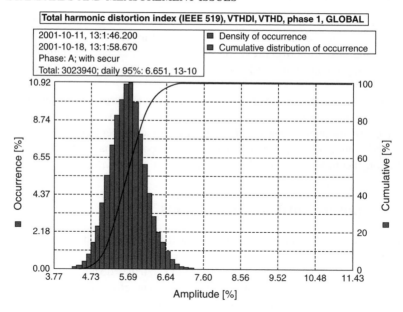

Figure 10.3   Histogram and cumulative distribution of THD levels for a week long measurement

address upper limits on distortion levels, or the time duration of individual bursts of high distortion. In order to further explore the impacts of harmonic bursts, it is of interest to study the impact of these bursts on AC power capacitors.

10.5   Capacitor aging

Capacitors are known to be sensitive to excessive harmonic levels. Partial discharge and over-temperature are two primary agents which lead to accelerated aging of power capacitors.

IEC standards call for harmonic measurements taken with approximately 200 ms windows (exactly 10 cycles on 50 Hz or 12 cycles on 60 Hz systems). These measurements are taken continuously, and the rms values of the harmonic components are then computed over 3 s intervals and over 10 min intervals. The established total harmonic distortion limits cannot be exceeded in any 10 min interval. The 10 min time frame is of the order of magnitude of power capacitor thermal time constants, so it is appropriate that steady-state limits should be enforced in this time frame, from the capacitor perspective. From a thermal point of view, shorter-term bursts of harmonics could be acceptable.

10.6   Partial discharge

Instantaneous voltages above the partial discharge inception voltage cause rapid aging of the capacitor. This aging is the primary reason for capacitor voltage limits, which are 120% of rated peak voltage in IEEE Standard 18, IEEE Standard for Shunt Power Capacitors. This standard also limits rms voltage to 110% of rated rms voltage. Therefore, for capacitors which are not derated and experience fundamental frequency overvoltages, it is important that the harmonics do not increase the peak instantaneous voltage by more than 10%. Montanari and Fabiani [3] investigated the effect of voltages above the partial discharge inception limit. They showed that capacitors exhibit rapid aging when the voltage is sufficient to cause partial discharge to occur. It is clear from this study that harmonic voltages above the partial discharge inception voltage should be avoided for the 3 s measurement as well as the 10 min measurement.

A 5% THD limit will not necessarily limit the instantaneous harmonic voltage to less than 10% – THD is an rms type measure, while voltage magnitudes add algebraically – experience shows that it must be assumed that harmonic voltage peaks and the fundamental voltage peak coincide. Therefore, the present THD limits on harmonic voltage levels are not directly compatible with the capacitor voltage standard. In many practical installations, however, the majority of the harmonic content is centered at one or two harmonic frequencies. Comparison of THD limits and the capacitor voltage limits shows that capacitor voltage limit is met if the fundamental voltage is within its limit, the THD limit is maintained and the harmonics are limited to a relatively low number of significant frequencies.

IEEE Standard 18 also provides allowance for transient overvoltage. These do require the capacitor to tolerate a limited exposure to voltages above 120% of the capacitor rated peak voltage. This transient allowance does not appear to have any relevance to harmonic bursts measured over a 3 s window. Another complicating factor is that it is not uncommon to derate capacitors in applications where harmonic levels are expected to be significant. If this practice is uniformly followed, relatively high distortion levels would not necessarily lead to overvoltage stress in the capacitor.

10.7   Thermal heating

Harmonic flows contribute losses in addition to those experienced under purely sinusoidal conditions. Steady harmonic levels are allowed for in capacitors through rms current ratings which are somewhat higher than would be experienced by the capacitor with rated sinusoidal voltage at the rated frequency.

The thermal time constant of a capacitor has been defined as the time it takes for the capacitor core temperature to reach 65% of its final value in response to a step change in ambient temperature. Harmonic variations which are slower than the thermal time constant of a capacitor will cause significant variations in the capacitor temperature. Conversely, harmonic over-currents which last less than 10% of the thermal time constant of the capacitor will cause relatively little temperature rise.

With regard to the IEC time limits, the 10 min harmonic window can be expected to be of the order of the thermal time constant of typical power factor correction capacitors. Significant over-temperature can be expected for over-currents lasting 10 min, therefore it is appropriate to apply steady-state limits in this time frame. Shorter-term harmonic over-currents, however, can occur without excessive over-temperature. In particular, a doubling of the harmonic current over a 3 s period should not lead to reduced life when the current rating is maintained over a 10 min measurement.

The IEC 3 s and 10 min limits refer to harmonic voltages, not currents. The capacitor will experience loss components proportional both to current and to voltage. The low capacitor impedance at the harmonic frequencies, however, makes capacitor current a primary concern.

It is not possible to specify a general voltage THD limit for a broad range of harmonic frequencies, and guarantee that the capacitor current will remain within the device rating. These ratings, however, can be sufficient in systems with typical harmonic voltage profiles, where the harmonic content is primarily limited to the lower orders.

10.8   Conclusions

The development of limits for short-term harmonic levels is clearly a difficult topic. These limits must provide an economical compromise between operational efficiency and equipment lifetimes. It is clear that there will be no answer which avoids some level of compromise.

These limits must involve both magnitude and duration. The correspondence of duration limits and equipment thermal limits is apparent, and any duration limit should be a small fraction of power equipment thermal time constants. The IEC's 3 s measurement certainly qualifies in this regard, and the requirement that 10 min measurements meet harmonic limits also seems reasonable. Middle ground may exist between the two for an intermediate limit.

The magnitude of an allowable burst poses more difficulties. In cases where a single harmonic is dominant (which is not unusual), a single harmonic limited to 3% in the steady state could be doubled to 6% in short-term situations without problem. If, however, there is a THD level of 3% consisting of two equal harmonics of 2.12%, in the worst case (which again is not unusual), these could sum and add 4.24% to the peak in the steady state. If this level is allowed to double in the short term, the peak voltage would increase by 8.5%, which is approaching the limit imposed by the standard (if the capacitor is operating at its rms limit, 10% above nominal voltage).

This difficulty of applying a short-term THD limit on capacitors is apparent. It appears that a relaxation of short-term THD limits should not allow levels to go more than double the steady-state limits, based on the existing levels for general distribution systems. An allowance of short-term variations of 50% above the steady-state limits appears to be a conservative level which should not lead to problems in most cases.

References

[1] IEC 61000-4-7 Ed. 2.0, Electromagnetic compatibility (EMC) - Part 4: Testing and measurement techniques - Section 7: General guide on harmonics and interharmonics measurements and instrumentation, for power supply systems and equipment connected thereto. International Electrotechnical Commission, 2002.

[2] IEEE Standard 18-2002, *IEEE Standard for Shunt Power Capacitors*. Institute of Electrical and Electronic Engineers. ISBN: 0-7381-3243-8, 2002.

[3] G. C. Montanari and D. Fabiani. 'The Effect of Non-Sinusoidal Voltage on Intrinsic Aging of Cable and Capacitor Insulating Materials', *IEEE Trans. on Dielectrics and Insulation*, **6**, 1999, 798–802.

11

Probabilistic harmonic indices

P. Caramia, G. Carpinelli, A. Russo, P. Verde and P. Varilone

11.1 Introduction

When the time-varying nature of harmonics is considered, the most adequate indices to represent it have to be chosen. The indices can refer either to a single customer point-of-common coupling or to a segment of the distribution system or, more generally, to the utility's entire distribution system.

While the single site index probabilistic treatment requires the collection of different observations in time, for system indices observations need to be collected at different sites. These last indices – well known to characterize such other aspects of service quality as sustained interruptions – have been considered only recently for harmonic purposes [1]; the system indices can serve as a metrics only and are of particular interest in the new liberalized market frame, because they can be used as a benchmark against which the quality level of different distribution systems, or of parts of the same distribution system, can be compared.

In this chapter, some single site and system probabilistic harmonic indices are analyzed and considerations about their statistical characteristics are presented, also on the basis of measurements on actual distribution systems; particular attention is paid to ascertain the statistical measures adequate to describe synthetically their statistical behavior (mean value, standard deviation, percentiles, etc.) and to be included in a harmonic standard or in a power quality contract.

On the basis of the statistical characteristics of the harmonic index probability density functions, it is shown that the expected value of the peak factor on a single site is a suitable statistical measure to be selected for inclusion in Standards or in a power quality contract. Furthermore, different system indices can be introduced in qualifying a distribution system or a part of it, particularly in the frame of the new liberalized market. Proper selection from among those proposed here depends on the particular aim of the analysis.

Time-Varying Waveform Distortions in Power Systems   Edited by Paulo F. Ribeiro
© 2009 John Wiley & Sons, Ltd

11.2   Single-site indices

The harmonic distortion can be described by several indices (individual harmonic distortion, total harmonic distortion, the peak factor, and so on). In this section, without loss of generality, we will refer to the total harmonic distortion (THD) and the peak factor (k_p); on the other hand, they are very useful indices [2–6] and also easily transferable in the probabilistic field, since only the marginal probability density function is generally required to describe their random properties fully. In any case, considerations about the probabilistic behavior of individual harmonic distortions are reported in [7, 8]. Moreover, the Joint Working Group CIGRE C4.07/CIRED in [9] has collected available measurement data and existing probabilistic indices for MV, HV and EHV systems and recommends a set of internationally relevant harmonic indices and objectives.

The total harmonic distortion of Y (current or voltage) is given by:

$$THDY = \frac{\sqrt{\sum_{h=2}^{\infty} Y_h^2}}{Y_1}, \tag{11.1}$$

with Y_h a harmonic superimposed onto the fundamental Y_1.

The peak factor is defined as:

$$k_{pY} = \frac{Y_p}{Y_{p1n}}, \tag{11.2}$$

where Y_p is the peak value of the distorted waveform Y, and Y_{p1n} is the rated value of the fundamental.

The peak factor can be separated into two components k_{psY} and k_{phY}, the first related to the effects of fundamental variations and the second one related to the effects of harmonics superimposed to fundamental Y_{p1}; it results:

$$k_{pY} = \frac{Y_{p1}}{Y_{p1n}} \frac{Y_p}{Y_{p1}} = k_{psY} k_{phY}. \tag{11.3}$$

Both total harmonic distortion and peak value can be applied to voltage and current distortions. The peak factor of voltage given by Equation (11.2) and its components given by Equation (11.3) have been considered in [5, 10] for translation in the probabilistic field while the current indices have been considered in [6].

The probability density functions (pdfs) of the above mentioned indices exhibit different statistical characteristics, as is clearly revealed by the analysis of several on site measurements and simulation results. First of all, the pdfs of all above indices can have several shapes and there is not a general rule for the choice of the probability density function which fits one of them best; in fact, various basic pdfs (Gaussian, uniform, linear increasing, and so on) have been tested to ascertain the one which approximates to the harmonic index pdf behavior best, but none of them can be generally applied.

From the analysis of several daily or weekly on-site measurements and of numerical simulations on low voltage and medium voltage distribution systems, the following considerations on their probabilistic behavior can be provided. The pdfs of the total harmonic distortion of the voltage (*THDV*) and current (*THDI*) can have both monomodal and multimodal (usually bimodal) patterns. Moreover, mainly when the harmonic data refer to a week, they are generally characterized by significant values of standard deviations, due to the large range of harmonics.

As an example, Figure 11.1(a) and (b) report the pdfs of (a) the *THDV* and (b) the *THDI* obtained processing recorded data over one week at a medium voltage distribution system. Starting from the considerations about both the pdf shapes and statistical measures of *THDV* and *THDI*, it follows that to impose reasonable probabilistic limits on these indices is very difficult. In fact, to set limits directly on the whole pdfs is neither realistic nor a reasonable solution, as a consequence of the obvious difficulties in the successive implementation in the Standards. On the other hand, to impose limits on one or more statistical measure (minimum value, maximum value, mean value, standard deviation, 95th percentile, and so on) can be practical but can lead to biassed decisions. In fact, if we refer to the 95th percentile or maximum value, as proposed by the IEC 61000-3-6 [3] in its assessment procedure, and we account for all the harmonic effects, disappointing consequences can follow, since *THDV* and *THDI* pdfs with the same 95th percentile and maximum value can have very different effects on electrical component behavior. As a very easy example, we can refer to the joule losses in a conductor (as a percentage of the losses at fundamental frequency) associated with the two different *THDI* pdfs of Figure 11.2, characterized by the same 95th percentile and maximum. Assuming the conductor resistance is independent of the frequency, the expected joule loss values in the two cases are 0.095 (Figure 11.2(a)) and 0.035 (Figure 11.2(b)), which are quite different.

The pdfs of the voltage peak factor (k_{pV}) are characterized, in the most general case, by very low values of standard deviations. In particular, the standard deviation usually shows values of at least very few percent, because the voltage harmonics have values lower than the fundamental voltage that ranges in not particularly large intervals for system operation safety. Due to the very low values of standard deviations, it is useless to talk about the shape of the pdf (monomodal or multimodal). A similar behavior has been experimented with reference to the component k_{phV}, relating to the effects of harmonics superimposed to the fundamental. As an example, Figure 11.3 shows the pdfs of (a) the k_{pV} and (b) the k_{phV} obtained from processing recorded data over one week at a medium voltage distribution system.

From the analysis of Figure 11.3(b), and from all the measurements on distribution systems we analyzed, it was also noted that the voltage harmonics in several cases caused an increase of the peak voltage with respect to the value that it would have been in the presence of only the fundamental component. However, in [10] a reduction in the peak voltage was discovered.

Starting from the considerations about the k_{pV} and k_{phV} pdf shapes and their statistical measures, it follows that to impose probabilistic limits on these indices is now possible and can be a reasonable choice. In fact, the very low values of standard deviations allow us to say that it can be enough to assign a limit on only the mean value. Moreover, in practice only the mean value of the voltage peak factor[1] k_{pV} seems to be suitable for inclusion in Standards or in power quality contracts. The component k_{phV} does not allow the computation of all the harmonic effects on distribution system components, because to distinguish the effects between fundamental and harmonics is a very difficult matter.[2]

---

[1] As shown in [4], the voltage peak has the dominant effect in accelerating the aging of distribution system components in a distorted regime and so the peak factor is the index most suited to take into account the harmonic effect.

[2] The component k_{phV} seems most appropriate for component designers as a valuable complementary criterion for choosing between different design solutions.

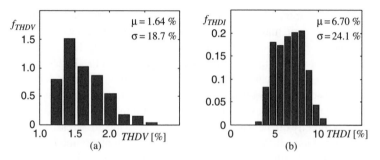

Figure 11.1   Probability density functions: (a) THDV and (b) THDI

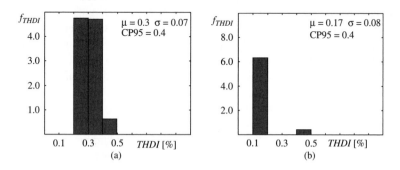

Figure 11.2   Different THDI pdfs with the same 95th percentile and maximum values

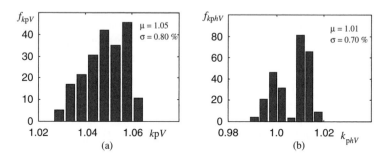

Figure 11.3   Probability density functions: (a) k_{pV} and (b) k_{phV}

As an example of setting limits on the mean value of the peak factor, let us consider the case of low-voltage industrial systems. The limits can be assigned making use of the diagram in Figure 11.4 [5], where the normalized expected value of life is plotted versus the mean value of the peak factor; each point of Figure 11.4 is obtained averaging the expected values of normalized useful life for different distribution system components (cables, transformers, capacitors and AC motors).

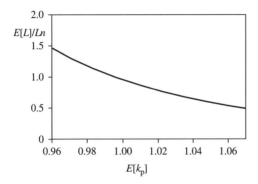

Figure 11.4   Expected value of normalized useful life versus expected value of peak factor averaged for cables, capacitors, transformers and AC motors

The pdfs of the current peak factor (k_{pI}) and of its component k_{phI}, related to the effects of harmonics superimposed onto the fundamental can have both monomodal and multimodal patterns and once again they are usually bimodal. The variance of the k_{pI} pdf is characterized, in the most general case, by significant values of standard deviations; on the contrary, reduced variances of k_{phI} pdfs were investigated.

As an example, Figure 11.5(a) and (b) show the pdfs of (a) the k_{pI} and (b) the k_{phI} obtained from processing data recorded over one week at an actual medium voltage distribution system.

From the analysis of Figure 11.5(a) and from all the measurements we analyzed on distribution systems, it was also noted that the high variability of current peak factor k_{pI} is due to the wide variation range of the fundamental component, as clearly highlighted in Figure 11.6, which shows the pdf of the component k_{psI} corresponding to the same measurements as those of Figure 11.5. This is a reasonable result, for the very large variability of connected loads.

From the analysis of Figure 11.5(b), it appears that the current harmonics usually cause a decrease in the peak current with respect to the value that would have been in the presence of only the fundamental component. However, this is not a general rule, because in other on-site

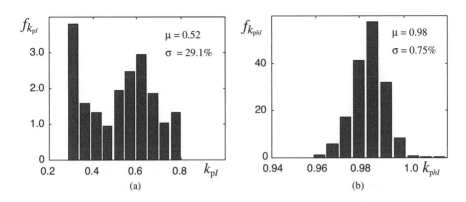

Figure 11.5   Probability density functions: (a) k_{pI} and (b) k_{phI}

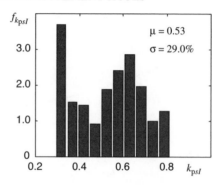

Figure 11.6   Probability density function of k_{psI}

measurements at distribution networks an increase – in some cases significant – of the peak current was experienced.

Finally, starting from the considerations about both the k_{pI} pdf shapes and their statistical measures, it follows that to impose reasonable probabilistic limits on this index is a difficult task, for the same reasons as those in the case of *THDV*.

On the contrary, if the trend shown by our measurements were to be confirmed, it would be possible to impose probabilistic limits on the k_{phI} index and also a reasonable choice. In fact, the very low value of the standard deviation allows us to say, once again, that it is enough to assign a limit only on the mean value. A possible criterion for assigning limits to the mean value of k_{phI} could be to set the fundamental current at the rated value and derive the limits on the basis of the component effects.

11.3   System indices

As is clearly pointed out in [1], the system indices serve as metrics and are not intended as an exact representation of the quality of service provided to each customer served from the assessed system. They can be used as a benchmark against which quality levels of parts of the distribution system and of different distribution systems can be compared.

The system indices taken into account here only refer to voltage distortion. They are:

- the system total harmonic distortion 95th percentile, *STHD95*;

- the system average total harmonic distortion (peak factor) *SATHD* (SAk_p);

- the system average excessive total harmonic distortion (peak factor) ratio index, $SAETHDRI_{THD^*}$, ($SAEk_pRI_{KP^*}$).

The system indices referred to as total harmonic distortion have been defined in [1], the ones concerning the peak factor are introduced in [6].

Let us refer to a distribution system with N busbars. Even if the system indices can also refer to small segments of the entire distribution system, such as a single feeder or a single busbar, in the following paragraphs, without loss of generality, we consider indices referred to the entire distribution system.

The *STHD95* index is defined as the 95th percentile value of a weighted distribution; this weighted distribution is obtained collecting the 95th percentile values of the *M* individual index distributions, each distribution obtained with the measurements recorded at a monitoring site. The weights can be linked to the connected powers or the number of customers served from the area that the monitored data represents, and so on; in the following, without loss of generality, we assume the connected powers as weights. Under these assumptions, the following relations compute *STHD95*:

$$\frac{\sum_{-\infty}^{STHD95} f_t(CP95_{THDs})L_s}{\sum_{-\infty}^{+\infty} f_t(CP95_{THDs})L_s} = 0.95, \tag{11.4}$$

with $CP95_{THDs}$ calculated via the following expression:

$$\frac{\sum_{-\infty}^{CP95_{THD,s}} f_{THD,s}(X_{THD_{s,i}})}{\sum_{-\infty}^{+\infty} f_{THD,s}(X_{THD_{s,i}})} = 0.95 \tag{11.5}$$

where:

$s =$ monitoring site;

$X_{THDs,i} =$ *i*th steady-state *THD* measurement at the site *s*;

$L_s =$ connected kVA served from the system segment where the monitoring site *s* is;

$f_{THDs}(X_{THDs,i}) =$ the probability distribution function of the sampled *THD* values for monitoring site *s*;

$CP95_{THDs} =$ the 95th cumulative probability value of the *THD* for the monitoring site *s*;

$f_t(CP95_{THDs}) =$ the probability distribution function of the individual monitoring site $CP95_{THDs}$ values.

The system index *STHD95* allows the measurements to be summarized both temporally and spatially handling measurements at *M* sites of the system in a defined time period, assumed as significant for the characterization of system service conditions. As an example, the value of *STHD95* obtained from processing data recorded over one week, referring to the same medium voltage distribution system considered in Section 11.2, is 2.2%.

The system average total harmonic distortion (peak factor) is based on the mean value rather than the *CP95* value:

$$SATHD = \frac{\sum_{s=1}^{M} L_s \mu(THD_s)}{L_T} \tag{11.6}$$

$$SAk_p = \frac{\sum_{s=1}^{M} L_s \mu(k_{p,s})}{L_T} \tag{11.7}$$

where:

$\mu(THD_s)$ = mean value of sampled THD values for monitoring site s;

$\mu(k_{p,s})$ = mean value of sampled k_p values for monitoring site s;

L_T = total connected kVA served from the system.

$SATHD$ and SAk_p give average indications of the system voltage quality and, thanks to the introduced weights, allow different levels of importance to be assigned to the various sections of the entire distribution system.

The values of $SATHD$ and SAk_p, again computed on the same distribution system referred to before, are 1.16% and 1.05, respectively. It is worthwhile noting that the $SATHD$ is very different from the $STHD95$ due to the not negligible value of standard deviation.

The system average excessive total harmonic distortion (peak factor) ratio index, $SAETHDRI_{THD^*}$ ($SAEk_pRI_{kp^*}$) is related to the number of steady-state measurements that exhibit a THD (k_p) value exceeding the THD^* (k_p^*) threshold. It is computed, for each monitoring site of the system, counting the measurements exceeding the THD^* (k_p^*) value and normalizing this number to the total number of the measurements effected at site s. They are defined by the following relations:

$$SAETHDRI_{THD^*} = \frac{\sum_{s=1}^{M} L_s \left(\frac{N_{THD^*s}}{N_{Tot,s}} \right)}{L_T} \tag{11.8}$$

$$SAEk_pRI_{k_p^*} = \frac{\sum_{s=1}^{M} L_s \left(\frac{N_{k_p^*s}}{N_{Tot,s}} \right)}{L_T} \tag{11.9}$$

where:

N_{THD^*s} (N_{kp^*s}) = the number of steady-state measurements at monitoring site s that exhibit a THD (k_p) value which exceeds the specified threshold THD^* (k_p^*);

$N_{Tot,s}$ = the total number of steady-state measurements effected at monitoring site s over the assessed period of time.

$SAETHDRI_{THD^*}$ and $SAEk_pRI_{kp^*}$ are linked to the measure of the total portion of the time that the system exceeds the specified THD^* and k_p^* thresholds, respectively.

Table 11.1 shows the values of $SAETHDRI_{THD^*}$ and $SAEk_pRI_{kp^*}$ computed on the same distribution system already referred to and with two thresholds.

Table 11.1   Values of $SAETHDRI_{THD^*}$ and $SAEk_pRI_{KP^*}$

$SAETHDRI_{5\%}$	0
$SAETHDRI_{2\%}$	0.016
$SAEk_pRI_{1.03}$	0.97
$SAEk_pRI_{1.05}$	0.41

Summarizing, the indices introduced provide different global measures to qualify the system in terms of service quality. The availability of several indices for a phenomenon can be useful to give evidence to different aspects of the same disturbance. No general rule can be established to suggest the most adequate index to be used; it will depend on which aspect is judged to be more relevant.

For example, in the considered real distribution system, the *SATHD* equal to 1.16% indicates not only the average level of distortion, but also takes into account the sites serving larger loads. This can be more evident if we consider that the average value of the *THD* on the same distribution system, without any weights, is greater of about 25%.

Further indications on the same system and on the same measurements concern the highest levels of distortion. The *STHD95* value indicates the weighted 95th percentile of the 95th percentiles of the system sites and the $SAETHDRI_{THD^*}$ values give more details about the weighted relative frequencies of the records above specified thresholds. Analogous considerations can be developed concerning the system indices referred to the peak factor.

An alternative way to compute the aforementioned indices concerns the statistical analysis of all the measurements effected at the *M* sites of the system. Assuming that the centralized handling of all the measurements of the *M* sites is not hard, this method seems to be interesting, since any statistical measure is estimated with an accuracy that depends on the number of processed data. This extensive procedure gives an indication of the global behavior of the system; on the contrary, the first recalled procedure furnishes the indication on the behavior of the sites.

For example, with reference to the *THD*, for each site *s* we compute the relative frequencies of the measured *THD* and we weight them with the ratio L_s/L_T. Starting from the weighted relative frequencies at each site *s*, we evaluate the system weighted relative frequencies WF_{THD}, and then compute the mean value and the 95th percentile that coincide with the indices *SATHD* and *STHD95*, respectively. The system weighted relative frequencies of the *THD* can also be used to evaluate the indices $SAETHDRI_{THD^*}$.

An analogous procedure can be implemented to calculate the indices SAk_p and $SAEk_pRI_{kp^*}$ referring to the peak factor. Figures 11.7 and 11.8 show the histograms of the system weighted relative frequencies of the *THD* and k_p (WF_{THD} and WF_{kp}), respectively, with reference to the

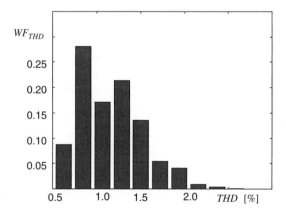

Figure 11.7   Histogram of the system weighted relative frequencies for the THD

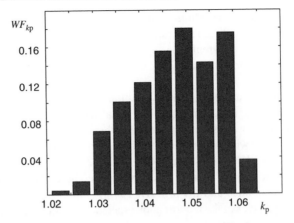

Figure 11.8   Histogram of the system weighted relative frequencies for k_p

distribution system under consideration. Table 11.2 shows the results in terms of the system indices obtained with the aforementioned procedure.

Comparing the results of the indices computed via the two procedures, it is important to note that the obtained values of $SATHD$, $SAETHDRI_{THD^*}$, SAk_p and $SAEk_pRI_{kp^*}$ coincide, as expected.

Conversely, the values of $STHD95$ are different. This is due to the fact that in the first procedure, the weighted distribution cannot take into account the values greater than the 95th percentiles. These values are instead considered, for the second procedure, in the construction of the system weighted relative frequency histogram, so that they can determine a lower value of the 95th percentile.

11.4   Conclusions

The problem of the random behavior of some harmonic indices has been taken into account. Both single site and system harmonic indices have been analyzed and considerations about their statistical characteristics have been given. The main conclusions are:

- With reference to single site indices, the expected value of the voltage peak factor seems to be a suitable harmonic index for translation to the probabilistic field. In fact, the very low value of its standard deviation allow us to say that it is enough to assign a limit on only the mean value; moreover, the index objective can easily be assigned considering the effects on the behavior of the distribution system components.

Table 11.2   Values of system indices

$STHD95$	1.94%
$SATHD$	1.16%
SAk_p	1.05
$SAETHDRI_{2\%}$	0.016
$SAEk_pRI_{1.03}$	0.97

- With reference to the system indices, it is worth noting that the availability of all the system indices proposed in this chapter is valuable because the analysis of their values allows a better understanding of the actual operating conditions of the system.

References

[1] D. D. Sabin, 'Analysis of Harmonic Measurements Data', IEEE Power Engineering Society Summer Meeting, Volume 2, Chicago, Illinois, USA, July 2002, pp. 941–945.

[2] European Standards EN50160, *Voltage Characteristics of Electricity Supplied by Public Distribution Systems*, 1999.

[3] IEC 61000-3-6, *Assessment of Emission Limits for Distorting Loads in LV and HV Power Systems*, 1996.

[4] P. Caramia, G. Carpinelli, P. Verde, G. Mazzanti, A. Cavallini, and G. C. Montanari, 'An Approach to Life Estimation of Electrical Plant Components in the Presence of Harmonic Distortions', Ninth International Conference on Harmonics and Quality of Power, Orlando, Florida, USA, October 2000, pp. 887–891.

[5] P. Caramia, G. Carpinelli, A. Russo, P. Varilone and P. Verde, 'An Integrated Probabilistic Harmonic Index', IEEE Power Engineering Society Winter Meeting, Volume 2, New York, USA, January 2002, pp. 1084–1089.

[6] P. Caramia, G. Carpinelli, A. Russo and P. Verde, 'Some Considerations on Single Site and System Probabilistic Harmonic Indices for Distribution Networks', IEEE Power Engineering Society General Meeting, Volume 2, Toronto, Canada, July 2003, pp. 1160–1165.

[7] D. D. Sabin, D.L. Brooks and A. Sundaram, 'Indices for Assessing Harmonic Distortion from Power Quality Measurements: Definitions and Benchmark Data', *IEEE Trans. on Power Delivery*, **14**, 1999, 489–496.

[8] A. E. Emanuel and S. R. Kaprielian, 'An Improved Real-time Data Acquisition Method for Estimation of the Thermal Effect of Harmonics', *IEEE Trans. on Power Delivery*, **9**, 1994, 668–674.

[9] Joint Working Group Cigrè C4.07/Cired, *Power quality indices and objectives,* January 2004, Rev. March 2004.

[10] D. Gallo, R. Langella and A. Testa, 'On the Effects on MV/LV Component Expected Life of Slow Voltage Variations and Harmonic Distortion', Tenth International Conference on Harmonics and Quality of Power, Vol. 2, Rio de Janeiro, Brazil, October 2002, pp. 737–742.

12

Measurement techniques and benchmarking

J. Driesen and J. Van den Keybus

12.1 Introduction

This chapter contains a brief description of measurement techniques used to obtain the harmonic content of current or voltage signals. First, the front-end signal chain is outlined, the output of which is assumed to be a digital data stream. Then three methods are discussed to produce the harmonic signal content based on this digital, sampled data stream.

The first method is the well-known fast Fourier transform (FFT). The second one is a model-based technique. These two methods are essentially finite impulse response (FIR) and infinite impulse response (IIR) filters, important properties to consider when applying them to non-stationary signals (i.e. signals in which the harmonic content varies with time). The third method uses wavelets but it is still under development and therefore only briefly mentioned at the end of this chapter.

The measurement of both current and voltage is usually performed in four steps as follows.

1. *Scaling*. This step converts the physical voltage or current into a value that is small enough to be transported through signal cables, but not too small to avoid noise pickup. Scaling devices include current and voltage transformers and capacitive dividers, with a precision of 0.1–5% at 50 or 60 Hz. Typically, the analog output rms value of a scaling operation lies in the range of 60–600 V for the voltage and 0.1–5 A for the current measurement.

2. *Isolation*. Galvanic isolation is mandatory for safety reasons and is often combined with the scaling operation. The construction of a precision isolation barrier for analog

Time-Varying Waveform Distortions in Power Systems   Edited by Paulo F. Ribeiro
© 2009 John Wiley & Sons, Ltd

signals is not easy. As an alternative, the barrier may be inserted after Step 4, when the signal is digital. The entire measurement system is then allowed to float at the potential of the power system.

3. *Conditioning.* Analog to digital converters (ADCs) require a voltage input signal. Precision shunt resistances convert a current input signal to a small voltage, which is subsequently amplified. A voltage input signal, on the other hand, is further scaled down and buffered. Conditioning also includes filtering to prevent aliasing errors. The ADC resolution largely dictates the precision of the conditioning stage.

4. *Analog to digital conversion.* The ADC is a single component converting the input signal to a digital value. Monolithic ADCs with resolutions of up to 24 bits are currently available and often reported, although 20 bits is rather a realistic limiting value if sampling rates of more than 100 Hz are desired.

Two comments must be made about this overview. First, the isolation barrier is not only advantageous for protection reasons; it also avoids ground loops and interference, especially if the power system contains noise at high frequencies, which is often the case if switching converters are present nearby. Some researchers [1] also decided on using fully floating measurement heads after being plagued by interference.

Secondly, the overview applies to sampling ADCs, which take a sample of their input, convert it somehow, and deliver the output value in one cycle. Other types exist, such as delta-sigma and pipelined ADCs. The pipelined ADC uses several conversion stages in series, leading to sample rates of more than 100 MSa/s at 14 bits, but with a nine sample delay. The delta-sigma ADC is basically a single bit converter, which produces a high-frequency bitstream that needs to be filtered digitally. As a result, these converters have a reduced input filtering requirement, facilitating the design of the conditioning electronics. They are excellent for 50 or 60 Hz measurements, which explains why most electronic wattmeter chipsets are based on them.

12.2   Measurement and time dependency of phasors

Strictly speaking, the specification of the harmonic signal content (also known as the phasor description) also implies that the signal is periodic. Because of this requirement, one fundamental frequency period is sufficient to determine all phasors fully. Excluding noise, the analysis of another equally sized time frame would invariably lead to the same result.

In practice, signals are seldom strictly periodic (e.g. any physical system must be powered up at some point in time) and frequency domain analysis is theoretically impossible. On the other hand, prolonged periods exist in which – apart from noise – signals are perfectly periodic, and in which a frequency domain analysis is actually a sensible thing to carry out. Thus, the phasors determined from experiments performed at two distinct points in time might differ, indicating a time dependency of the phasors obtained in such a way.

The following sections deal with techniques to determine voltage or current phasors based on a frame of consecutive current and voltage samples. First, the well-known Fourier transform (FT) is highlighted. Next, model based techniques are described and compared with the FT. Finally, reference is made to wavelet-based techniques.

12.3   Fourier transform

The Fourier transform is widely used and covered by many textbooks and other publications. Therefore, the discussion of this integral transform is kept short and the reader must refer to [2], for example, for an in-depth treatment.

The continuous time Fourier transform provides the full frequency domain description of any arbitrary signal, but is impossible to apply in practical sampled data systems. Its discrete time counterpart, the discrete Fourier transform (DFT), is given by:

$$\underline{C}_n = \sum_{k=0}^{N-1} y_k e^{-2\pi jnk/N}. \tag{12.1}$$

It is very popular due to its computationally efficient implementation, the fast Fourier transform (FFT). Furthermore, if the N samples of y_k cover one grid cycle, it readily provides the phasors.

A few assumptions are intrinsically made when applying the DFT for power phasor measurements:

- the signal is strictly periodic; the sampling frequency is an integer multiple of the fundamental grid angular frequency ω_G; the sample frequency is at least twice the highest frequency in the signal to be analyzed; each frequency in the signal is an integer multiple of the fundamental frequency.

Additionally, when using an FFT it is advantageous to have a number of samples that have an integer power of two.

When the above assumptions are satisfied, the results of the DFT are accurate, but if they are not, generally, three problems are encountered [3]: aliasing, leakage and the picket-fence effect.

1. *Aliasing.* Aliasing originates from the presence of frequencies in the signal above $f_S/2$, the Nyquist frequency. Although an anti-aliasing filter may be present in the data acquisition system, special attention must be paid to the noise introduced in the measurement system after this filter.

2. *Leakage.* The DFT must be performed on a finite series of signal samples. Leakage is caused by truncation of the (theoretically) infinite measurement series y_k. It is unavoidable and can be mitigated by the use of window functions. If the signal is periodic and an integer number of periods are used for analysis, there is no leakage.

3. *Picket-fence effect.* Picket fencing occurs when the signal contains a frequency that is not an integer multiple of the fundamental frequency (interharmonic), increasing the Fourier coefficients of neighbouring frequencies. Again, if the signal is periodic and well synchronized to the signal samples, no interharmonics exist.

12.3.1   Synchronization errors

If the sampling frequency f_S and the grid fundamental frequency are not properly synchronized, the leakage and the picket fence effect occur simultaneously.

It can be shown that the phasor value of the kth harmonic has an error proportional to the synchronization frequency error, and inversely proportional to the order of the harmonic. Phasors that should be zero deviate in roughly the same manner, but are also proportional to f_S. Therefore, if the measurement is badly synchronized, the presence of a strong harmonic impedes the accurate assessment of other small harmonics.

In islanding grids, the accuracy of the grid frequency may very well be of the order of 1%. In this case, synchronization using, for example, a phase locked loop (PLL) is indispensable. The DFT may also be used to analyze more than one period of the fundamental frequency. The leakage effect is less dominant in this case. This method is used in the IEC 61000-4-7 Standard, at the expense of a longer observation interval.

12.3.2   Filter properties and transient response

1. Observing the DFT can be considered a finite impulse response (FIR) filter, having the properties if the filter length is N, any input before the N last samples no longer contributes to the output;

2. a FIR filter never becomes unstable.

The transition from a periodic signal with amplitude A_n and phase φ_n to another periodic signal with amplitude A'_n and phase φ'_n causes the signal temporarily not to meet the periodicity requirement and due to Property 1, one must wait at least N cycles before the phasor calculation of the signal becomes valid again.

12.3.3   Computational effort

The computational effort to compute the $N/2$ Fourier coefficients from N samples is proportional to $N \log_2(N)$. If the calculation is done in sliding mode, that is the FFT is repeatedly applied to a frame of N elements consisting of the last $(N-1)$ shifted elements of the previous frame and a single new element, $N^2 \log_2(N)$ calculations are needed.

12.4   Model based techniques

Model based harmonic estimation is quite different from the Fourier transform approach. The technique is based on a mathematical model of the system to be analyzed and uses voltage or current measurements to correct the state of the model. At any time, phasors can be derived from the model state, which is updated every time a measurement becomes available. This is a clear advantage over the DFT that processes entire frames of data and cannot provide in-between data.

12.4.1   Linear discrete time Kalman filter

The discrete Kalman filter is as much an algorithm as it is a filter. For completeness, the main equations are repeated here, but the reader must refer to the many existing, excellent handbooks (e.g. [4]) covering Kalman filtering in depth.

System modeling is done using the state variable approach, that is. the system is modeled in the discrete time domain:

$$\begin{aligned} x_{k+1} &= Ax_k + Bu_k \\ y_k &= Cx_k + Du_k \end{aligned}$$
(12.2)

where x_k is the state vector, y_k is the output vector and u_k is the input vector of the system at time point k. The A and B matrices entirely describe the system's behavior.

Suppose the system to be observed can be described in discrete state space by:

$$x_{k+1} = \Phi_k x_k + w_k.$$
(12.3)

It is assumed that a number of process states (elements of x_k) are measured by:

$$y_k = H_k x_k + v_k.$$
(12.4)

Note that both Φ_k and H_k can be time-dependent. In these equations, w_k and v_k are white noise sequences with a known covariance structure. They represent the inability of the model to track the observed system on one hand and the measurement errors on the other. The stochastic properties of v_k and w_k can be mathematically described by:

$$\begin{aligned} E[w_k w_i^T] &= Q_k(i = k), 0(i \neq k) \\ E[v_k v_i^T] &= R_k(i = k), 0(i \neq k) \\ E[v_k w_i^T] &= 0 \end{aligned}$$
(12.5)

in which $E[]$ represents the expectancy operator. In other words, the noises described by the elements of v and w are white. Next, the estimation error is defined as:

$$e_k^- = x_k - \hat{x}_k^-$$
(12.6)

with the associated covariance matrix:

$$P_k^- = E[e_k^- e_k^{-T}]$$
(12.7)

where x_k denotes the real 'correct' state which is immeasurable, and $\hat{x}_k^-$ is the estimate of x_k. The circumflex '^' indicates that the variable is an estimation and '−' means that the estimation is an a priori one, that is based on all previous information but the last measurement, z_k. The a posteriori estimate is obtained, based on the a priori estimate and the difference between the new measurement and its expected value:

$$\hat{x}_k = \hat{x}_k^- + K_k(z_k - H_k \hat{x}_k^-)$$
(12.8)

where K_k is the Kalman gain for time instant k. It shows that blending occurs between the previous estimate and the measurement's deviation from the expected one, which indicates a state estimation mismatch. The calculation of the successive Kalman gains, K_k, can be conveniently summarized by a loop (Figure 12.1) clearly demonstrating the algorithmic nature of the Kalman filter [4].

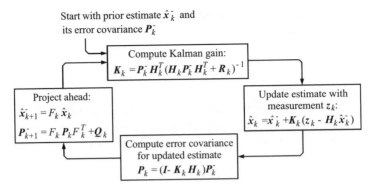

Start with prior estimate $\hat{x}_k^-$ and its error covariance P_k^-

Compute Kalman gain:
$$K_k = P_k^- H_k^T (H_k P_k^- H_k^T + R_k)^{-1}$$

Project ahead:
$$\hat{x}_{k+1}^- = F_k \hat{x}_k$$
$$P_{k+1}^- = F_k P_k F_k^T + Q_k$$

Update estimate with measurement z_k:
$$\hat{x}_k = \hat{x}_k^- + K_k (z_k - H_k \hat{x}_k^-)$$

Compute error covariance for updated estimate
$$P_k = (I - K_k H_k) P_k^-$$

Figure 12.1   Kalman filter loop [4]

12.4.2   Time invariant linear discrete Kalman filter

In many cases, Φ_k, H_k, Q_k and R_k are all time invariant and, therefore, do not depend on k. In that case, K_k and P_k calculated in the algorithm shown in Figure 12.1 converge to a time invariant gain K and an associated covariance matrix P.

The calculation of K and P can be done offline (e.g. using the MATLAB kalman command) and the use of the estimator then reduces to the evaluation of Equation (12.8), with a minimum of computational effort.

12.4.3   State variable description of a sinusoidal signal

There are two methods to represent a sinusoidally varying signal in the state space [5]. The *first method (Model 1)* uses a harmonic oscillator.

$$\begin{bmatrix} x_{\text{Re}} \\ x_{\text{Im}} \end{bmatrix}_{k+1} = x_{k+1} = A_{\text{OSC1}} x_k = \begin{bmatrix} \cos(\omega\Delta t) & \sin(\omega\Delta t) \\ -\sin(\omega\Delta t) & \cos(\omega\Delta t) \end{bmatrix} \begin{bmatrix} x_{\text{Re}} \\ x_{\text{Im}} \end{bmatrix}_k \qquad (12.9)$$

Note that the state transition matrix A_{OSC1} is constant. x is a vector containing the real (in-phase) and imaginary (quadrature) part of the harmonic oscillator. ω is the natural pulsation and Δt represents the model's sample time.

A_{OSC1} can also be considered a transformation matrix, rotating the vector x over an angle $\varphi = \omega\Delta t$ in every time step whilst preserving its amplitude. The output of the system, y_k, is given by the in-phase component, thus:

$$C_{\text{OSC1}} = \begin{bmatrix} 1 & 0 \end{bmatrix} \qquad (12.10)$$
$$D_{\text{OSC1}} = 0.$$

The *second method (Model 2)* uses a state vector representing a phasor instead of the actual voltages, which, ideally, does not change. Therefore:

$$x_{k+1} = A_{\text{OSC2}} x_k = \begin{bmatrix} 1 & 0 \\ 0 & 1 \end{bmatrix} x_k. \qquad (12.11)$$

The measurement equation may be expressed as:

$$C_{OSC2} = [\sin(\omega\, t_k)\ \cos(\omega\, t_k)]$$
$$D_{OSC2} = 0 \tag{12.12}$$

In this case, C_{OSC2} is no longer constant in time. Both systems have no input ($B_{OSC1} = B_{OSC2} = 0$). Once started with an initial state vector x_0, they output the same sine waveform forever. The associated states are uncontrollable, but observable.

12.4.4   Grid voltage modeling

In what follows, the model based grid voltage estimation is highlighted. The grid voltage is not stationary and contains noise of various origins. The measurement taken to improve the grid voltage estimation may also contain noise.

To some extent, the nonstationary property of the grid voltage and its noise can be modeled as a single white noise source $w(t)$. A second, white noise source $v(t)$ represents the measurement errors. Including these noises, the grid voltage model is:

$$x_{k+1} = Ax_k + w_k$$
$$y = Cx_k + v_k \tag{12.13}$$

in which $v_k = v(k\Delta t)$ and $w_k = w(k\Delta t)$ are the sampled data versions of $v(t)$ and $w(t)$ and the A and C matrices must be replaced by A_{OSC1} and C_{OSC1}, or A_{OSC2} and C_{OSC2}, depending on which grid modeling technique is used. Also note that w_k is a vector containing two independent white noise sources.

In the form of Equation (12.2) and provided that Q and R are available, the model can be processed by the Kalman filter. Note that C_{OSC2} in Model 2 is time-dependent and the algorithm of Figure 12.1 must be executed for every time step to obtain the – also time-dependent – Kalman gain K_k.

12.4.5   Transient response

The dynamic behaviour of the first model based grid voltage estimator (Model 1) is studied by observing the response to a state mismatch $e_k = x_k - \hat{x}_k$. Figure 12.2 shows the data flow schematically. In this section, for notational convenience, A and C are used to represent A_{OSC1} and C_{OSC1}, respectively.

Under the assumption that the model is perfect, i.e. $A = A'$ and $C = C'$, the state estimation error e_k can be rewritten as:

$$e_{k+1} = x_{k+1} - \hat{x}_{k+1}$$
$$= A'x_k - A\hat{x}_k - L(C'x_k - C\hat{x}_k) \tag{12.14}$$
$$= (A - LC)\, e_k.$$

The dynamic behavior of the state estimation error e_{k+1} at $t_{k+1} = (k+1)\Delta t$ is determined by $A - LC$ and depends on the preceding error e_k. Thus, in general, the model based grid voltage estimator is an infinite impulse response (IIR) filter. This is fundamentally different from the Fourier series based methods, which are essentially FIR filters.

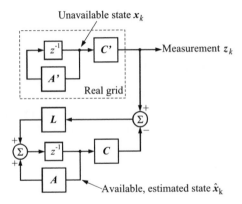

Figure 12.2   Model based grid voltage estimation

If $L = 0$, e_k tends to zero if all eigenvalues of A lie within the unit circle. Since A is a rotation matrix, its eigenvalues lie exactly on the unit circle and, without some feedback, it is impossible to reduce the state estimation error.

Using pole placement techniques to determine $L \neq 0$, the eigenvalues of $A - LC$ can be moved towards the center of the unit circle, leading to a (faster) decay of e_k, assuming that the system is controllable. In has been shown that e_k in Figure 12.2 is controllable if:

$$\text{Rank}[(A^T)^{n-1}C, \ \ldots, A^T C, C] = n \tag{12.15}$$

with n the number of states in the system or the number of elements in e_k. Since A is a rotation matrix, this condition is usually fulfilled unless n is so large that the $(A^T)^{n-1}$ start to overlap.

L can be chosen in such way that $e_k = 0$ in at most n cycles, leading to a deadbeat response. In that case, however, the sensitivity is high to process and measurement noises, $w(t)$ and $v(t)$, which have a detrimental effect on the estimation. By moving the system poles away from zero, one obtains a reduced sensitivity to $w(t)$ and $v(t)$ at the expense of an increased settling time. If $w(t)$ and $v(t)$ are white and a minimal squared estimation error (MSEE) is desired, the Kalman filter provides an optimum for L, further referred to as K.

12.4.6   Remark concerning the model errors and Q

A remark concerning the modeling error is in place here. In many cases, $w(t)$ or its spectrum is not well known. Sometimes, the errors are systematic and cannot be represented by a white noise sequence.

If the noise is still stochastic but nonwhite, the system model can be extended to include a filter in order to generate a colored noise signal, on which the model can depend subsequently. This technique is called 'state augmentation'. In some cases, the real system even contains an unknown deterministic input. For example, consider the 'process' of the sinusoidal grid voltage. An unknown 'load signal' at its unavailable 'input' varies the output voltage and phase. While monitoring this process, one often considers the unknown input signal as a noise contribution and includes it in the Q_k matrix. This is a questionable practice as the load signal is probably nonwhite, that is loads are usually switched on or off for an extended period.

During transitions, the Kalman gain is no longer an optimum blending factor in the MSEE sense, but the estimator response may still be acceptable in practice.[1]

In the sensorless active filter described in Chapter 6 the model based grid voltage estimator is used with $Q = I \cdot 0.016$, being in the same order of magnitude as used in [5] and [6]. Methods exist to determine Q_k during Kalman filter operation and the papers of [7] and [8] may provide a useful starting point for the interested reader.

12.4.7   Computational effort

In the case where multiple harmonic frequencies are introduced in the model, A_{OSC1} becomes semidiagonal and the computational effort is $(2 + k)N$, k indicating the number of measurements, that is the number of rows in C_{OSC1}. A frame containing N data samples therefore requires $(2 + k)N^2$ operations, which is slightly less than the FFT.

12.5   Wavelet based technique

The use of wavelets [12] introduces a totally different approach to the problem of determining phasors. They offer a trade off between frequency and time resolution, something the DFT is incapable of. The technique is mentioned here only for completeness. An introduction to the application of wavelets in power analysis can be found in [9]; [10] discusses the use of the wavelet transform as a harmonic analysis tool in general. The use of the Morlet wavelet in power system analysis is interesting, as it strongly resembles the (windowed) DFT, the window being a variable width Gauss bell curve [11]. In that case, the wavelet coefficients have a strong correspondence to phasors. This is further discussed in Chapter 16.

References

[1] I. A. Robinson, 'An optical-fibre ring interface bus for precise electrical measurements', *Measurements Science Technology*, **2**, 1991, 949–956.

[2] A. Papoulis, *The Fourier Integral and its applications*, McGraw-Hill, New York, 1962.

[3] G. D. Bergland, 'A Guided Tour of the Fast Fourier Transform', *IEEE Spectrum*, **6**, 1969, 41–52.

[4] R. G. Brown and Y. C. Hwang, *Introduction to Random Signal Analysis and Applied Kalman Filtering*, John Wiley & Sons, Inc., New York, USA, 1985.

[5] A. Girgis, W. Bin Chang and E. B. Makram, 'A Digital Recursive Measurement Scheme for On-Line Tracking of Power System Harmonics', *Trans. IEEE Power Delivery*, **6**, 1991, 1153–1159.

[6] M. S. Sachdev and M. Nagpal, 'A Recursive Least Error Squares Algorithm for Power System Relaying and Measurement Applications', *Trans. IEEE Power Delivery*, **6**, 1991, 1008–1015.

[7] R. K. Mehra, 'On the Identification of Variances and Adaptive Kalman Filtering', *Trans. IEEE Automatic Control*, **15**, 1970, 175–184.

[8] B. Carew and P. B. Bélanger, 'Identification of Optimum Filter Steady-State Gain for Systems with Unknown Noise Covariances', *Trans. IEEE Automatic Control*, **18**, 1973, 582–587.

---

[1] Theoretically, the estimation would vastly improve if the unknown 'load signal' were made available to the observer.

[9] W.-K. Yoon and M. J. Devaney, 'Power Measurement Using the Wavelet Transform', *Trans. IEEE Instrumentation. and Measurement*, **47**, 1998, 1205–1210.

[10] D. E. Newland, 'Harmonic wavelet analysis', *Proc. R. Soc., London*, **A443**, 1993, 203–225.

[11] H. Shyh, H. Cheng and H. Ching, 'Application of Morlet Wavelets to Supervise Power System Disturbances', *Trans. IEEE Power Delivery*, **14**, 1999, 235–243.

[12] S. Mallat, *A Wavelet Tour of Signal Processing*, Academic Press, San Diego, 1999.

Part V

APPLICATIONS AND CASE STUDIES

This part covers some applications and case studies intended to test the concepts already presented. Chapter 13 examines existing harmonic summation methods defined in IEC 61000-3-6 and illustrates the application of these methods based on measured data from various EAF sites. Chapter 14 presents a methodology for the treatment of harmonic currents measured in a third and fifth harmonic filter located at the Foz do Iguaçu Converter Station of the High Voltage DC System of Itaipu, Brazil. The methodology has been based on the statistical treatment applied to field measurements by comparing the amplitude average value and the dispersion of the harmonic currents for different configurations and loading conditions of the HVDC system.

13

Harmonic summation for multiple arc furnaces

J. Wikston

13.1 Introduction

The IEEE Standard 519–1992 establishes limits for harmonic voltage and current levels at the point of common coupling (PCC). When estimating compliance to these limits for new or expanding facilities with multiple time-varying harmonic loads such as electric arc furnaces (EAFs) the summation of harmonics from the time-varying loads becomes an important issue. This chapter examines existing harmonic summation methods defined in IEC 61000-3-6 and illustrates the application of these methods based on measured data from various EAF sites.

The science of harmonic simulation has advanced a great deal in the past two to three decades. The problem of how to represent multiple harmonic sources and background harmonics so that future levels can be predicted when expanding an existing facility or building a new facility still represents a difficult task. When the loads are time varying loads such as EAFs the representation becomes even more complex.

There are a few methods for combining multiple sources:

- linear vector sum,

- root sum of squares (RSS),

- vector sum randomly varying phase,

- linear randomly varying magnitude, and

- vector sum random phase and magnitude.

A linear combination gives the most pessimistic results of voltage distortion since it is assumed that the current injections from multiple sources are all in phase. A vector sum is the

Time-Varying Waveform Distortions in Power Systems   Edited by Paulo F. Ribeiro
© 2009 John Wiley & Sons, Ltd

ideal method but requires information on the phase of the harmonics with respect to each other, which is usually difficult to obtain. The RSS only requires magnitude information but assumes the square law correctly defines the combination due to phase angle differences, which is sometimes conservative and sometimes pessimistic. The other three methods are statistical methods that try to account for the phase angle and magnitude variations when not all the information is known.

In the IEEE Standard 519-1992 *Recommended Practices and Requirements for Harmonic Control in Electrical Power Systems* [1] there are tables for harmonic voltage and current limits for recommended practices for utilities and consumers. However, there is little guidance on how to apply these limits in particular with regards to the method used for determining the harmonic levels. If, for example, the loads did not vary with time and the system background harmonics were constant then this would be a nonissue; however, that is far from practical. Most loads vary with time and background harmonics also vary but to a lesser degree than some of the more severe loads like EAFs. In the IEEE Standard 519-1992 a reference is made to 'the values in the table maybe exceeded for a one-hour period' even though it is not stated it can be assumed that this is one hour per day. With this assumption the tables can then be viewed as values not to be exceeded 95% of the time ($23/24 \approx 95\%$). We round to 95% because IEC standards also refer to values not to be exceeded 95% of the time to evaluate harmonic emissions. Also in Standard 519, there is a cumulative probability function (cpf) illustrating the time-varying nature of harmonic sources. This cpf example introduced the concept of having to apply the tables against a value not to exceed a certain percentage of the time. The other complication in applying the Standard 519 tables is how to do the measurements: by measuring one cycle at a time or by measuring multiple cycles? In [2] a measurement philosophy is described which overcomes this problem. The measurement philosophy uses multiple cycles which allows interharmonics to be identified. The technique has four different measurement intervals on which to evaluate harmonics: 200 ms, 3 s, 10 min or 2 hours, with each interval being related by the following equations:

$$H_{3\,s} = \sqrt{[(1/15)*(\Sigma H^2_{200\,ms})]}$$
$$H_{10\,min} = \sqrt{[(1/200)*(\Sigma H^2_{3\,s})]}$$
$$H_{2\,hour} = \sqrt{[(1/12)*(\Sigma H^2_{10\,min})]}.$$

The IEC Standards use the 10 min interval for comparing the 95% cpf value as well as the 3 s interval. The use of harmonic snapshots whether 200 ms or 1 cycle to compare to the cpf value is also commonly applied. In this chapter we focus on the combination of harmonic currents from EAFs for instantaneous and the 10 min intervals.

13.2   Harmonic limits used by IEC standards

Figure 13.1 illustrates a disturbance immunity planning and compatibility level graph. In the graph all disturbances are represented by a probability density function. Another, slightly overlapping, probability density function represents equipment/user immunity levels. As illustrated in the graph, in a generic sense, some disturbances exceed the defined compatibility level and some of the environment (equipment, people, etc.) may not be immune even at the defined planning level. The defined planning and compatibility levels are chosen such that the system is not overdesigned and to make best use of the economical investment in the system.

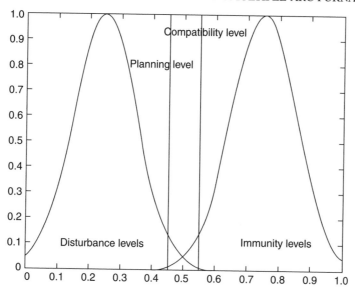

Figure 13.1   Planning levels illustrated with disturbance and immunity statistics

Compatibility levels for voltage harmonics in [3] are defined to be a THD of 5%, 8% and 10% for equipment Class 1, 2 and 3 respectively. Class 1 applies to protected supplies and has compatibility levels lower than the public network. Class 2 applies to public networks. Class 3 applies to a point of common couple inside a facility, which meets certain requirements. These compatibility values refer to continuous levels; time-varying components may have higher values. Reference [4] gives the following guidelines on how to apply the compatibility levels to time varying harmonics.

- The greatest 95% probability daily value of $V_{h,\mathrm{vs}}$ (rms value of the individual harmonic component over 'very short' 3 s periods) should not exceed the planning level.

- The maximum weekly value of $V_{h,\mathrm{sh}}$ (rms value of the individual harmonic component over 'short' 10 min periods) should not exceed the planning level.

- The maximum weekly value of $V_{h,\mathrm{vs}}$ should not exceed 1.5 to 2 times the planning level.

13.3   Harmonic summation techniques used by IEC standards

In [4] two methods for harmonic summation are presented. The first is a simple linear combination of the magnitudes with a diversity factor. This is used for the new voltage distortion when connecting a new load and the existing background distortion is known. The diversity factors depend on:

- the type of load,
- the harmonic order,

Table 13.1   Summation exponents for harmonics

m	Harmonic order
1	$h < 5$
1.4	$5 < h < 10$
2	$h > 10$

Note – when it is known that the harmonics are likely to be in phase (i.e. phase angle difference less than 90°), then an exponent $m = 1$ should be used for order 5 and above.

- the size of the load versus system strength, and

- the phase angle of background and new load

The second more generic law for both harmonic voltages and currents is:

$$H_{\text{TOTAL}} = \left(\sum H_{\text{individual}}^{m} \right)^{(1/m)}.$$

This equation is applied to each harmonic component of interest.

The values for m are given in [4] (Table 13.1) for 95% nonexceeding probability values. The exponent m depends mainly on two factors:

- the chosen value of the probability for the actual value not to exceed, and

- the degree to which individual harmonic voltages/currents vary randomly in terms of magnitude

13.4   Empirical exponent values for multiply EAFs

The second law of harmonic summation for currents from EAF loads is the focus of this chapter. The following tables illustrate the exponent, ratio of exponents and ratio of values for the 95% and 99% cpf values for snapshots and the 10 min evaluation interval.

From Tables 13.2 and 13.3 it can be seen that the recommended exponent value of 1.0 for orders less than the fifth will result in too high an estimate for harmonic current combination from multiple EAFs.

The values in the tables show that even harmonic components are less likely to combine with an average exponent of 1.5, 1.7, 2.0 and 1.9 as compared with the odd-order harmonic components which had an average exponent of 1.4, 1.4, 1.6 and 1.4 for the 95% snapshot, 95% 10 min, 99% snapshot and 99% 10 min respectively.

For the even harmonic components at 95% the 10 min values are less likely to combine than the snapshot values, since the 10 min exponents have a higher average. For the odd harmonic components the snapshot and 10 min values have approximately the same average exponent value.

For the even harmonic components at 99% the 10 min and the snapshot have approximately the same exponent value. For the odd harmonic components at 99% the

Table 13.2   Exponents and ratios for harmonic snapshot

Harmonic order		2	4	6	3	5	7
Exponent 95%	Site 1	1.4	1.4	1.5	1.5	1.5	1.4
	Site 2	1.5	1.4	1.5	1.5	1.3	1.3
	Site 3	1.7	1.6	1.5	1.5	1.3	1.2
	Site 4	1.6	1.5	1.7	1.7	1.5	1.6
	Average	1.5	1.5	1.5	1.5	1.4	1.4
Exponent 99%	Site 1	1.6	1.7	1.8	1.7	1.9	1.6
	Site 2	1.8	1.9	2.0	1.8	1.3	1.4
	Site 3	2.0	1.9	1.6	1.7	1.5	1.4
	Site 4	2.2	2.7	2.5	2.2	1.4	1.4
	Average	1.9	2.0	2.0	1.8	1.5	1.4
Ratio 99% to 95%	Site 1	1.3	1.3	1.4	1.4	1.4	1.4
	Site 2	1.5	1.7	1.7	1.4	1.2	1.2
	Site 3	1.4	1.4	1.6	1.3	1.2	1.2
	Site 4	1.6	1.6	1.6	1.4	1.3	1.3
	Average	1.5	1.5	1.6	1.4	1.3	1.3
Ratio exponent 99% to 95%	Site 1	1.1	1.2	1.2	1.1	1.3	1.2
	Site 2	1.3	1.4	1.5	1.2	1.0	1.0
	Site 3	1.2	1.1	1.0	1.1	1.2	1.1
	Site 4	1.3	1.8	1.7	1.3	0.9	0.9
	Average	1.2	1.3	1.3	1.2	1.1	1.1

Table 13.3   Exponents and ratios for harmonic 10 min interval

Harmonic order		2	4	6	3	5	7
Exponent 95%	Site 1	1.4	1.4	1.6	1.4	1.5	1.3
	Site 2	1.8	1.6	1.7	1.4	1.2	1.2
	Site 3	1.5	1.4	2.0	1.4	1.1	1.4
	Site 4	2.4	2.0	1.8	1.7	1.3	1.4
	Average	1.8	1.6	1.8	1.5	1.3	1.3
Exponent 99%	Site 1	1.4	1.4	1.6	1.7	1.9	1.5
	Site 2	1.8	2.2	1.8	1.4	1.1	1.2
	Site 3	1.7	1.7	2.0	1.2	1.2	1.5
	Site 4	1.8	2.9	2.4	1.5	1.1	1.2
	Average	1.7	2.0	1.9	1.4	1.3	1.3
Ratio 99% to 95%	Site 1	1.2	1.1	1.1	1.2	1.2	1.2
	Site 2	1.2	1.3	1.3	1.1	1.1	1.1
	Site 3	1.0	1.1	1.1	1.1	1.0	1.0
	Site 4	1.3	1.2	1.2	1.3	1.2	1.2
	Average	1.2	1.2	1.2	1.2	1.1	1.1
Ratio exponent 99% to 95%	Site 1	1.0	1.0	1.0	1.2	1.2	1.2
	Site 2	1.0	1.4	1.1	1.0	0.9	1.0
	Site 3	1.1	1.2	1.0	0.9	1.0	1.0
	Site 4	0.8	1.5	1.3	0.9	0.9	0.8
	Average	1.0	1.3	1.1	1.0	1.0	1.0

snapshot values are less likely to combine than the 10 min values since the average exponent value is higher.

The ratio of 99% to 95% values is high for even harmonic components as compared with odd harmonic components with an average of 1.5 compared with 1.3 for the snapshot values and 1.2 compared with 1.1 for the 10 min values.

13.5   Conclusions

The generic summation law is extremely useful since information about harmonic phase angles is not required. The calibration of the exponents for different percentiles, measurement intervals, odd/even components and component order is required for loads with unique profiles such as EAFs.

References

[1] IEEE Standard 519-1992, *Recommended Practices and Requirements for Harmonic Control in Electrical Power Systems*, June 1992.

[2] IEC 61000-4-7, 'Electromagnetic compatibility (EMC) – Part 4: Testing and measurement techniques – Section 7: General guide on harmonics and interharmonics measurements and instrumentation, for power supply systems and equipment connect thereto', (Work in Progress IEC 61000-4-7 Ed. 2.0 Stage Code CCDV), August 1991.

[3] IEC 61000-2-4, 'Environment – Section 4: Compatibility levels in industrial plants for low frequency conducted disturbances', February, 1994.

[4] IEC 61000-3-6, 'Limits – Section 6: Assessment of emission limits for distorting loads in MV and HV power systems – Basic EMC publication', October, 1996.

14

Treatment of measured harmonic currents in filters of an HVDC system

S. Carneiro Jr and A. C. de Freitas Marotti

14.1   Introduction

This chapter presents a methodology for the treatment of harmonic currents measured in a third/fifth harmonic filter located at the Foz do Iguaçu Converter Station of the High Voltage DC System of Itaipu, Brazil. The methodology has been based on the statistical treatment applied to field measurements. By comparing the amplitude average value and the dispersion of the harmonic currents for different configurations and loading conditions of the HVDC System, the methodology has been of assistance with the identification of some factors that may affect the levels of the harmonics. Experience obtained from this work indicates the importance of three-phase representation to study the harmonics in the system studied as the level varies substantially between the phases.

The Itaipu Hydro Plant is a binational venture situated in the Paraná River between Brazil and Paraguay. Due to the different frequencies used in the two countries, half of the 18 generators are 60 Hz and the other half are 50 Hz. However, most of the generated power has to be transmitted to the Brazilian grid, and for this reason a 6300 MW, $+/-600$ kV DC link was constructed. The DC line connects the Converter Station at Foz do Iguaçu to a major substation near São Paulo, 800 km away, where the power is converted to 60 Hz. In November 1999, there was an emergency which caused the disconnection of the third/fifth harmonic filters at the Foz do Iguaçu Converter Station. Even though the harmonic overload protection did not indicate any anomaly, it was decided that the filters should be tripped once it was observed that some filter reactors were actually burning.

Time-Varying Waveform Distortions in Power Systems   Edited by Paulo F. Ribeiro
© 2009 John Wiley & Sons, Ltd

In order to monitor the levels of harmonics, it was decided to install meters in the affected reactor of one of the third/fifth harmonic filters. The equipment was installed and remained in operation for several months, generating vast quantities of data. The objective of this chapter is to describe the methodology adopted to analyze the measured data using statistical concepts.

14.2   Field measurements

A modern digital meter was installed to measure the currents flowing through the three phases of the reactor. The meter was connected to the secondary of the reactor CTs. The instrument processed the data using FFT with a fixed two-cycle window which was synchronized through the detection of zero crossing. The sampling rate was 32 points per cycle with 1% accuracy [1], and the measurements were updated at 5 s intervals.

The meter remained installed for several months and this allowed the monitoring of the harmonic levels under different operating conditions of the DC link. As an example, Figures 14.1 and 14.2 show respectively the third and the fifth harmonics as a percentage of the fundamental for a 24 hour period.

It is apparent from these figures that the level of harmonics varies considerably between phases. The influence of the system load condition on the harmonics can also be observed, particularly in Figure 14.2 where it can be seen that the level of the fifth harmonic rises substantially during the peak load period, from 19:00 to 23:00 hours.

Considering the vast quantity of data generated by the metering campaign, as well as the fairly wide variations in harmonic levels as observed above, it was decided to introduce a statistical methodology to assist with the analyses of the harmonic behavior. Experience with

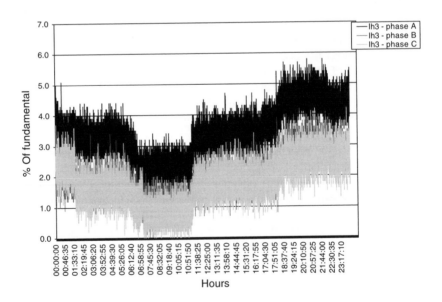

Figure 14.1   Third harmonic components of measured currents on phases A, B and C on the third/fifth harmonic filter – 24 hour period

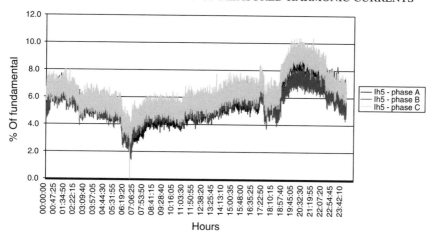

Figure 14.2   Fifth harmonic components of measured currents on phases A, B and C on the third/fifth harmonic filter – 24 hour period

the operation of the DC link over the years since its installation has shown that for any given load condition, steady-state operation can be considered to remain fixed during a period of about 7.5 min. Since the meter registers quantities at 5 s intervals, this would correspond to 90 samples.

Figures 14.3 and 14.4 show the third and fifth components, respectively, over a given 7.5 min operating condition. This condition was chosen to start at 18:57:00 hours, that is, just before the beginning of the peak system load period. It is clear from these figures that, although the system is at steady state, there is a random variation of the harmonic levels.

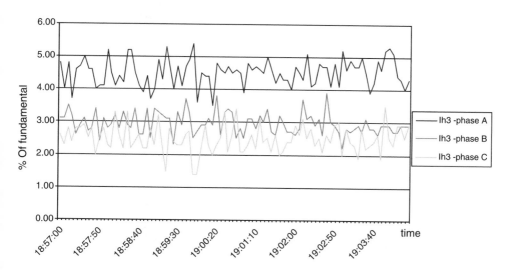

Figure 14.3   Third harmonic components of measured currents on phases A, B and C – 7.5 min of a steady-state operating condition

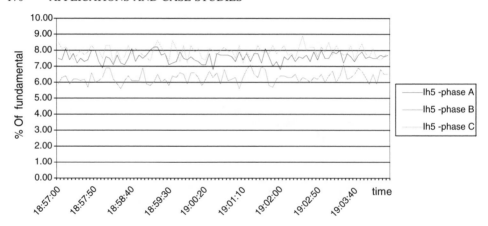

Figure 14.4   Fifth Harmonic Components of Measured Currents on phases A, B and C – 7.5 min of a steady-state operating condition

14.3   Statistical treatment

Student's t-distribution was selected to process the data. This choice was based on observations of the behavior of the harmonic variations and on the fact that it was desired to attain a confidence interval for each harmonic level having a confidence coefficient of 90%. There was no prior knowledge of the standard variation, or of the σ^2 value for the variable distribution [2–4].

For each operating condition, the following calculations were performed.
Calculation of the sample mean:

$$\bar{I} = \left(\frac{1}{n}\right)\sum_{i=1}^{n}Ii, \quad \text{where } n = 90. \tag{14.1}$$

Standard deviation:

$$S = \sqrt{\frac{1}{n-1}\sum_{i=1}^{n}(Ii-\bar{I})^2}. \tag{14.2}$$

The coefficient for Student's t-distribution, for a confidence interval of 90%, and 89 degrees of freedom (90 samples) is $t_{89;0.95} = 1.662$. Thus the confidence interval can be calculated from:

$$L = n^{-1/2}.S.t_{n-1,1-\alpha/2}. \tag{14.3}$$

The above calculations were performed for a wide range of operating conditions and it was found that the results were consistent with the chosen distribution, as will be described below.

14.4   Case studies

From the available data measurements, the following four different DC link power operating conditions were selected for discussion here. In all cases, the two bipoles were operating and only one of the third/fifth filters was connected.

Table 14.1   Statistical treatment – data measurements

Power DC link (MW)	Ih3 A	Ih3 B	Ih3 C	Ih5 A	Ih5 B	Ih5 C
	[% fundamental]					
2330	2.44	1.13	1.06	2.67	2.46	3.52
2790	2.94	1.70	2.03	3.84	4.17	4.38
3600	3.79	2.42	2.37	6.33	6.09	6.41
5590	4.51	2.96	2.49	7.55	6.29	7.91

Case 1 - Day: 23/04/2000, 6:45 a.m. – 2330 MW,

Case 2 - Day: 29/01/2000, 6:45 a.m. – 2790 MW,

Case 3 - Day: 29/01/2000, 7:00 p.m. – 3600 MW,

Case 4 - Day: 23/04/2000, 7:00 p.m. – 5590 MW.

For space reasons, the most relevant information is included here. The complete set of results can be obtained in [5]. The results for the four cases are summarized in Table 14.1.

It can be seen that the fundamental component of the filter current is almost constant for all the conditions. This is due to the voltage control at the busbar. However, it is apparent that the levels of harmonics in the filters are very much dependent on the operating conditions. It is also clear that there is substantial unbalance in the phase values, and this stresses the importance of three-phase measurements when dealing with harmonic problems [6].

14.5   Conclusions

This work has shown that Student's t-distribution is a method suitable to describe the statistical behavior of harmonic levels in a power system. The fairly wide variations in the harmonic levels observed in the three phases of the filters has been useful to stress the importance of three-phase modeling to examine this phenomenon. Simulations using the EMTP program were also carried out, and the results are summarized in [5]. The proposed methodology has been applied to the study of harmonic behavior in filters connected to an HVDC system, but it could be readily applied to other kinds of equipment.

References

[1] Power Measurement 3720 ACM, *Installation and Operation Manual*, Power Measurement Ltd., Canada, May 1997.

[2] P. L. Meyer, *Introduction Probability and Statistical Application*, Addison-Wesley Publishing Company, Inc., Massachusetts, USA, 1965.

[3] S. Ehrenfeld and S. B. Littauer, *Introduction to Statistical Method*, McGraw-Hill Book Company, New York, USA, 1964.

[4] I. Miller and J. E. Freund, *Probability and Statistics for Engineers*, PrenticeHall, Inc., Englewood Cliffs, New Jersey, USA, 1965.

[5]   A. C. F. Marottti, 'Treatment of Measured and Calculated Harmonic Currents in Filters of the Itaipu HVDC System', M.Sc. Thesis (in Portuguese), UFRJ/COPPE, March 2001.

[6] S. Carneiro Jr and A. C. D. Marotti, 'Treatment of Measured and Calculated Harmonic Currents in Filters of the Itaipu HVDC System', Proceedings of the International Power Systems Transients Conference, New Orleans, USA, September 2003 (Paper 11b-2, available for download at http://www.ipst.org).

Part VI
ADVANCED TECHNIQUES

This part presents new and advanced methods for characterizing time-varying waveform distortions. While in the past probabilistic analysis was the most recommended technique, which was based on the fact that harmonics could only be properly visualized as steady-state components, this part presents a number of techniques which give a deeper insight into our understanding of time-varying harmonic distortions.

Chapter 15 shows how wavelet multiresolution analysis can be used to help in the visualization and physical understanding of time-varying waveform distortions. The approach is then applied to waveforms generated by high-fidelity simulation using real-time digital simulation on a shipboard power system.

Chapter 16 utilizes wavelets to characterize several power quality deviation parameters such as harmonics, flicker, transients and voltage sags, while Chapter 17 proposes the utilization of fuzzy logic to analyze, compare and diagnose time-varying harmonic distortion indices in a power system.

Chapter 18 deals with a real-time digital simulation as a new tool to model and simulate time-varying phenomena which can include hardware in the loop for greater accuracy of studies.

In Chapter 19 a statistical signal processing technique, known as independent component analysis (ICA), is used for harmonic source identification and estimation. If the harmonic currents are statistically independent, ICA is able to estimate the currents using a limited number of harmonic voltage measurements and without any knowledge of the system admittances or topology.

In Chapter 20 enhancements and modifications to the Hilbert–Huang method are presented for signal processing applications in power systems. The rationale for the modifications is that the original empirical mode decomposition (EMD) method is unable to separate modes with instantaneous frequencies lying within one octave. The proposed modifications significantly improve the resolution capabilities of the EMD method, in terms of detecting weak higher-frequency modes, as well as separating closely spaced modal frequencies.

In Chapter 21 reference is made to ideal supply conditions which allow recognizing the frequencies and the origins of those interharmonics which are generated by the interaction between the rectifier and the inverter inside the ASD. Formulas to forecast the interharmonic component frequencies are developed firstly for LCI drives and then for synchronous sinusoidal PWM drives. Afterwards, a proper symbolism is proposed to make it possible to recognize the interharmonic origins. Finally, numerical analyses, performed for both ASDs considered in a wide range of output frequencies, give a comprehensive insight into the complex behavior of interharmonic component frequencies; also some characteristic aspects, such as the degeneration in harmonics or the overlapping of an interharmonic couple of different origins, are described.

Although it is well known that Fourier analysis is, in reality, only accurately applicable to steady-state waveforms, it is a tool which is widely used to study and monitor time-varying signals, which are commonplace in electrical power systems. The disadvantages of Fourier analysis, such as frequency spillover or problems due to sampling (data window) truncation, can often be minimized by various windowing techniques. These last two chapters present two newly developed methods which allow for the previous mentioned limitations to be overcome in a more effective way thus allowing a greater understanding and characterization of time-varying waveforms.

Chapter 22 presents a method to separate the odd and the even harmonic components, until the 15th. It uses selected digital filters and down-sampling to obtain the equivalent band-pass filters centered at each harmonic. After the signal is decomposed by the analysis bank, each harmonic is reconstructed using a nonconventional synthesis bank structure. This structure is made up of filters and up-sampling that reconstructs each harmonic to its original sampling rate. This new tool allows for a clear visualization of time-varying harmonics which can lead to better ways to track harmonic distortion and understand time-dependent power quality parameters. It also has the potential to assist with control and protection applications. While estimation technique is concerned with the process used to extract useful information from the signal – such as amplitude, phase and frequency – signal decomposition is concerned with the way that the original signal can be split into other components, such as harmonic, inter-harmonics, sub-harmonics, etc.

Chapter 23 presents a time-varying harmonic decomposition using a sliding-window discrete Fourier transform. Despite the fact that the frequency response of the method is similar to the short time Fourier transform, with the same inherent limitation for asynchronous sampling rate and interharmonic presence, the proposed implementation is very efficient and helpful to track time-varying power harmonics. The new tool allows a clear visualization of time-varying harmonics, which can lead to better ways of tracking harmonic distortion and understand time-dependent power quality parameters. It also has the potential to assist with control and protection applications.

Chapter 24 describes a phase-locked loop (PLL) based harmonic estimation system which makes use of an analysis filter bank and multirate processing. The filter bank is composed of bandpass adaptive filters. The initial center frequency of each filter is purposely chosen to be equal to the harmonic frequencies. The adaptation makes it possible to track time-varying frequencies as well as interharmonic components. A down-sampler device follows the filtering stage which reduces the computational burden, specially, because an undersampling operation is realized.

Chapter 25 deals with Proni analysis which does not require prior knowledge of existing harmonics. It works by detecting frequencies, magnitudes, phases and especially damping factors of exponential decaying or growing transient harmonics.

15

Visualization of time-varying waveform distortions with wavelets

P. M. Silveira and P. F. Ribeiro

15.1   Introduction

Harmonic distortion assumes steady-state conditions and is consequently inadequate to deal with time-varying waveforms. Even the Fourier transform is limited in conveying information about the nature of time-varying signals. The objective of this chapter is to demonstrate and encourage the use of wavelets as an alternative to the inadequate traditional harmonic analysis and still maintain some of the physical interpretation of harmonic distortion viewed from a time-varying perspective. The chapter shows how wavelet multiresolution analysis (MRA) can be used to help in the visualization and physical understanding of time-varying waveform distortions. The approach is then applied to waveforms generated by high-fidelity simulation using a real time digital simulator of an isolated power system.

In recent years, utilities and industries have focused much attention on methods of analysis to determine the state of health of electrical systems. The ability to obtain a prognosis for a system is very useful, because attention can then be brought to any problem a system may exhibit before it causes the system to fail. Considering the increased use of power electronic devices, utilities have experienced – in some cases – a higher level of voltage and current harmonic distortions. A high level of harmonic distortions may lead to failures in equipment and systems, which can be inconvenient and expensive.

Time-Varying Waveform Distortions in Power Systems   Edited by Paulo F. Ribeiro
© 2009 John Wiley & Sons, Ltd

Traditionally, harmonic analysis of time-varying harmonics have been done using a probabilistic approach and assuming that harmonic components vary too slowly to affect the accuracy of the analytical process [1–3]. Another paper has suggested a combination of probabilistic and spectral methods, also referred to as evolutionary spectrum [4]. The techniques applied rely on Fourier transform methods that implicitly assume stationarity and linearity of the signal components. However, in reality distorted waveforms are varying continuously and in some cases (during transients, notches, etc.) too fast for the traditional probabilistic approach.

The ability to give a correct assessment of time-varying waveform/harmonic distortions becomes crucial for the control and proper diagnosis of possible problems. The issue has been analyzed before and a number time–frequency techniques have been used [5]. Also, the use of the wavelet transform as a harmonic analysis tool in general has been discussed [6]. The techniques tend to concentrate on determining equivalent coefficients and do not seem to quite satisfy the engineer's physical understanding given by the concept of harmonic distortion.

In general, harmonic analysis can be considered a trivial problem when the signals are in a steady state. However, it is not simple when the waveforms are nonstationary signals whose characteristics make Fourier methods unsuitable for analysis.

To address this concern, this chapter reviews the concept of time-varying waveform distortions, which are caused by different operating conditions of the loads, sources and other system events, and relates them to the concept of harmonics (that implicitly imply the stationary nature of a signal for the duration of the appropriate time period). Also, this chapter shows how multiresolution analysis with wavelet transform can be useful to analyze and visualize voltage and current waveforms and unambiguously show graphically the harmonic components varying with time. Finally, the authors emphasize the need for additional investigations and applications to further demonstrate the usefulness of the technique.

15.2   Steady-state and time-varying waveform distortions and Fourier analysis

To illustrate the concept of a time-varying waveform Figure 15.1 shows two signals. The first is a steady-state distorted waveform, whose harmonic content (in this case third, fifth and seventh) is constant with time or, in other words, the signal is a periodic one. The second signal represents a time-varying waveform distortion in which the magnitude and phase of each harmonic vary during the observed period of time.

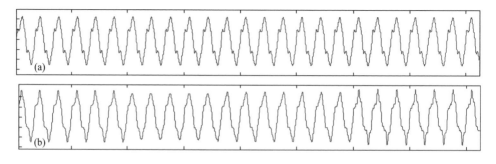

Figure 15.1   (a) Steady-state distorted waveform; (b) time-varying waveform distortion

Power systems, independent of the nature of the signal (stationary or not), need to be constantly measured and analyzed for reasons of control, protection and supervision. Many of these tasks need specialized tools to extract information in time, in frequency or both.

The most well-known signal analysis tool used to obtain the frequency representation is Fourier analysis which breaks down a signal into constituent sinusoids of different frequencies. Traditionally it is very popular, mainly because of its ability to translate a signal in the time domain for its frequency content. As a consequence of periodicity these sinusoids are very well localized in frequency, but not in time, since their support has an infinite length. In other words, the frequency spectrum essentially shows which frequencies are contained in the signal, as well as their corresponding amplitudes and phases, but does not show at which times these frequencies occur.

Using the Fourier transform one can perform a global representation of a time-varying signal but it is not possible to analyze the time localization of frequency contents. In other words, when nonstationary information is transformed into the frequency domain, most of the information about the nonperiodic events contained on the signal is lost.

In order to demonstrate the lack of ability of FFT in dealing with time-varying signals, let us consider the hypothetical signal represented by Equation (15.1) in which, during some time interval, the harmonic content assumes variable amplitude.

$$f = \begin{cases} \sin(2\pi 60t) + 0.2 \sin(10\pi 60t) => 0 < t \leq 0.2\,\text{s} \\ \sin(2\pi 60t) + 0.2.t.\sin(10\pi 60t) => 0.2 < t \leq 2\,\text{s} \\ \sin(2\pi 60t) + 0.2 \sin(10\pi 60t) => 2 < t \leq 5\,\text{s} \end{cases} \tag{15.1}$$

Fourier transform has been used to analyze this signal and the result is presented in Figure 15.2. Unfortunately, as can be seen, this classical tool is not enough to extract features from this kind of signal, first because the information in time is lost and, secondly, the harmonic magnitude and phase will be incorrect when the entire data window is analyzed. In this example the magnitude of the fifth harmonic has been indicated as 0.152 p.u.

Considering the simplicity of the case, the result may be adequate for some simple application; however, large errors will result when detailed information of each frequency is required.

15.3   Dealing with time-varying waveform distortions

Frequently a particular spectral component occurring at a certain instant can be of particular interest. In these cases it may be very beneficial to know the behavior of those components during a given interval of time. Time–frequency analysis, thus, plays a central role in signal processing analysis. Harmonics or high-frequency bursts, for instance, cannot be identified. Transient signals, which are evolving in time in an unpredictable way (like time-varying harmonics) need the notion of frequency analysis that is local in time.

Over the last 40 years, a large effort has been made to deal with the drawbacks previously cited efficiently and represent a signal jointly in time and frequency. As a result a wide variety of possible time–frequency representations can be found in specialized literature, for example in [5]. The most traditional approaches are the short time Fourier transform (STFT) and the Wigner–Ville distribution [7]. For the first case, STFT, whose computational effort is smaller, the signal is divided into short pseudo-stationary segments by means of a window function and, for each portion of the signal, the Fourier transform is found.

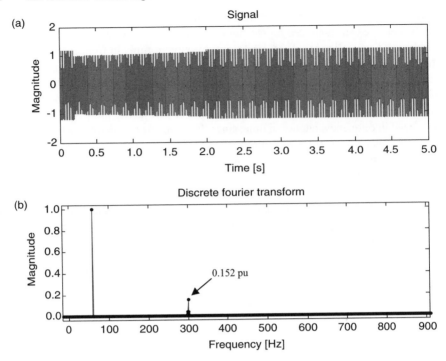

Figure 15.2   (a) Time-varying harmonic and (b) its Fourier (FFT) analysis

However, even these techniques are not suitable for the analysis of signals with complex time–frequency characteristics. For the STFT, the main reason is the width of the fixed data window. If the time-domain analysis window is made too short, frequency resolution will suffer, and lengthening it could invalidate the assumption of a stationary signal within the window.

15.4   Wavelet multiresolution decomposition

Wavelet transforms provide a way to overcome the problems cited previously by means of short width windows at high frequencies and long width windows at low frequencies. In doing so, the use of wavelet transform is particularly appropriate since it gives information about the signal both in the frequency and the time domains.

The continuous wavelet transform of a signal $f(t)$ is then defined as

$$CWT_f^{\psi}(a,b) = \int_{-\infty}^{\infty} f(t)\, \psi_{ab}^{*}(t)dt \tag{15.2}$$

where

$$\psi_{ab}(t) = \frac{1}{\sqrt{a}}\psi\left(\frac{t-b}{a}\right) \quad a,b \in \Re;\ a \neq 0 \tag{15.3}$$

ψ being the mother wavelet with two characteristic parameters, namely dilation (a) and translation (b), which vary continuously.

The results of Equation (15.2) give many wavelet coefficients, which are a function of a and b. Multiplying each coefficient by the appropriately scaled and shifted wavelet yields the constituent wavelets of the original signal.

Just as a discrete Fourier transform can be derived from a continuous Fourier transform, so a discrete wavelet transform can be derived from a continuous wavelet transform. The scales and positions are discretized based on powers of two while the signal is also discretized. The resulting expression is shown in Equation (15.4)

$$DWT_f^{\psi}(j,k) = \frac{1}{\sqrt{a_0^j}} \sum_{n=-\infty}^{\infty} f(n)\, \psi\left[\frac{n-a_0^j k b_0}{a_0^j}\right] \tag{15.4}$$

where j, k, $n \in Z$ and $a_0 > 1$.

The simpler choice is to make $a = 2$ and $b = 1$. In this case the wavelet transform is called dyadic-orthonormal. With this approach the DWT can be easily and quickly implemented by filter bank techniques normally known as multiresolution analysis (MRA) [8]. The multiresolution property of wavelet analysis allows for both good time resolution at high frequencies and good frequency resolution at low frequencies.

Figure 15.3 shows an MRA diagram, which is built and performed by means of two filters: a high-pass filter with impulse response $h(n)$ and its low-pass mirror version with impulse response $g(n)$. These filters are related to the type of mother wavelet (ψ) and can be chosen

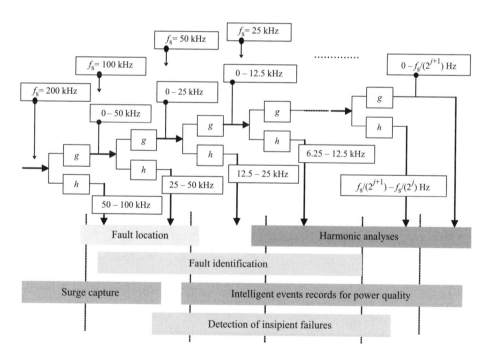

Figure 15.3   Multiresolution analysis and applications in power systems

according to the application. The relation between $h(n)$ and $g(n)$ is given by:

$$h(n) = (-1)^n g(k-1-n) \tag{15.5}$$

where k is the filter length and $n \in \mathbf{Z}$. Each high-pass filter produces a detailed version of the original signal and the low-pass a smoothed version.

The same Figure 15.3 summarizes several kinds of power system applications using MRA, which have been published in the last decade [9–13]. The sampling rate (f_s) showed in this figure represents just a typical value and can be modified according to the application with the faster time-varying events requiring higher sampling rates.

It is important to notice that several of these applications have not been comprehensively explored yet. This is the case of harmonic analysis, including sub-harmonic, inter-harmonic and time-varying harmonic. And the reason for that is the difficulty to physically understand and analytically express the nature of time-varying harmonic distortions from a Fourier perspective. The other aspect, from the wavelet perspective, is that not all wavelet mothers generate physically meaningful decomposition.

Another important consideration, mainly for protection applications, is the computational speed. The time of the algorithm is essentially a function of the phenomena (the sort of information needed to be extracted), the sampling rate and the processing time. Applications which require the detection of fast transients – like traveling waves [10, 11] – normally have a very short processing time. Another aspect to be considered is the sampling rate and the frequency response of the conventional current transformers (CTs) and voltage transformers (VTs).

15.5   The selection of the mother wavelet

Unlike the case of a Fourier transform, there exists a large selection of wavelet families depending on the choice of the mother wavelet. However, not all wavelet mothers are suitable for assisting with the visualization of time-varying (harmonic) frequency components.

For example, the celebrated Daubechies' wavelets (Figure 15.4(a)) are orthogonal and have compact support, but they do not have a closed analytic form and the lowest-order families do not have continuous derivatives everywhere. On the other hand, wavelets like a modulated Gaussian function or harmonic waveform are particularly useful for harmonic analysis due to their smoothness. This is the case of Morlet and Meyer wavelets (Figure 15.4(b)) which are able to show amplitude information [14].

The 'optimal' choice of the wavelet basis will depend on the application. For discrete computations the Meyer wavelet is a good option for visualization of time-varying frequency components because the MRA can clearly indicate the oscillatory nature of time-varying frequency components or harmonics in the Fourier sense of the word.

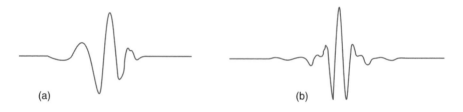

(a)   (b)

Figure 15.4   (a) Daubechies-10 coefficients (db5) and (b) Meyer wavelets

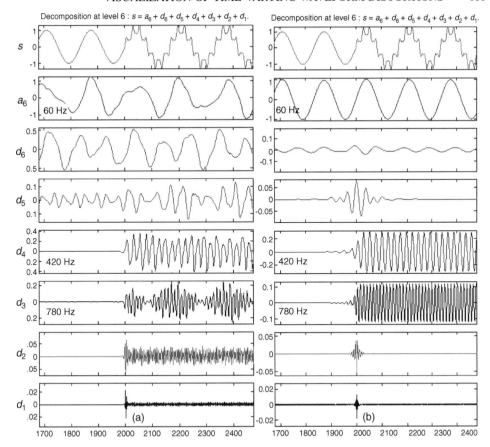

Figure 15.5   An MRA with six levels of decomposition using (a) a Daubechies five wavelet and (b) a Meyer wavelet

In order to exemplify such an application the MRA has been performed to decompose and visualize a signal composed of 1 p.u. of 60 Hz, 0.3 p.u. of seventh harmonic, 0.12 p.u. of 13th harmonic and some noise. The original signal has a sampling rate of 10 kHz and the harmonic content has not been present all the time.

Figure 15.5 shows the results of an MRA with six levels of decomposition, first using Daubechies length five (Db5) and next the Meyer wavelet (dmey). It is clear that even for a high-order wavelet (Db5 – 10 coefficients) the output signals will be distorted when using Daubechies wavelet and, otherwise, will be perfect sinusoids with a Meyer wavelet as it can be seen at level three (13th) and at level four (7th). Some of the detail levels are not the concern because they represent only noise and transition states.

15.6   Impact of sampling rate and filter characteristic

It is important to recognize that the sampling rate and the characteristic of filters in the frequency domain, will affect the ability of the MRA to separate the frequency components and avoid frequency crossing in two different detail levels as previously recognized [15].

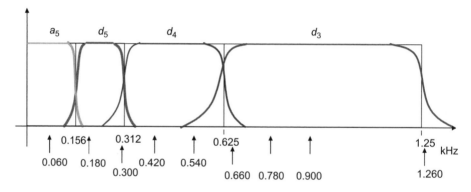

Figure 15.6   MRA filters: frequency support

This problem can be clarified more with the aid of Figure 15.6. The pass-band filters' location is defined by the sampling rate and the frequency support of each filter ($g[n]$ and $h[n]$) by the mother wavelet. If a certain frequency component of interest is positioned inside the crossing range of the filters, this component will be impacted by the adjacent filters. As a consequence, the frequency component will appear distorted in two different levels of decomposition.

In order to illustrate this question let us consider a 60 Hz signal in which a fifth harmonic is also present and whose sampling rate is 10 kHz. An MRA is performed with a 'dmey' filter. According to Figure 15.6 the fifth harmonic is located in the crossing between detail levels d_4 and d_5. The result can be seen in Figure 15.7 where the fifth harmonic appears as a beat frequency in levels d_4 and d_5.

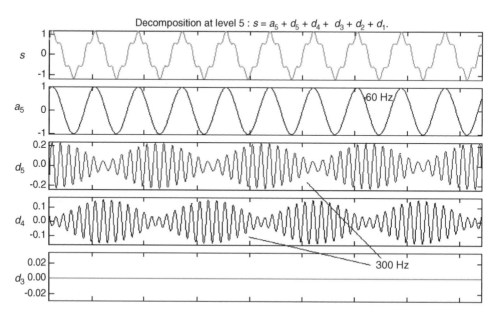

Figure 15.7   MRA with five decomposition levels; the fifth harmonic is revealed on level d_4 and d_5

This problem previously cited may reduce the ability of the technique to track the behavior of a particular frequency in time. However, artificial techniques can be used to minimize this problem. For example, the simple algebraic sum of the two signals ($d_4 + d_5$) will result in the fifth harmonic. Of course, if other components are present in the same level, a more complex technique must be used.

As a matter of practicality the problem can sometimes be easily overcome as shown in Figure 15.5 where the fifth and seventh harmonics (two of the most common components found in power system distortions) can easily be separated if an adequate sampling rate and number of scales/detail coefficients are used.

15.7 Time-varying waveform distortions with wavelets

Let us consider the same signal given by Equation (15.2), whose variable amplitude of the fifth harmonic is not revealed by a Fourier transform. By performing the MRA with a Meyer wavelet in six decomposition levels the fifth harmonic has been revealed in d_5 (fifth detailed level) during all the time of analysis, including the correct amplitude of the contents.

As can be seen from Figure 15.8 this decomposition can be very helpful to visualize time-varying waveform distortions in which both the frequency/magnitude (harmonics) and

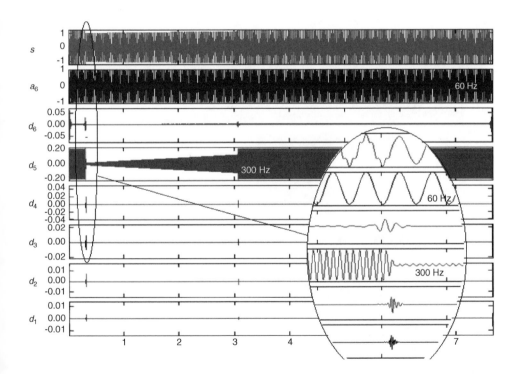

Figure 15.8   *MRA of the signal (Equation (15.1)) showing a time-varying fifth harmonic in level d_5*

time information is clearly seen. This can be very helpful for understanding the behavior of distortions during transient phenomena as well as being used for possible control and protection action.

It is important to remark that other information, such as rms value and phase, can be extracted from the detailed and the approximation levels. For the previous example the rms value during the interval from 0.2 to 2 s has been easily achieved.

15.8   Application to an isolated power system time-varying distortions

In order to apply the presented concept a time-varying voltage waveform, resulting from a pulse in an isolated power system [16], is decomposed by MRA using a Meyer mother wavelet with five levels (only detail levels 5 and 4 and the approximation coefficients are shown). This is shown in Figure 15.9. The approximation coefficient a_5 shows the behavior of the fundamental frequency whereas d_5 and d_4 show the time-varying behavior of the fifth and seventh harmonics respectively. The time-varying behavior of the fundamental, the fifth and the seventh 'harmonics can easily be followed.

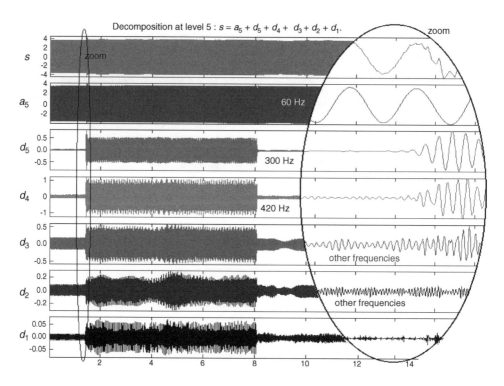

Figure 15.9   Time-varying voltage waveform caused by a pulsed load in an isolated power system: MRA decomposition using a Meyer mother wavelet

15.9   Conclusions

This chapter has attempted to demonstrate the usefulness of wavelet MRA to visualize time-varying waveform distortions and track independent frequency component variations. This application of MRA can be used to further the understanding of time-varying waveform distortions without losing the physical meaning of frequency components (harmonics) variation with time. It is also possible that this approach could be used in control and protection applications.

The chapter recognizes that the successful tracking of the filters is dependent on the adequate selection of the sampling rate and location of the filters. Also the wavelet mother type and the number of detail levels of decomposition may have an impact on the clarity of the information provided.

An isolated power system simulation with a pulsed load was then used to verify the usefulness of the method. The MRA decomposition applying a Meyer mother wavelet was used and the transient behavior of the fundamental, fifth and seventh harmonic were clearly visualized and properly tracked from the corresponding MRA.

References

[1] IEEE Task Force on Harmonics Modeling and Simulation, 'Modeling and Simulation of the Propagation of Harmonics in Electric Power Networks – Part I: Concepts, Models and Simulation Techniques', *IEEE Trans. on Power Delivery*, **11**, 1996, 452–465.

[2] Probabilistic Aspects Task Force of Harmonics Working Group, 'Time-Varying Harmonics: Part II Harmonic Summation and Propagation', *IEEE Trans. on Power Delivery*, **17**, 2002, 279–285.

[3] R. E. Morrison, 'Probabilistic Representation of Harmonic Currents in AC Traction Systems', *IEE Proceedings*, **131**, 1984, 181–189.

[4] P. F. Ribeiro, 'A novel way for dealing with time-varying harmonic distortions: the concept of evolutionary spectra', IEEE Power Engineering Society General Meeting, 2003, Volume 2, July 2003, pp. 1151–1153.

[5] P. Flandrin, *Time-Frequency/Time-Scale Analysis*, Academic Press, London, UK, 1999.

[6] D. E. Newland, 'Harmonic Wavelet Analysis', *Proc. R. Soc., London*, **A443**, 1993, 203–225.

[7] G. Matz and F. Hlawatsch, 'Wigner Distributions (nearly) everywhere: Time-frequency Analysis of Signals, Systems, Random Processes, Signal Spaces, and Frames', *Signal Process.*, **83**, 2003, 1355–1378.

[8] C. S. Burrus, R. A. Gopinath and H. Guo, *Introduction to Wavelets and Wavelet Transforms - A Primer*, 10th edition, Prentice-Hall Inc., New Jersey, USA, 1998.

[9] O. Chaari, M. Meunier and F. Brouaye, 'Wavelets: A New Tool for the Resonant Grounded Power Distribution Systems Relaying', *IEEE Trans. on Power System Delivery*, **11**, 1996, 1301–1038.

[10] F. H. Magnago and A. Abur, 'Fault Location Using Wavelets', *IEEE Trans. on Power Delivery*, **13**, 1998, 1475–1480.

[11] P. M. Silveira, R. Seara and H. H. Zürn, 'An Approach Using Wavelet Transform for Fault Type Identification in Digital Relaying', IEEE PES Summer Meeting, Edmonton, Canada, 1999, 937–942.

[12] V. L. Pham and K. P. Wong, 'Wavelet-transform-based Algorithm for Harmonic Analysis of Power System Waveforms', *IEE Proc. – Gener. Transm. Distrib.*, **146**, 1999, 249–254.

[13] O. A. S. Youssef, 'A wavelet-based technique for discrimination between faults and magnetizing inrush currents in transformers', *IEEE Trans. Power Delivery*, **18**, 2003, 170–176.

[14] N. C. F. Tse, 'Practical Application of Wavelet to Power Quality Analysis', IEEE 1-4244-0493-2/06/2006.

[15] M. H. J. Bollen and I. Y. H. Gu, *Signal Processing of Power Quality Disturbances*, IEEE Press, New York, USA, 2006.

[16] M. Steurer, S. Woodruff, M. Andrus, J. Langston, L. Qi, S. Suryanarayanan and P. F. Ribeiro, 'Investigating the Impact of Pulsed Power Charging Demands on Shipboard Power Quality', The IEEE Electric Ship Technologies Symposium, Arlington, Virginia, USA, 2007.

16

Wavelets for the measurement of electrical power signals

J. Driesen

16.1 Introduction

In modern electrical energy systems, voltages and especially currents become very irregular due to large numbers of nonlinear loads and generators in the grid. In particular, power electronic based systems such as adjustable speed drives, power supplies for IT equipment, high efficiency lighting and inverters in systems generating electricity from distributed renewable energy sources are many sources of disturbances. Distortions encountered are, for instance, harmonics, rapid amplitude variations (flicker) and transients, which are all elements of 'power quality' problems.

In this situation, the registration of electrical energy signals and power related quantities may become problematic, due to differences in power definitions, extended for nonfundamental frequency content, and nonstandardized measurement procedures, which are all based on sinusoidal voltages and currents, and equipment originally designed for undistorted signals [1]. In general, voltage and current signals have become less stationary and periodicity is sometimes completely lost. This fact poses a problem for the correct application of Fourier-based frequency domain power quantities. Practical measurements use the FFT algorithm for this purpose, implicitly assuming infinite periodicity of the signal to be transformed. Therefore, the registered power quantities are often unreliable, and it is not possible to localize distortions with a higher resolution than the time slot.

Time-Varying Waveform Distortions in Power Systems   Edited by Paulo F. Ribeiro
© 2009 John Wiley & Sons, Ltd

16.2   Real wavelet-based power measurement approaches

16.2.1   Basic principles

A wavelet transform maps the time-domain signals of voltage(s) and current(s) in a real-valued time–frequency domain (using a notation similar to [2]), where the signals are described by the wavelet coefficients:

$$i(t) = \sum_{k=0}^{2^{j_0}-1} c_{j_0,k}\phi_{j_0,k}(t) + \sum_{j \geq j_0}^{N-1} \sum_{k=0}^{2^{j-1}} d_{j,k}\psi_{j,k}(t) \tag{16.1}$$

$$v(t) = \sum_{k=0}^{2^{j_0}-1} c'_{j_0,k}\phi_{j_0,k}(t) + \sum_{j \geq j_0}^{N-1} \sum_{k=0}^{2^{j-1}} d'_{j,k}\psi_{j,k}(t) \tag{16.2}$$

with

$$c_{j_0,k} = \langle i(t), \phi_{j_0,k}(t) \rangle \quad \text{and} \quad d_{j,k} = \langle i(t), \psi_{j,k}(t) \rangle \tag{16.3}$$

$$c'_{j_0,k} = \langle v(t), \phi_{j_0,k}(t) \rangle \quad \text{and} \quad d'_{j,k} = \langle v(t), \psi_{j,k}(t) \rangle, \tag{16.4}$$

where j is the wavelet frequency scales, k is the wavelet time scale and c and d are the wavelet coefficients – the accent indicates the voltage coefficients.

16.2.2   Active power

The active power is computed as an averaged sum of the physical power transfer over a certain time interval (T is the averaging time interval and N the highest scale):

$$P = \frac{1}{T} \int_0^T i(t)u(t)dt = P_{j_0} + \sum_{j \geq j_0}^{N-1} P_j$$

$$= \frac{1}{2^N} \left(\sum_{k=0}^{2^{j_0}-1} c_{j_0,k}c_{j_0,k} + \sum_{j \geq j_0}^{N-1} \sum_{k=0}^{2^{j-1}} d_{j,k}d_{j,k} \right). \tag{16.5}$$

In this way it is possible to distinguish the power over the different wavelet scales, to be interpreted as frequency bands. Most interest usually goes to the base band j_0 containing the fundamental power frequency coefficients.

16.2.3   Reactive power

Reactive power is computed in a similar way, but the voltage signal first has to pass a phase-shifting filter network, producing a delay or phase shift of 90° in the approach of [3]:

$$Q = \frac{1}{T} \int_0^T i(t)u(t-90°)dt = Q_{j_0} + \sum_{j \geq j_0}^{N-1} Q_j$$

$$= \frac{1}{2^N} \left(\sum_{k=0}^{2^{j_0}-1} c_{j_0,k}c''_{j_0,k} + \sum_{j \geq j_0}^{N-1} \sum_{k=0}^{2^{j-1}} d_{j,k}d''_{j,k} \right). \tag{16.6}$$

For purely sinusoidal voltages and currents this results in a correct value, but in the presence of distortions it is just one of a number of possible definitions of reactive power, still being a topic of discussion. In the phase shifting implementation approach for reactive powers, often used in traditional measuring equipment, only the 'baseband' value Q_{j0} is associated with the accepted fundamental reactive power [4].

The implementation of the 90° phase shift, actually a time delay, in the time–frequency domain, after the wavelet transform instead of in the time domain as in Equation (16.6), forms an interesting, yet simple alternative for a shift in the time domain. Such a delay in the wavelet domain means a shift of the time-bound wavelet coefficient as calculated in Equation (16.4). Since there are far fewer coefficients involved than time samples, this comes down to a rather simple memory operation.

Hence, no special filter is required, considerably simplifying the computation procedure. Only an extra vector of memory elements to store the previous wavelet coefficients is required. The shift M of the coefficients depends on the wavelet transform parameters, in particular the number of coefficients describing a fundamental period in the base band:

$$Q_{j0} = \frac{1}{2^N} \sum_{k=0}^{2^{j_0}-1} c_{j_0,k} c''_{j_0,k-M} \tag{16.7}$$

with the double accent indicating the delayed coefficients, here the voltage coefficients.

An alternative is found in a projection operation in which the current is split into a component responsible for the active power transfer and a complementary component, the 'reactive' current component. Actually, the current is projected on the voltage wave shape, resulting in a proportional waveform that only transfers active power. The active current component is:

$$P_{j0} = U_{j0} \cdot I_{\text{act}} \tag{16.8}$$

with the rms value of the voltage U (in the base band j_0) filled in, this yields:

$$I_{\text{act}} = \frac{P}{\sqrt{\dfrac{1}{2^N} \sum_{k=0}^{2^{j_0}-1} (c'_{j_0,k})^2}} \tag{16.9}$$

and P already calculated, for example by means of Equation (16.5).

This current rms value is used as a scaling factor for the unity series of voltage wavelet coefficients:

$$\{c_{P,j_0,k}\} = I_{\text{act}} \frac{\{c'_{j_0,k}\}}{\{c'_{j_0,k}\}} . \tag{16.10}$$

Hence, the wavelet transform coefficients of the current proportional to the voltage (also in the time domain), transmitting the same active power, are obtained. The current split is obtained for the base band in the time–frequency domain by computing the complementary 'reactive' series of wavelet coefficients:

$$\{c_{Q,j_o,k}\} = \{c_{j_o,k}\} - \{c_{P,j_o,k}\}. \tag{16.11}$$

In this way, the rms value of the reactive current component can be defined:

$$I_{\text{act}} = \sqrt{\frac{1}{2^N} \sum_{k=0}^{2^{j_o}-1} (c_{Q,j_o,k})^2}. \tag{16.12}$$

This is then used to compute the reactive power value:

$$Q_{j_0} = U_{j_0} \cdot I_{\text{react}}. \tag{16.13}$$

16.3   Application of real wavelet methods

As an example, the two approaches mentioned above are applied on a voltage and current containing harmonic distortion, mainly fifth-order harmonics (Figure 16.1).

These waveforms are sampled at 128 samples per period and are wavelet transformed using a 'symmlet' wavelet [5]. Both resulting coefficients are plotted in Figure 16.2.

Note that the base band is described by only four coefficients. Hence buffering in order to obtain the 90 ° shift represents a limited operation, compared to a delay or phase shift in the time domain in which all 128 samples have to be processed.

The active and reactive powers are calculated and compared to analytically derived reference quantities in Table 16.1.

It can be noticed that the accuracy of the wavelet approach is rather good: the processed values are within 0.5% of the analytical values. The difference is due to numerical errors and 'leakage' between frequency bands.

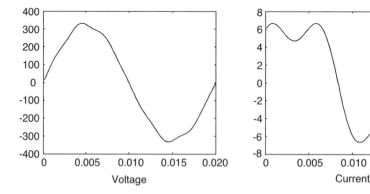

Figure 16.1   Voltage and current period used for the example calculation

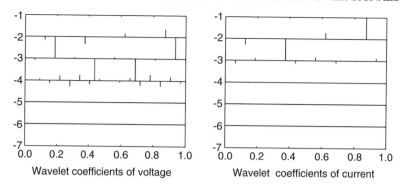

Figure 16.2   Wavelet coefficients at different scales: top scale is associated with the base band

Table 16.1   Comparison of computed power quantities

	P	Q
Analytical reference	995.93	575.00
Delay approach	991.80	575.24
Splitting approach	991.80	575.24

16.4   Concept using complex wavelet based power definitions

16.4.1   Introduction

The power calculation procedure using real wavelets is less subject to errors due to the irregularity of signals, but is still based on classical averaging over a designated time interval (Equation (16.5)), an idea also present in the derivation of regular frequency domain quantities, where a sampled interval is covered. Complex wavelets seem to have been used only rarely in power system analysis, except for [6].

To be able to calculate any power quantities, one needs to be able to analyze amplitudes and phase differences between the related voltage(s) and current(s). A (complex) Fourier transform yields the appropriate phasors, but a (real) wavelet transformed signal does not readily deliver phase information. The 'hidden' phase-related information, present in the time localization property of the wavelet coefficient, influences the average power values as in [2, 3].

A complex wavelet, however, does yield phase information. It is even possible to retrieve an instantaneous phase $\varphi(t)$ using the polar representation of the complex coefficients [5]. It is therefore a candidate to serve as a basis in novel power definitions having better time localization properties. Therefore, this would result in true time–frequency domain power quantities.

16.4.2   Complex wavelet based power definitions

The relevant voltages and currents are transformed to the time–frequency domain using a well-chosen complex wavelet $\psi(t)$, with scaling parameter s (setting the frequency range) and translation parameter t (determining the time localization) [5]:

$$f_W(t,s) = \langle f, \underline{\psi}_{t,s} \rangle = \int_{-\infty}^{+\infty} f(t') \cdot \frac{1}{\sqrt{s}} \underline{\psi}\left(\frac{t'-t}{s}\right) dt'. \tag{16.14}$$

In a single-phase system this yields two series of complex wavelet coefficients for the voltage U and for the current I, indicated by a subscript w: U_w and I_w. From these coefficients, instantaneous amplitude and phase values are derived for the different subbands.

$$\underline{U}_W(t,s) = U_W(t,s) \angle \varphi_{U,W}(t,s) \tag{16.15}$$

$$\underline{I}_W(t,s) = I_W(t,s) \angle \varphi_{I,W}(t,s). \tag{16.16}$$

For most power measurement applications, the most interesting subband is the one covering the fundamental frequency, here indicated as s_f. A complex-wavelet based power quantity is then defined in a way analogous to the Fourier-based active power definition, now using the instantaneous voltage and current amplitude and the instantaneous phase difference between voltage and current:

$$p_W(t, s_f) = U_W(t, s_f) \cdot I_W(t, s_f) \cdot \cos(\varphi_W(t, s_f)) \tag{16.17}$$

with the phase difference, based on the difference of the instantaneous phases:

$$\varphi_W(t, s_f) = \varphi_{U,W}(t, s_f) - \varphi_{I,W}(t, s_f). \tag{16.18}$$

Similarly, a complex-wavelet based reactive power quantity can be defined as well:

$$q_W(t, s_f) = U_W(t, s_f) \cdot I_W(t, s_f) \cdot \sin(\varphi_W(t, s_f)). \tag{16.19}$$

Both p_W and q_W can be obtained immediately by

$$\underline{p}_W(t, s_f) = p_W + jq_W = \underline{U}_W(t, s_f) \text{conj}(\underline{I}_W(t, s_f)). \tag{16.20}$$

A 'momentary' power factor can be defined using the instantaneous phases:

$$\text{dPF}(t) = \cos(\varphi_W(t, s_f)). \tag{16.21}$$

An apparent power-like quantity is obtained as well:

$$s_W(t, s_f) = \sqrt{p_W(t, s_f) + q_W(t, s_f)}. \tag{16.22}$$

16.4.3   Wavelet choice

The complex wavelet (16.14), has to be appropriate for the analysis of power signals. Therefore, a smooth oscillating function is preferred. The following functions are candidate [5]: the complex Gaussian wavelet (C_p is a scaling parameter):

$$\psi_G(x) = C_p \cdot e^{-x^2} \cdot e^{-jx} \tag{16.23}$$

and the complex Morlet wavelet (f_b: a bandwidth parameter, f_c: the center frequency):

$$\psi_M(x) = \sqrt{\pi f_b} \cdot e^{-\frac{x^2}{f_b}} \cdot e^{2j\pi f_c x}. \tag{16.24}$$

The real and imaginary part of this wavelet, with $f_b = 50\,\text{Hz}$ are plotted in Figure 16.3.

16.4.4   Discussion: physical interpretation

The physical interpretation of power definitions is always debatable. The only physically correct value is the time domain based power $p(t) = u(t)i(t)$, containing a DC part as well as higher frequency oscillations. Its average, global or for a certain harmonic frequency in the case of periodic signals, is expressed by the frequency domain active powers P or P_h [4]. Reactive power can be regarded in a similar way, as a measure for the power oscillating between source and user.

The complex-wavelet based power rather yields a sort of sliding average power, as the time window over which it averages is limited to the length of the wavelet function and determined by the dilation parameter [5]. Due to the decaying shape of the wavelet, this average is weighted. Thus, this power can be localized in time. The frequency localization is broader than just a sharp single harmonic as the complex wavelet covers a certain frequency band.

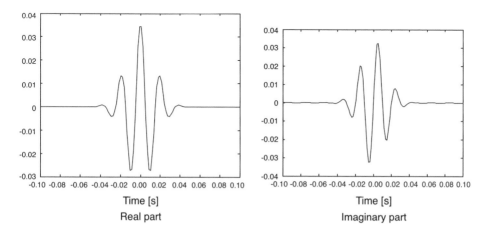

Figure 16.3   Real and imaginary part of the Morlet wavelet

16.4.5   Comparison with fourier-based approach

A traditional Fourier-based power computation starts with calculating an FFT of a frame of current or voltage samples. Then adequate frequencies are selected and processed, yielding one value per power quantity over the whole frame, resulting in a poor time resolution. Alternatively, to provide a better time resolution, one can use overlapping frames, causing one sample to be used more than once. The frequency resolution is equidistant.

A wavelet transform is typically calculated using filters [5]. This yields several transformed coefficients over the same time frame, depending on the scaling of the wavelet basis. This provides a better time localization. In return, the frequency resolution is not as sharp, especially not for the higher frequencies. This does not pose as many problems in practical power systems as one is mainly interested in a sharp distinction of the fundamental frequency behavior, a relatively good distinction of the low-frequency harmonics, which in practice are found in pairs (e.g. fifth/seventh) and a rough idea about high-frequency phenomena. The typical wavelet frequency resolution pattern is well suited for such requirements.

16.5   Examples

16.5.1   Practical implementation aspects

The wavelet function providing the best results, in the sense that similar results for stationary signals were obtained as with Fourier analysis, was the Morlet function (Equation (16.24)), with a pseudo-frequency of 50 Hz, the fundamental power system frequency in the examples ($f_b = 4 \times 10^{-4}$ Hz, $f_c = 50$ Hz). This wavelet function is limited to eight periods of the fundamental frequency. One period contains 128 samples.

In the following examples, the (continuous) wavelet transform is implemented as a convolution. In a practical implementation a faster filter algorithm is required. To eliminate boundary effects, an appropriate number of transformed samples is omitted.

16.5.2   Example 1: voltage dip

The first example shows a typical nonperiodic distortion, a dip in the voltage (Figure 16.4). This power quality problem may occur due to a sudden change in the load, for example the start

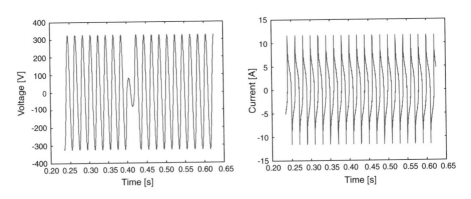

Figure 16.4   Voltage with a dip and current with harmonics

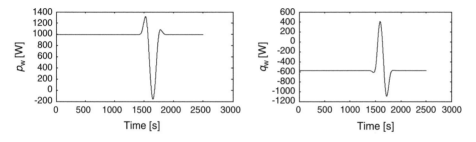

Figure 16.5 p_W and q_W associated with the waveforms in Figure 16.4

of a large motor or an arc furnace. In an FFT, this results in the presence of subharmonic frequencies and a lower fundamental, without knowing when the dip occurred. The current is identical to the current in the previous example.

The resulting powers p_W and q_W are computed and plotted in Figure 16.5. They clearly show the sudden drop in transferred power and change in reactive power due to the local change in phase difference. The small overshoots at the beginning of the power curves are due to the discontinuous jump in phase angles in this simulated voltage. These will be smaller in reality as most voltage dips settle in more slowly. Even more, the reaction of the load, usually a lower current, is not taken into account.

16.5.3   Example 2: dynamically changing current

The second example contains a continuously rising distorted current, for instance due to an adjustable speed drive with an increasing mechanical output, for instance due to increasing speed (Figure 16.6).

The resulting powers p_W and q_W are computed and plotted in Figure 16.7. They follow the smooth rate of change of the current. In a Fourier based analysis, the spectrum of this nonperiodic current would result again in some sort of subharmonic occurrences.

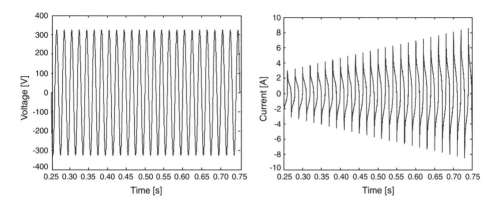

Figure 16.6   *Clean voltage and transient harmonically distorted current*

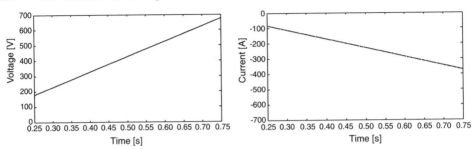

Figure 16.7 p_W and q_W associated with the waveforms in Figure 16.6

References

[1] J. Driesen, T. Van Craenenbroek and D. Van Dommelen, 'The registration of harmonic power by analog and digital power meters', *IEEE Trans. on Instrumentation and Measurement*, **47**, 1998, 195–198.

[2] W.-K. Yoon and M. J. Devaney, 'Power Measurement Using the Wavelet Transform', *IEEE Trans. on Instrumentation and Measurement*, **47**, 1998, 1205–1210.

[3] W.-K. Yoon and M. J. Devaney, 'Reactive Power Measurement Using the Wavelet Transform', *IEEE Trans. on Instrumentation and Measurement*, **49**, 2000, 246–252.

[4] IEEE Working Group on Nonsinusoidal Situations: Effects on Meter Performance and Definitions of Power, 'Practical Definitions for Powers in Systems with Nonsinusoidal Waveforms and Unbalanced Loads: A Discussion', *IEEE Trans. on Power Delivery*, **11**, 1996, 79–101.

[5] S. Mallat, *A Wavelet Tour of Signal Processing*, Academic Press, San Diego, USA, 1999.

[6] M. Meunier and F. Brouaye, 'Fourier transform, Wavelets, Prony analysis: Tools for Harmonics and Quality of Power', Proceedings of the 8th International Conference on Harmonics and Quality of Power, Athens, Greece, 1998, pp. 71–76.

17

Fuzzy logic application for time-varying harmonics

B. R. Klingenberg and P. F. Ribeiro

17.1 Introduction

Harmonic distortion in power systems continues to grow in importance due to the proliferation of nonlinear loads and sensitive electronic devices. Due to the inherently time-varying nature of harmonics, it is difficult to predict the exact level of harmonics in the system. The use of traditional tools such as fast Fourier transforms (FFT) may not be appropriate for the analysis of time-varying harmonics. Hence, more advanced techniques are required to quantify their impact properly. This chapter proposes the utilization of fuzzy logic to analyze, compare and diagnose time-varying harmonic distortion indices in a power system.

When nonlinear loads are connected to an electric power system they tend to draw nonlinear currents and consequently distort the system voltage [1]. Typically the harmonics are assumed to be periodical/time-invariant. However, harmonic components are continually changing with time [2]. It is important then to look at the harmonics from a time-varying perspective. Harmonic distortions can adversely affect many systems by causing erratic behavior in microcontrollers, breakers and relays. The most substantial effect of harmonic distortions within a system is the increase in the temperature which results in increased losses, transformer derating and possible equipment failure [1, 2]. When techniques such as FFT are applied to quantify the spectrum of time-varying harmonics, the method suffers a break down. Hence, in order to analyze, compare and quantify the time-varying nature of harmonic

Time-Varying Waveform Distortions in Power Systems   Edited by Paulo F. Ribeiro
© 2009 John Wiley & Sons, Ltd

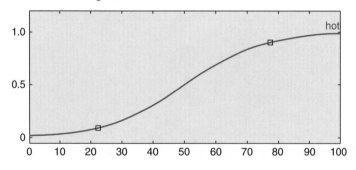

Figure 17.1   Fuzzy membership function for hotness

distortions, the framework of a conceptual application of a fuzzy logic technique is presented in this chapter. Synthetic data are used for the purpose of demonstrating the concept.

17.2   Fuzzy logic

In a traditional bivalent logic system an object either is or is not a member of a set. The idea of fuzzy sets is that the members are not restricted to true or false definitions. A member in a fuzzy set has a degree of membership to the set. For example, the set of temperature values can be classified using a bivalent set as either hot or not hot. This would require some cut-off value where any temperature greater than that cut-off value is 'hot' and any temperature less than that value is 'not hot'. If the cut-off point is at 50 °C then this set does not differentiate between a temperature that is 20 °C and a temperature of 49 °C, they are both 'not hot' [3–5].

If a fuzzy set were to be used in this situation each member of the set, or each temperature, would have a degree of membership to the set of temperature. The function that determines this degree of membership is called the fuzzy membership function as shown in Figure 17.1.

There are different membership function topologies that can be used; the most common are triangular, Gaussian and sigmoidal. The function in Figure 17.1 is a sigmoidal function. The attributes of the membership function can be modified based on the desired input [6]. If the relevant temperature range was between 20 and 60 °C, for example, and more weight was needed for higher temperatures, then an appropriate membership function can be determined. The determination of this function is dependent on the desired weighting of the input.

17.3   Fuzzy logic control

The basic fuzzy logic control system is composed of a set of input membership functions, a rule-based controller and a defuzzification process. The fuzzy logic input uses member functions to determine the fuzzy value of the input. There can be any number of inputs to a fuzzy system and each one of these inputs can have several membership functions. The set of membership functions for each input can be manipulated to add weight to different inputs. The output also has a set of membership functions. These membership functions define the possible responses and outputs of the system [6].

The fuzzy inference engine is the heart of the fuzzy logic control system. It is a rule-based controller that uses 'If–Then' statements to relate the input to the desired output [6]. The fuzzy inputs are combined based on these rules and the degree of membership in each function set. The output membership functions are then manipulated based on the controller for each rule. Several different rules will usually be used since the inputs will usually be in more than one membership function. All of the output member functions are then combined into one aggregate topology. The defuzzification process then chooses the desired finite output from this aggregate fuzzy set. There are several ways to do this such as weighted averages, centroids or bisectors. This produces the desired result for the output.

17.4   Fuzzy logic in power systems

There are relatively few implemented systems using fuzzy logic in the power industry at this time [4]. This is due to the fact that most of the focus with fuzzy systems has been in research and not in implementation. The application of fuzzy logic in power systems has been mainly focused on controllers and system stabilizers [5]. There are other applications in areas such as prediction, optimization and diagnostics [5]. The diagnosis area of application is particularly attractive because it is difficult to develop precise numerical models for failure modes [4]. Understanding the failure modes of a system is usually only done by approximations at best; therefore, the prediction of a failure is difficult to do because of this inherent imprecision. There is rarely a single measurement that can indicate impeding failure and so several measurements need to be taken and compared based on the specific system [5]. A generic diagnostic tool is difficult to develop since it needs to be tuned to a specific system to obtain a reasonable performance [4].

17.5   The harmonic distortion fuzzy model

The fuzzy model for the harmonic distortion diagnostic tool was implemented in MATLAB using the fuzzy logic toolbox. This toolbox allows for the creation of input membership functions, fuzzy control rules and output membership functions [7]. To implement this system in Simulink the system will need to have two different inputs: the harmonic voltage and the temperature. The temperature is used in the analysis because the temperature of a piece of electrical equipment will increase as the harmonic distortion content increases [2]. These two inputs will then be processed by a fuzzy logic controller that will output a degree of caution. This degree of caution is then decoded into one of four possible outputs: no problem, caution, possible problem and imminent problem. A simple (two-variable example) diagnostic system was created as shown in Figure 17.2.

This diagnostic system uses random number inputs for the harmonic voltage and temperature inputs. The harmonic voltage input (shown in Figure 17.3) is a random distribution in the range of 0 to 10. The temperature input (shown in Figure 17.4) is a random distribution in the range of 30 to 100 °C.

These input function ranges can now be used to determine the fuzzy membership sets. The fuzzy system will have these two inputs and one indicating output as shown in Figure 17.5. The fuzzy system used will be a Mamdani system [6] with the centroid method for defuzzification [6]. The input membership function for harmonic voltage (shown in Figure 17.6) will have five membership functions: very low, low, medium, high and very high. The range of this function is 0 to 10, these are the possible input values. The very low and very high membership functions continue on to infinity in either direction to include any voltage value out of range.

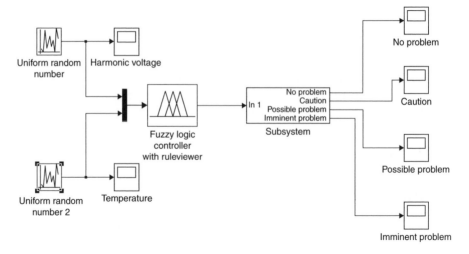

Figure 17.2   Harmonic distortion diagnostic Simulink model

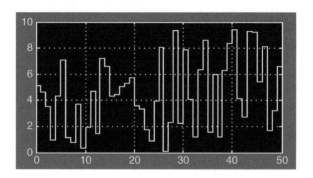

Figure 17.3   Harmonic voltage input

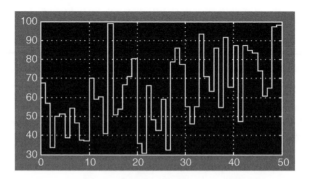

Figure 17.4   System temperature input

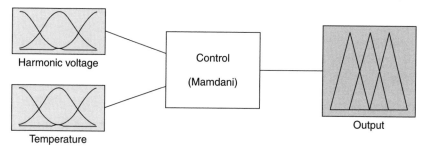

Figure 17.5   The fuzzy logic diagnostic controller

The harmonic voltage membership functions define anything from 0 to 5 as low, using a triangular function; anything from 2.5 to 7.5 is medium and anything from 5 to 10 is high. An input with a harmonic voltage of 3 will have about an 80% membership in the low function and about a 20% membership in the medium function. The total membership in this case will add up to be 100% but this is not required in a fuzzy set.

There are four temperature input membership functions as shown in Figure 17.7. The below normal function is a triangular function centered at 30 °C that extends up to 53 °C. The normal triangular function is centered at 53 °C, extending from 30 to 76 °C at its limits. The overheating triangular membership function is centered at 76 °C with the same magnitude of range as the normal function. The very hot function begins at 76 °C and peaks at 100 °C where it extends on past the set maximum input of 100 °C to cover out of limit values.

The output has four membership functions, no problem, caution, possible problem and imminent problem (shown in Figure 17.8). These membership functions are all triangular and are spread evenly over a range from 0 to 1.

Once all of the input and output membership functions have been defined the heart of the control can now be defined – the rules. The fuzzy rules are in the form of 'if–then' statements. These statements look at both inputs and determine the desired output. In this system increasing voltage and increasing temperature will lead to an imminent problem. A low temperature with a relatively high voltage will not necessarily be an imminent problem though. The rules defined for this system are in listed in Table 17.1.

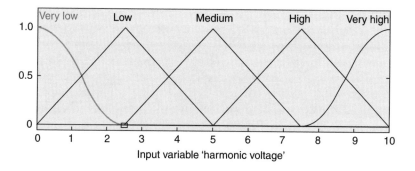

Figure 17.6   The harmonic voltage input membership functions

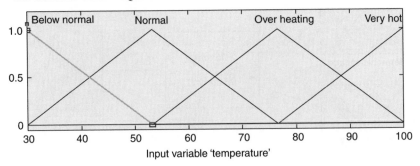

Figure 17.7   The system temperature input membership functions

These rules are the defining elements of this system. They determine the output based on the input. These rules can be seen graphically as a rule map (shown in Figure 17.9). This rule map illustrates the response of the system to different inputs. On the map the dark grey represents no problem and the light yellow represents an imminent problem. The intermediate colors show the mix of fuzzy options in between.

Now that the fuzzy control system has been entirely defined it is exported into the Simulink model. The model includes some decoding logic that will output different discrete levels for each of the possible outputs, similar to the one shown in Figure 17.2. This could serve as an input to some other system.

17.6   Expanded model

The inputs for this example system have been shown before; they are randomly generated data within a valid range. The system can be simulated using these data. There are four different output scopes that will indicate the output signal of the fuzzy controller. These results could then be used to compute probability distribution functions and/or send alarm notes to a central controller.

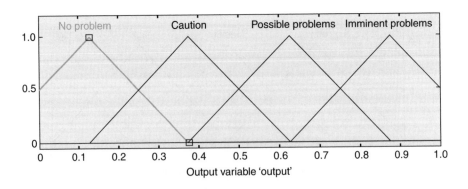

Figure 17.8   The output membership function

Table 17.1   Membership rules

If harmonic voltage is:	And the temperature is:	Then the output is:
Very_low	Below_normal	No_problem
Very_low	Normal	No_problem
Very_low	Overheating	No_problem
Very_low	Very_hot	Caution
Low	Below_normal	No_problem
Low	Normal	No_problem
Low	Overheating	Caution
Low	Very_hot	Possible_problems
Medium	Below_normal	No_problem
Medium	Normal	Caution
Medium	Overheating	Possible_problems
Medium	Very_hot	Possible_problems
High	Below_normal	Caution
High	Normal	Possible_problems
High	Overheating	Possible_problems
High	Very_hot	Imminent_problems
Very_high	Below_normal	Possible_problems
Very_high	Normal	Possible_problems
Very_high	Overheating	Imminent_problems
Very_high	Very_hot	Imminent_problems

A better diagnostic tool can be developed that takes more data into account and provides a single output [5]. This can be done using the previous model and setting up a more involved case. This case will look at the fundamental voltage variation as well as the variation of the third, fifth and seventh harmonics. The following variations, shown in Table 17.2, will be used in this case.

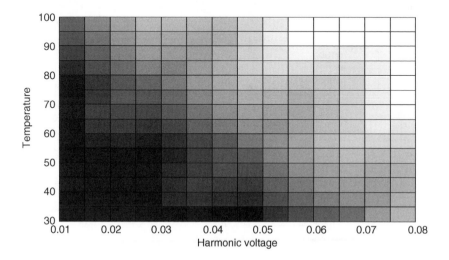

Figure 17.9   Fuzzy control rule map

Table 17.2   Input variations

Fundamental	+/− 10%
V3	1%–8%
V5	1%–8%
V7	1%–8%
VTHD	1%–13%
Temperature	30 °C–100 °C

These inputs were chosen based on the recommended harmonic voltage limits from [1]. The fundamental and the harmonics will have uniform random function generators as inputs. These function generators will generate a uniform distribution of inputs within the variances given in Table 17.2. The total harmonic distortion will be calculated depending on these inputs and will be in the range of 1–13%, which are the best and worst case scenarios. The temperature variation will remain the same as in the previous case.

Using these input data and the basic model developed in the first case, a Simulink model can be developed that processes all the input data and gives an appropriate indication for each harmonic, the fundamental and THD. These indications will remain the same as in the previous case. Each indicator will have a fuzzy logic controller that implements one of three control topologies: one for the fundamental, THD and the harmonics. The final model can be seen in Figure 17.10.

The first fuzzy logic controller will use the fuzzy inference system that has a rule similar to the one shown in Figure 17.10 except that the scale is now from 0.9 to 1.1. This represents the percentage of the ideal. The total harmonic distortion rule surface and the harmonic voltage rule surfaces will be essentially the same except with different scales again based on the variations given in Table 17.2. The second fuzzy controller in Figure 17.10 uses the THD fuzzy

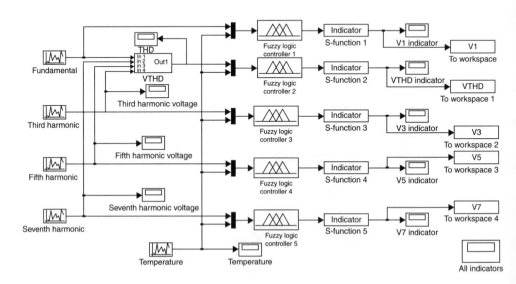

Figure 17.10   Final Simulink model

```
if  u < 0.25
        sys= 0;
elseif   u < 0.5
        sys= 1;
elseif   u < 0.75
        sys= 2;
else
        sys= 3;
end;
```

Figure 17.11   S-function code

inference system and the remaining three fuzzy controllers use the harmonic fuzzy inference system. The S-functions in the model are simple MATLAB files that process the fuzzy logic controller output and determine the output level using the code shown in Figure 17.11.

This code will split the fuzzy output into four ranges and then output either 0, 1, 2 or 3 corresponding to no problem, caution, possible problems and imminent problems. This provides a simple way to visualize the output data for this proof of concept case. After this final decoding step the output is sent to scopes and also to workspace variables. These workspace variables will allow for statistical analysis after simulation and eliminate the need for having four separate scopes, one for each fuzzy controller.

17.7   Simulation results

The inputs for the simulation are all from uniform random number generators that will generate random numbers within the previously defined ranges. The system will be modeled for a 24 hour period with the number generators producing a new number every 2 min. This is what the input from a typical power system monitoring device would be [1].

Each random number generator has a different seed value to produce a different set of numbers. All of the inputs can be seen in Figure 17.12. From top to bottom the inputs are the fundamental, third harmonic, fifth harmonic, seventh harmonic and the temperature in percentages. The data used in this system are purely fictitious since this is a proof of concept experiment. This model provides a base for future applications which would use real data. In any future application the system would have to be retuned to meet the requirements of that system. This retuning process would involve the membership functions and the fuzzy rules and is a common aspect of fuzzy logic applications [5].

The outputs of the system will be one of four options: 0, 1, 2 or 3 which represent the possible warning indicators. These outputs can be seen in Figure 17.13. In this figure the outputs, from top to bottom, are the fundamental, third, fifth, seventh harmonics and finally the total harmonic distortion (THD).

Most of the outputs appear to be in the 'caution' or 'possible problems' state by looking at the output plot. The outputs were exported into MATLAB where they could be plotted in a histogram so that the distribution of outputs could easily be seen. This histogram can be seen in Figure 17.14. From this we can tell that most of the results are the first three states,

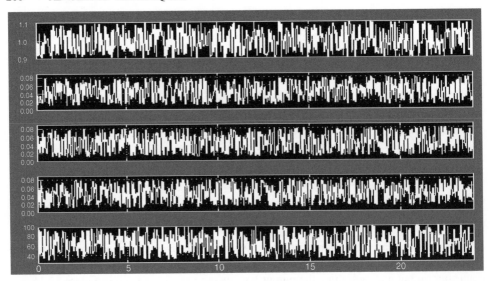

Figure 17.12   All system inputs (top down): fundamental, third, fifth and seventh harmonics and temperature

except with the total harmonic distortion where most of the results are either 'caution' or 'imminent problems'.

The simulation was repeated with different input distributions as well. Figure 17.15 shows the output histogram when the inputs are all Gaussian distributions. The output histograms

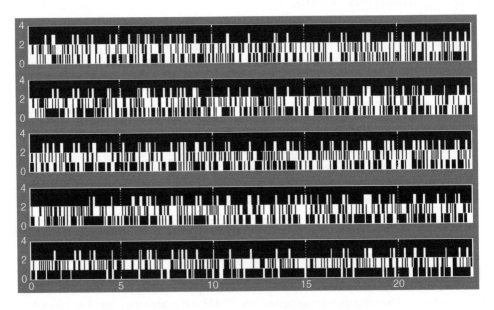

Figure 17.13   Output indicators (top down): fundamental, third, fifth and seventh harmonics and THD

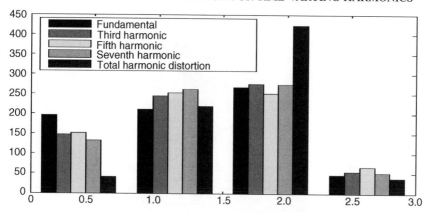

Figure 17.14   Output histogram: uniform distribution

are broken up into four sections as a result of the filtering that occurs in the S-function code, shown in Figure 17.11, which reduces the fuzzy controller output to only four states.

The output distribution of the fuzzy controllers can be plotted with higher resolution; for an example the output of the fundamental fuzzy controller can be seen in Figure 17.16. This output is the unprocessed output of the fuzzy logic controller.

The simulation is based on possible typical data modeled by random number generators. Any statistical trends in the system are a result of the nature of the input for this simulation. The inputs for this simulation do not represent actual harmonic levels and so the statistical analysis is performed simply to show how it would be done if real data inputs were used. Other parameters such as voltage unbalance and mechanical vibration may be studied as outputs as shown in Figure 17.17.

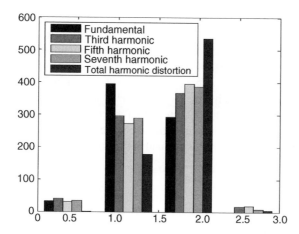

Figure 17.15   Output histogram: Gaussian distribution

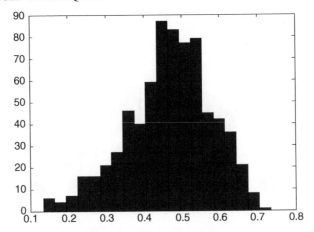

Figure 17.16   Fundamental fuzzy logic controller output histogram

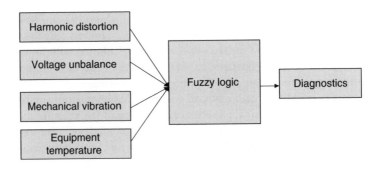

Figure 17.17   Schematic of an expanded fuzzy logic based diagnosis tool

17.8   Conclusions

This chapter presented a methodology to analyze harmonic distortion impact on system equipment using a fuzzy logic based system. The examples simulated indicate the potential for using such a procedure for studying complex systems and performing a meaningful evaluation and/or analysis of the impact of time-varying harmonic distortion on a power system. The use of advanced techniques such as fuzzy logic may help in overcoming the challenges faced with traditional tools such as FFT for the diagnosis and analysis of time-varying harmonics. Other intelligent techniques such as agents and artificial neural networks may also serve as suitable candidates for the analysis and diagnosis of time-varying harmonic distortions.

References

[1] IEEE Standard 519-1992, *IEEE Recommended Practices and Requirements for Harmonic Control in Electrical Power Systems*, April 1993.

[2] S. M. Halpin, 'Harmonics in Power Systems', in *The Electric Power Engineering Handbook*, (L. L. Grigsby, ed.) CRC Press, Boca Raton, Florida, USA, 2001.

[3] Bart Kosko, *Fuzzy Thinking*. Hyperion, New York, USA, 1993.

[4] T. Hiyama and K. Tomsovic, 'Current Status of Fuzzy System Applications in Power Systems', Proceedings of the IEEE SMC99, Tokyo, Japan, October, 1999, pp. VI 527–532.

[5] K. Tomsovic, 'Fuzzy Systems Applications to Power Systems', Chapter IV-Short Course, International Conference on Intelligent System Application to Power Systems, Rio de Janeiro, Brazil, April 1999.

[6] H. T. Nguyen, R. P. Nadipuram, C. L. Walker and E. A. Walker. *A First Course in Fuzzy and Neural Control*. Chapman & Hall/CRC, Boca Raton, Florida, USA, 2003.

[7] The MathWorks. [Online]. Fuzzy Logic Toolbox. Available: http://www.mathworks.com/products/fuzzylogic/ [Dec 16, 2005].

18

Real-time simulation of time-varying harmonics

Y. Liu, M. Steurer and P. F. Ribeiro

18.1 Introduction

Emulating conditions existing on a real power system has always been a critical requirement for myriad applications, including testing new control and protection equipment. Historically, this requirement has been met by Transient Network Analyzers (TNA) which were built using scaled-down analog models of power system equipment and which were interconnected in their original configuration. Although TNAs were inherently real-time in nature, they soon ran into problems of complexity, inadequate scaling, accuracy, and cost. These limitations motivated the introduction of off-line digital simulators such as the Electromagnetic Transient Programs (EMTP), which model the system mathematically and solve it numerically in the time-domain. Nevertheless, their main drawback is the lack of real-time interaction with the equipment being tested. Rapid advances in modern computers and digital signal processing hardware have finally led to the development of fully digital, real-time simulators that are capable of simulating adequately detailed power system models with sufficient speed to meet the output bandwidth requirement for representing real network conditions. The chief characteristic of such simulators is their ability to interact in real-time with actual hardware connected in closed-loop. Therefore, many tests that cannot be performed on a real system can be safely and efficiently done on a real-time digital simulator.

In the power quality (PQ) and harmonics research area real-time digital simulators can play a vital role. Traditionally, time-varying harmonics were studied indiscriminately using statistical and probabilistic methods for periodic harmonics. However, the practice could not accurately describe the random characteristics of the time-varying processes, or capture the

Time-Varying Waveform Distortions in Power Systems   Edited by Paulo F. Ribeiro
© 2009 John Wiley & Sons, Ltd

reality of physical phenomena giving rise to such harmonics. To precisely interpret the time-varying processes, a time-dependent spectrum is needed to compute the local power-frequency distribution at each instant of time. With real-time digital simulators, this intense requirement on computational power can be easily satisfied.

The concern of PQ has also led to significant advances in the equipment development for PQ measurement, waveform generation, disturbance detection, and mitigation. Several PQ measurement and monitoring devices are currently being developed using both DSP and general purpose microprocessor technologies. There is a need for testing these fast acting apparatus and their controllers to evaluate their response to typical PQ disturbances such as voltage sags, swells, harmonics, impulses, transients, flicker, unbalanced operation and interruptions. A real-time digital simulator can be used to simulate such disturbances and apply them to the tested device under closed-loop conditions [1–6].

To achieve better accuracy in the power quality studies of large and complex power systems in an economic way, a novel power quality assessment method based on a real-time (RT) hardware-in-the-loop (HIL) simulator can be used. Hardware-in-the-loop is an idea of simultaneous use of simulation and real equipment. Generally, an HIL simulator is composed of a digital simulator, one or more hardware pieces under test, and their analog and digital signal interfaces (e.g. high performance A/D and D/A cards).

This chapter describes a sensitivity analysis of the power quality deviations of a variable speed drive controller card using real time simulation and hardware in the loop. The experiment has contributed to the design of a power quality test bed, which can be used to test the immunity of electric components and equipment and the consequent impact on AC distribution systems.

18.2   Description of the RT–HIL platform

Figure 18.1 shows the diagram of the RT–HIL platform. The platform is composed of a digital simulator, tested hardware, and their interface (e.g. power amplifiers and transducers). The

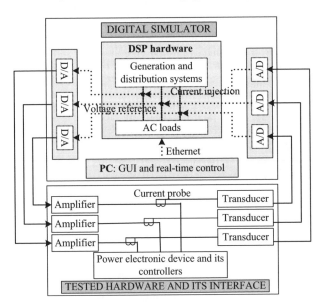

Figure 18.1   Diagram of the RT–HIL platform

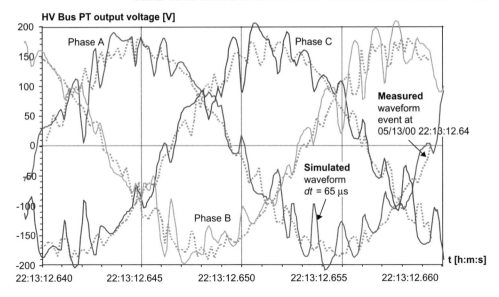

Figure 18.2   Measured AC bus voltage waveforms (broken gray lines) of a real shipboard distribution system (with only one bridge of each cycloconverter operating)

simulator can be used either as an independent simulation system (e.g. no hardware in the loop) or with tested hardware. In Figure 18.1, a real power electronic device is connected to the simulated power system through D/A adaptors and power amplifiers. The supply current of the AC/DC converter is measured and fed back into the system at the common coupling point through transducers and A/D adaptors. In fact, any component (e.g. controllers of power electronic devices as well as control and protection equipment) in power systems could be tested in the platform [1].

A real distribution system of a shipboard power system was used for demonstrating the simulator's capabilities. Figure 18.2 shows a good agreement between the complete software simulation results and the field measurements.

18.3   Sample case: testing the sensitivity of a thyristor firing board to poor quality power

A three-phase thyristor firing board was tested in the platform for its sensitivity to poor quality power. Also, the impact of its sensitivity on the DC load and the AC distribution systems were considered. The schematic of the simulated AC distribution system is shown in Figure 18.3.

The application was tested under extreme conditions as shown in Table 18.1.

Figure 18.4 and Figure 18.5 show the single-phase voltage sag, with and without any phase shift, and their impact on the DC output voltage of the rectifier. The sag with a phase shift resulted in the reboot of the firing board. The reboot then resulted in a 0.1 s DC blackout and about a 1.5 s transient on both DC and AC systems. However, the sag without phase shifts only resulted in a DC voltage drop. This result could not be discovered by using traditional laboratory tests.

Table 18.1   The RT–HIL simulation results for the firing board

PQ phenomena	Simulation results
THD	• Tolerate THD up to 14.8% and higher • Tolerance has no impact on distribution systems
Voltage sag	• Tolerance depends on not only the time duration and voltage reduction, but also on the phase shift • Tolerance results in DC voltage drop or blackout
Frequency change	• Tolerate system frequency from 30 Hz to 80 Hz

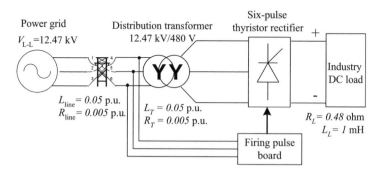

Figure 18.3   Diagram of the simulated industrial distribution system and rectifier load (60 Hz, power base = 833 kW)

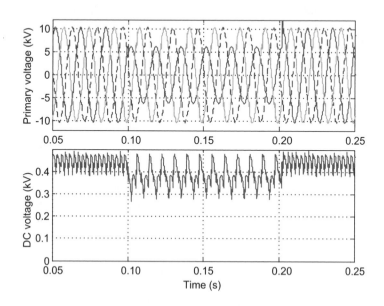

Figure 18.4   Single-phase voltage sag (0.1 s duration, 40% voltage reduction, no phase shift) and its impact on the rectifier DC output (delay angle $\alpha = 7°$)

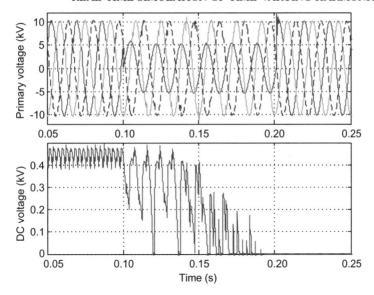

Figure 18.5   Phase-shifted single-phase voltage sag (0.1 s duration, 40% voltage reduction) and its impact on the rectifier DC output (delay angle $\alpha = 7°$)

Figure 18.6 shows the diagram of a power quality test bed. A universal interface is built to connect any firing board easily. Test systems and power quality phenomena can be selected from the existing ones in the digital simulator or self-designed for a special purpose.

18.4   Conclusions and future work

A power quality assessment method has been proposed. The method was applied in the RT–HIL platform to test an industry firing board. The successful initial test results showed that the tested board can tolerate highly distorted voltages, significant sudden frequency change

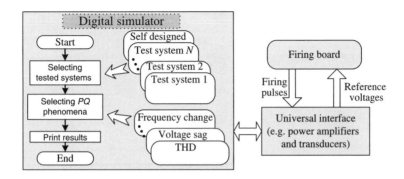

Figure 18.6   Diagram of universal power quality test bed

and three-phase voltage sags, but it cannot tolerate certain short-term phase-shifted single-phase voltage sags. This result, which could only have been revealed through the proposed RT–HIL method, is helpful for future product improvements. The successful experiment has contributed to the conceptual design of a power quality test bed, in which any kind sensitivity of power quality deviation could be revealed.

References

[1] A. J. Grono, 'Synchronizing Generators with HITL Simulation', *IEEE Computer Applications in Power*, **14**, 2001, 43–46.

[2] M. Steurer and S. Woodruff, 'Real Time Digital Harmonic Modeling and Simulation: An Advanced Tool for Understanding Power System Harmonics Mechanisms', IEEE PES General Meeting, Denver, Colorado, USA, June 2004.

[3] Lok-Fu Pak, Dinavahi, V., Gary Chang; Steurer, M., Ribeiro, P.F., 'Real-Time Digital Time-Varying Harmonic Modeling and Simulation Techniques', *IEEE, Transactions on Power Delivery*, v 22, n 2, April 2007, p 1218–27.

[4] J. Langston, S. Suryanarayanan, M. Steurer, M. Andris, S. Woodruff and P. Ribeiro, 'Experiences with the Simulation of a Notional All-Electric Ship Integrated Power System on a Large-Scale Electromagnetic Transient Simulator', 2006 IEEE Power Engineering Society General Meeting, Montreal, Quebec, Jun. 2006.

[5] W. Ren, M. Steurer, S. Woodruff and P.F. Ribeiro, 'Augmenting E-Ship Power System Evaluation and Converter Controller Design by Means of Real-Time Hardware-in-Loop Simulation', in Proc. of 2005 IEEE Electric Ship Technologies Symposium, July 25–27, Philadelphia, USA, pp. 171–175.

[6] S. Suryanarayanan, W. Ren, M. Steurer, P. Ribeiro and G. T. Heydt, 'A real-time controller concept demonstration for distributed generation interconnection', accepted for presentation at the 2006 IEEE Power Engineering Society General Meeting, Montreal, QC, Canada, Jun. 2006.

19

Independent component analysis for harmonic studies

E. Gursoy and D. Niebur

19.1   Introduction

Due to an increase in the amount of power electronic equipment and other harmonic sources, the identification and estimation of harmonic loads are of concern in electric power transmission and distribution systems. Conventional harmonic state estimation requires a redundant number of expensive harmonic measurements. In this chapter we explore the use of a statistical signal processing technique, known as independent component analysis (ICA) for harmonic source identification and estimation. If the harmonic currents are statistically independent, ICA is able to estimate the currents using a limited number of harmonic voltage measurements and without any knowledge of the system admittances or topology. Results are presented for the modified IEEE 30 bus system.

Identification and measurement of harmonic sources has become an important issue in electric power systems, since the increased use of power electronic devices and equipment sensitive to harmonics, has increased the number of adverse harmonic related events.

Harmonic distortion causes financial expense for customers and electric power companies. Companies are required to take necessary action to keep the harmonic distortion at levels defined by standards, for example IEEE Standard 519-1992. Marginal pricing of harmonic injections is addressed in [1] to determine the costs of mitigating harmonic distortion. Harmonic levels in the power system need to be known to solve these issues. However, in a deregulated network, it may be difficult to obtain sufficient measurements at substations owned by other companies.

Time-Varying Waveform Distortions in Power Systems   Edited by Paulo F. Ribeiro
© 2009 John Wiley & Sons, Ltd

Harmonic measurements are more sophisticated and costly than ordinary measurements because they require synchronization for phase measurements, which is achieved by global positioning systems (GPS). It is not easy and economical to obtain a large number of harmonic measurements because of instrumentation installation, maintenance and related measurement acquisition issues.

Harmonic state estimation (HSE) techniques have been developed to assess the harmonic levels and to identify the harmonic sources in electric power systems [2–9]. Using synchronized, partial, asymmetric harmonic measurements, harmonic levels can be estimated by system-wide HSE techniques [5]. An algorithm to estimate the harmonic state of the network is partially developed using a limited number of measurements [8]. The number and the location of harmonic measurements for HSE are determined from observability analysis [10]. Either for a fully observable or a partially observable network, the number of required harmonic measurements is much larger than the number of sources.

HSE techniques require detailed and accurate knowledge of network parameters and topology. Approximation of the system model and poor knowledge of network parameters may lead to large errors in the results. Measurement of harmonic impedances [11, 12] can be a solution, which again is impractical and expensive for large networks. It is therefore very desirable to estimate the harmonic sources without the knowledge of network topology and parameters, using only a small number of harmonic measurements.

In this chapter we present the estimation of harmonic load profiles of harmonic sources in the system using a blind source separation algorithm (BSS) which is commonly referred to as independent component analysis (ICA). The proposed approach is based on the statistical properties of loads. Both the linear loads and nonlinear loads are modeled as random variables.

19.2   Independent component analysis – ICA model

Blind source separation (BSS) algorithms [13] estimate the source signals from observed mixtures. The word 'blind' emphasizes that the source signals and the way the sources are mixed, that is the mixing model parameters, are unknown.

Independent component analysis is a BSS algorithm, which transforms the observed signals into mutually statistically independent signals [14]. The ICA algorithm has many technical applications including signal processing, brain imaging, telecommunications and audio signal separation [15].

The linear mixing model of ICA is given as

$$x(t_i) = As(t_i) + n(t_i) \tag{19.1}$$

where $s(t_i) = [s_1(t_i), \ldots, s_N(t_i)]$ is the N-dimensional vector of unknown source signals, $x(t_i) = [x_1(t_i), \ldots, x_M(t_i)]$ is the M-dimensional vector of observed signals, A is an $M \times N$ matrix called a mixing matrix and t_i is the time or sample index with $i = 1, 2, \ldots, T$. In Equation (19.1), $n(t_i)$ is a zero mean Gaussian noise vector of dimension M. Assuming no noise, the matrix representation of the mixing model in Equation (19.1) is

$$X = AS. \tag{19.2}$$

Here X and S are $M \times T$ and $N \times T$ matrices whose column vectors are observation vectors $\mathbf{x}(t_1), \ldots, \mathbf{x}(t_T)$ and sources $\mathbf{s}(t_1), \ldots, \mathbf{s}(t_T)$, A is an $M \times N$ full column rank matrix.

The objective of ICA is to find the separating matrix W which inverts the mixing process such that

$$Y = WX \qquad (19.3)$$

where Y is an estimate of the original source matrix S, and W is the (pseudo) inverse of the estimate of the matrix A. An estimate of the sources with ICA can be obtained up to a permutation and a scaling factor. Since ICA is based on the statistical properties of signals, the following assumptions for the mixing and demixing models need to be satisfied:

- The source signals $s(t_i)$ are statistically independent.

- At most one of the source signals is Gaussian distributed.

- The number of observations M is greater than or equal to the number of sources N ($M \geq N$).

There are different approaches for estimating the ICA model using the statistical properties of signals. Some of these methods are: ICA by maximization of nongaussianity, by minimization of mutual information, by maximum likelihood estimation, by tensorial methods [15, 16].

19.3   ICA by maximization of nongaussianity

In this chapter the ICA model is estimated by maximization of nongaussianity. A measure of nongaussianity is negentropy $J(y)$ (see Equation (19.4)) which is the normalized differential entropy. By maximizing the negentropy, the mutual information of the sources is minimized. Note that mutual information is a measure of the independence of random variables. Negentropy is always nonnegative and zero for Gaussian variables.

$$J(y) = H(y_{\text{gauss}}) - H(y). \qquad (19.4)$$

The differential entropy H of a random vector y with density $p_y(\eta)$ is defined as

$$H(y) = -\int p_y(\eta) \log p_y(\eta) d\eta. \qquad (19.5)$$

In Equations (19.4) and (19.5), the estimation of negentropy requires the estimation of probability functions of source signals which are unknown. Instead, the following approximation of negentropy is used:

$$J(y_i) = J(E(w_i^T x)) = \{E[G(w_i^T x)] - E[G(y_{\text{gauss}})]\}^2. \qquad (19.6)$$

Here E denotes the statistical expectation and G is chosen as a nonquadratic function [17]. This choice depends on assumptions of super- or subgaussianity of the underlying probability distribution of the independent sources.

The optimization problem using a single unit contrast function subject to the constraint of decorrelation, can be defined as

$$\text{maximize} \sum_{i=1}^{N} J(y_i) = \sum_{i=1}^{N} J(w_i^T x)$$

$$\text{under constraint } E[(w_k^T x)(w_j^T x)] = \delta_{jk} \tag{19.7}$$

where w_i, $i = 1, \ldots N$ are the rows of the matrix W. The optimization problem given in Equation (19.7) is single unit deflation algorithm, where independent components are estimated one by one. To estimate several independent components, this algorithm is executed using several units. After each iteration, vectors are decorrelated to prevent convergence to the same maxima. This algorithm, called FastICA, is based on a fixed-point iteration scheme [17]. In this chapter we used the FastICA algorithm for the estimation for harmonic sources.

To simplify the ICA algorithm, signals are preprocessed by centering and whitening. Centering transforms the observed signals to zero-mean variables and whitening linearly transforms the observed and centered signals, such that the transformed signals are uncorrelated, have zero mean and their variances equal unity.

19.4   Harmonic load profile estimation

In this section the harmonic load identification procedure is given. The system equations under nonsinusoidal conditions are given by the following linear equation:

$$I_h = Y_h V_h \tag{19.8}$$

where h is the harmonic order of the frequency, I_h is the bus current injection vector, V_h is the bus voltage vector and Y_h is the system admittance matrix at frequency h. The linear Equation (19.8) is solved for each frequency of interest.

As mentioned in the introduction, it may be difficult to obtain (a) accurate system parameters (Y_h), especially for higher harmonics, and (b) enough harmonic measurements for an observable system. Using time sequence data of available measurements, that is complex harmonic voltage measurement sequences on a limited number of busses, the ICA approach is able to estimate the load profiles of harmonic current sources without the knowledge of system parameters and topology using the statistical properties of time series data only.

The linear measurement model for the harmonic load flow given in Equation (19.8) can be defined as

$$V_h(t_i) = Z_h I_h(t_i) + n(t_i) \quad i = 1, 2 \ldots, T. \tag{19.9}$$

Here $V_h(t_i) \mathbb{C}^M$ are the known harmonic voltage measurement vectors, $I_h(t_i) \in \mathbb{C}^N$ are unknown harmonic current source vectors, $Z_h \in \mathbb{C}^{M \times N}$ is the unknown mixing matrix relating measurements to the sources, $n(t_i) \in \mathbb{C}^M$ is the Gaussian distributed measurement error vector, h is the harmonic order, t_i is the sample index and T is the number of samples. In the presence of only harmonic voltage measurements, Z_h is the system impedance matrix at harmonic order h.

The general representation of the linear system equations in (19.8) can be written as

$$
\begin{bmatrix} I_h^1 \\ I_h^2 \\ \vdots \\ I_h^N \end{bmatrix} = \begin{bmatrix} Y_h^{1,1} + Y_h^{L1} & Y_h^{1,2} & \cdots & Y_h^{1,N} \\ Y_h^{2,1} & Y_h^{2,2} + Y_h^{L2} & \cdots & Y_h^{2,N} \\ \vdots & \vdots & \ddots & \vdots \\ Y_h^{N,1} & Y_h^{N,2} & \cdots & Y_h^{N,N} + Y_h^{LN} \end{bmatrix} \begin{bmatrix} V_h^1 \\ V_h^2 \\ \vdots \\ V_h^N \end{bmatrix}.
\tag{19.10}
$$

Here $Y_h^{i,j}$ represents the equivalent admittance at frequency h between node i and j and Y_h^{Li} represents the admittance of linear loads connected to bus i which are modeled with impedance models. We can separate the admittance matrix into two parts:

$$
\begin{bmatrix} I_h^1 \\ I_h^2 \\ \vdots \\ I_h^N \end{bmatrix} = \begin{bmatrix} Y_h^{1,1} & Y_h^{1,2} & \cdots & Y_h^{1,N} \\ Y_h^{2,1} & Y_h^{2,2} & \cdots & Y_h^{2,N} \\ \vdots & \vdots & \ddots & \vdots \\ Y_h^{N,1} & Y_h^{N,2} & \cdots & Y_h^{N,N} \end{bmatrix} \begin{bmatrix} V_h^1 \\ V_h^2 \\ \vdots \\ V_h^N \end{bmatrix}
$$
$$
+ \begin{bmatrix} Y_h^{L1} & 0 & \cdots & 0 \\ 0 & Y_h^{L2} & \cdots & 0 \\ \vdots & \vdots & \ddots & \vdots \\ 0 & 0 & \cdots & Y_h^{LN} \end{bmatrix} \begin{bmatrix} V_h^1 \\ V_h^2 \\ \vdots \\ V_h^N \end{bmatrix}.
\tag{19.11}
$$

The second term on the right-hand side of Equation (19.11) represents the linear loads as a vector of harmonic current sources. Rewriting Equation (19.11), we obtain

$$
-I_h^L + I_h = Y_h V_h.
\tag{19.12}
$$

Here I_h^L is the harmonic current source vector corresponding to the second term on the right-hand side of Equation (19.11). In the ICA model, the mixing matrix A, which represents the admittance matrix Y_h in the harmonic domain, is required to be time-independent for all time steps t_i, $i = 1, 2, \ldots T$. Using this simple manipulation, the load model for linear loads changes from impedance model to current model and the admittance matrix is kept constant. Using current models for linear loads increases the number of sources to be estimated in addition to the nonlinear loads. However, considering there are no linear loads on harmonic source busses and combining the linear loads, which have similar load profiles, the number of sources to be estimated can be reduced.

19.5   Statistical properties of loads

Estimation with ICA requires the statistical independence and nongaussianity of sources. Load profiles of electric loads consist of two parts: a slow varying component and a fast varying component. The slow varying component represents the variation of loads depending on the temperature, weather, day of the week, time of day, and so on. The fast varying component can be modeled as a stochastic process, which represents temporal variation. Generally electrical loads are not statistically independent because of the slow varying component. This dependency can be removed by applying a linear filter to the observed data [18]. Fast varying components remain after filtering the slow varying part of the time series data. Fast fluctuations are assumed to be statistically independent and nongaussian distributed. It is shown in [18], for a particular load, that fast fluctuations are statistically independent and that they follow a supergaussian distribution. Linear filtering of data does not change the mixing matrix A. Therefore, ICA can be applied to the fast varying part of the data and the original sources can be recovered by the estimated demixing matrix. In studies of probabilistic analysis of harmonic loads, recorded signals are treated as the sum of a deterministic and a random component [19, 20]; furthermore, harmonic sources are assumed to be independent [21] similar to the general electric loads.

19.6   Harmonic load estimation

There are some ambiguities in estimation by ICA. Independent components can be estimated up to a scaling and a permutation factor. This is due to the fact that both the sources s and the mixing matrix A are unknown. A source can be multiplied by a factor k and the corresponding column of the mixing matrix can be divided by k, without changing the probability distribution and the measurement vector. Similarly, permuting two columns of A and the two corresponding rows of source s will not affect the measurement vector.

This indeterminacy can be eliminated if there is some prior knowledge of sources. In fact, in electric power systems, it is reasonable to assume that historical load data are available, which can be used to match the estimated sources to original sources; furthermore, it is assumed that forecasted peak loads are available which can be used to scale the load profiles [18].

For simplicity, we assume that the number of measurements is equal to the number of sources. In other words, A is a square matrix. If there is a redundancy in measurements, the dimension of the measurement vector can be reduced using principle component analysis (PCA).

In this chapter, we assume constant power factor for loads and statistical independence of loads. Using the FastICA algorithm, we are able to estimate the real and imaginary part of each harmonic component individually.

The harmonic load identification algorithm described above can be summarized as:

1. If the mixing matrix A is not square, use PCA to reduce the dimension of measurements.

2. Apply a linear filter to obtain the fast varying components of the measurement vector V_h.

3. Centralize and whiten the measurement data.

4. Apply FastICA algorithm to real and imaginary parts of the fast varying component of V_h.

5. Obtain the estimates of real and imaginary parts of the harmonic current sources at harmonic frequency h.

6. Perform steps 2 through 5 for each harmonic component of interest.

7. Reorder and scale the estimated sources using historical data.

19.7   Case studies

19.7.1   Case 1

For this chapter typical load profiles were downloaded from the website of Electric Reliability Council of Texas (ERCOT) [22]. To distinguish fast- and slow-varying components of the normalized and centered load profiles which thus have a zero mean and unity variance, a zero mean Laplace distributed random variable with 0.002 variance was added to the normalized and centered load profiles. Here the Laplace distributed random variable represents the fast-varying components, which are statistically independent, and the load profiles represent the slow-varying components mentioned in the previous part. The apparent power of each load was multiplied by one of these load profiles. Harmonic measurement vectors, that is the harmonic bus voltages, are simulated by harmonic power flow. The public-domain MATPOWER [23] program was modified to carry out both fundamental and harmonic power flow calculations. First, the fundamental frequency power flow solution was obtained. Harmonic sources were modeled as constant power loads in this step. Harmonic current source models were obtained for harmonic sources using the power flow solution. Next, we calculated the harmonic bus voltages by solving the linear system Equations (19.8) for each harmonic frequency of interest. In harmonic analysis, a transmission line, a generator and a transformer were modeled as a π-model, a subtransient reactance and a short-circuit impedance respectively. The impedance model #2 given in [24] was used to model linear loads in the harmonic domain. The harmonic measurement vector obtained by harmonic power flow was used in ICA for the estimation of harmonic sources.

The proposed harmonic load identification algorithm was tested on a modified IEEE 30-bus test system shown in Figure 19.1. The system was assumed to be balanced. We

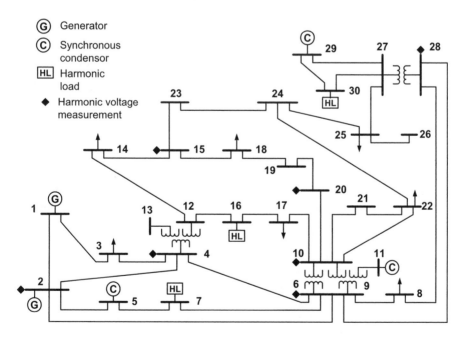

Figure 19.1   Modified IEEE 30-bus system

placed three harmonic producing loads at buses 7, 16 and 30. These buses were geographically and electrically far from each other, which is usually the case in power systems. These harmonic loads were modeled as harmonic current injection sources. Harmonic current spectrums given in [25] were used for these loads and simulations were obtained up to the 17th harmonic. Harmonic source power ratings were $45 + j20$ MVA, 25 MVar and $25 + j16$ MVA at bus 7, 16 and 30 respectively. In addition to harmonic loads, there were seven linear loads at buses 3, 8, 14, 17, 18, 22 and 25. Four of these linear loads had a power rating close to harmonic load power ratings. The remaining buses were no-load buses. Both the harmonic loads and linear loads had 1 min varying load shapes, which were obtained by adding Laplace distributed random variables to slow-varying load shapes normalized to unity and by multiplying them with the power rating of each load. For the measurements, harmonic voltage measurements were used, since in general these measurements were easier and more reliable than other harmonic measurements, such as power measurements. There were seven harmonic voltage measurements placed at buses 2, 4, 6, 10, 15, 20 and 28. In general it was easier and less expensive to obtain measurements at substations than at other buses. Therefore six of these measurements were located at substations. None of the measurements was on the load buses. This reflected the case of a deregulated network where generation and distribution were unbundled.

The proposed algorithm explained in Section 19.6 was used to estimate the load shapes of the harmonic sources. Estimates of the real and imaginary part of the harmonic current sources are shown in Figures 19.2 to 19.4.

Figure 19.2 shows the smoothed current profiles of the estimated harmonic current injection at bus 7. The solid line represents the actual load shapes and the dashed line represents the estimated ones. The real and imaginary parts of the current are estimated separately by applying the ICA algorithm to the real and imaginary parts of the observations, harmonic voltage measurements. Table 19.1 shows the error and the correlation coefficients between estimated and actual load shapes. The results from Figure 19.2 and Tables 19.2 and 19.3 show that the proposed algorithm is capable of estimating the load profiles of harmonic injection within a small error range. Correlation coefficients are close to one, indicating a matching of high accuracy between estimated and actual profiles.

In Figures 19.3 and 19.4, harmonic current injection profiles of the nonlinear loads at bus 16 and 30 are given respectively. The graphs show that there are some slight differences between estimates and actual shapes. However, estimates are tracking the actual load shapes. Estimates of the harmonic source at bus 30 are better than the estimates of bus 16. From the figures we can see that, the high percentage error occurs at small values of current magnitude. Estimates are more accurate when the magnitude of the current is high. For example, for the imaginary part of the fifth harmonic, the mean percentage errors are 1.98%, 6.90% and 2.21% at buses 7, 16 and 30 respectively. Error measures for nonlinear loads at bus 16 and 30 are given in Tables 19.2 and 19.3 respectively.

In our simulations, measurement noise is ignored. As investigated in [18], measurement noise increases the errors in estimates at lower load levels and additional measurements increase the performance of the algorithm. In the harmonic domain, the linear loads can be viewed as distracters adding some nonlinearity to the ICA model by changing the mixing matrix or additional sources as shown in Equations (19.9)–(19.11). To reduce the effect of the linear loads on the estimation of harmonic sources, we used more

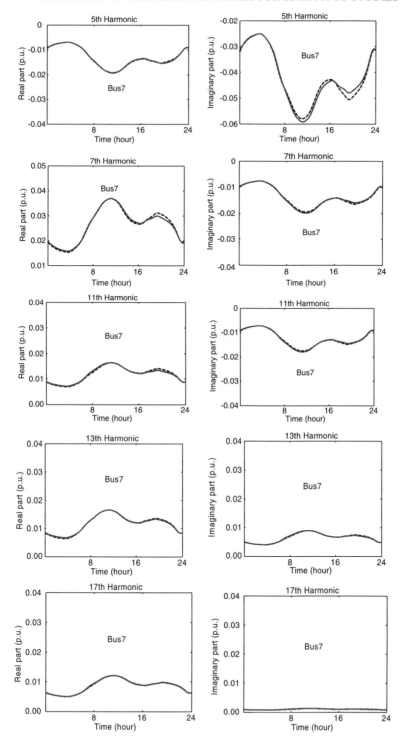

Figure 19.2   Harmonic components of current injection at bus 7

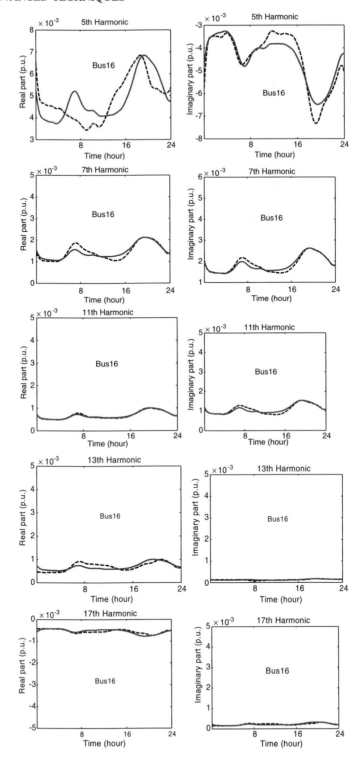

Figure 19.3   Harmonic components of current injection at bus 16

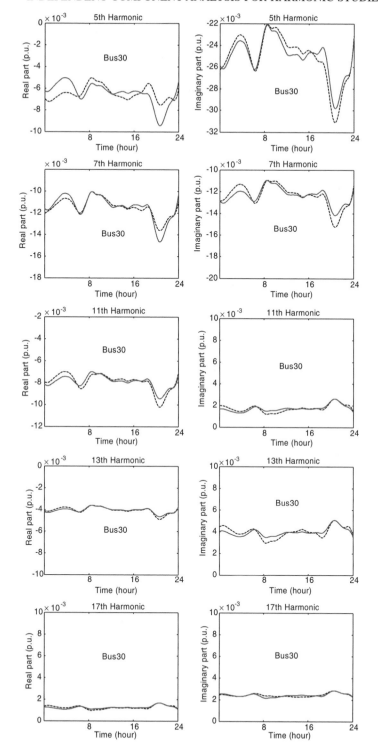

Figure 19.4   Harmonic components of current injection at bus 30

Table 19.1   Errors between estimated and actual smoothed harmonic current profiles at bus 7

	Harmonic order		Correlation coefficient	Maximum percentage error (%)	Mean percentage error (%)
BUS 7	5	Real	0.9988	2.73	0.73
		Imag	0.9932	5.30	1.98
	7	Real	0.9964	4.71	1.74
		Imag	0.9975	3.81	1.39
	11	Real	0.9933	5.43	2.31
		Imag	0.9974	3.20	1.47
	13	Real	0.9989	5.51	1.54
		Imag	0.9967	3.77	1.15
	17	Real	0.9992	2.71	0.67
		Imag	0.9713	9.38	2.29

Table 19.2   Errors between estimated and actual smoothed harmonic current profiles at bus 16

	Harmonic order		Correlation coefficient	Maximum percentage error (%)	Mean percentage error (%)
BUS 16	5	Real	0.6825	24.85	14.58
		Imag	0.9309	21.91	6.90
	7	Real	0.9122	21.44	7.70
		Imag	0.9461	12.30	4.38
	11	Real	0.9857	9.03	2.15
		Imag	0.9121	16.15	6.12
	13	Real	0.6780	36.34	14.96
		Imag	0.5429	42.42	11.48
	17	Real	0.7476	25.84	9.28
		Imag	0.6849	35.78	11.92

measurements (seven measurements) than the number of harmonic sources (three harmonic sources). However, the total number of loads, including the linear loads, is less than the number of measurements, indicating that there are some additional effects that contribute to the corruption of the load estimates.

Table 19.3   Errors between estimated and actual smoothed harmonic current profiles at bus 30

	Harmonic order		Correlation coefficient	Maximum percentage error (%)	Mean percentage error (%)
BUS 30	5	Real	0.6150	27.13	11.59
		Imag	0.9640	6.39	2.21
	7	Real	0.9592	7.21	2.15
		Imag	0.9525	7.55	2.66
	11	Real	0.9485	7.95	3.20
		Imag	0.8817	19.39	8.47
	13	Real	0.9669	5.43	1.81
		Imag	0.8787	15.07	6.02
	17	Real	0.8940	12.97	5.96
		Imag	0.9398	10.18	3.14

Table 19.4   Errors between estimated and actual smoothed harmonic current profiles at bus 7 with two harmonic measurements

	Harmonic order		Correlation coefficient	Maximum percentage error (%)	Mean percentage error (%)
BUS 7	5	Real	0.9982	3.29	1.00
		Imag	0.9916	6.31	2.26
	7	Real	0.9949	6.20	2.02
		Imag	0.9989	2.62	0.79
	11	Real	0.9989	2.09	0.90
		Imag	0.9993	2.17	0.54
	13	Real	0.9977	5.21	1.06
		Imag	0.9983	2.25	1.12
	17	Real	0.9417	16.68	6.48
		Imag	0.9593	8.95	2.38

19.7.2   Case 2

In this case, the estimation process was based on fewer measurements in order to test the effectiveness of the proposed algorithm for estimating the largest harmonic source in the network. Using the same test system as in Figure 19.1, two harmonic voltage measurements were taken from bus 2 and 4 instead of seven measurements as in Case 1.

Estimation results for the harmonic source at bus 7 are given in Table 19.4. Using the ICA algorithm, this harmonic load is estimated with a small error. The second output of the ICA estimation cannot be matched with the other two harmonic sources at bus 16 and bus 30, because the correlation coefficients are small, that is 0.3516 and 0.4587.

From the results given in Cases 1 and 2, we can see that the harmonic source with a harmonic current rating higher than the other system loads, can be estimated with a very good range of error. As future work, the impact of the location and the number of measurements, as well as the number of time steps per measurement, on the accuracy of the estimation needs further investigation.

19.8   Conclusion

Independent component analysis is used to estimate the load profiles of harmonic sources without any prior knowledge of network topology and parameters. This method is based on the statistical properties of loads. The statistical independence of loads is assured by separating the fast- and slow-varying components in load profiles using a linear filter and by using only the fast-varying component for the independent component analysis. The application of the proposed algorithm to the modified IEEE 30-bus test system shows that current profiles of harmonic sources can be estimated using only a small number of harmonic voltage measurements.

The proposed method is quite promising for the application in a deregulated network since the number of measurements is small and the measurements can be taken a long way from the sources. Also this algorithm can be extended to find the minimum set of measurements to reduce the measurements cost and increase estimation accuracy for the harmonic meter placement problem. Further analysis and research on the topic presented in this chapter is published in [26].

Acknowledgment

The authors would like to acknowledge Huaiwei Liaos help in the early stages of this work.

References

[1] P. J. Talacek and N. R. Watson, 'Marginal pricing of harmonic injections', *IEEE Transactions on Power Systems*, **17**, 2002, 50–56.

[2] G. T. Heydt, 'Identification of harmonic sources by a state estimation technique', *IEEE Transactions on Power Delivery*, **4**, 1989, 569–576.

[3] Z. P. Du, J. Arrillaga and N. Watson, 'Continuous harmonic state estimation of power systems', *IEE Proceedings-Generation Transmission and Distribution*, **143**, 1996, 329–336.

[4] A. P. S. Meliopoulos, F. Zhang and S. Zelingher, 'Power-System Harmonic State Estimation', *IEEE Transactions on Power Delivery*, **9**, 1994, 1701–1709.

[5] Z. P. Du, J. Arrillaga, N. R. Watson and S. Chen, 'Identification of harmonic sources of power systems using state estimation', *IEE Proceedings-Generation Transmission and Distribution*, **146**, 1999, 7–12.

[6] V. L. Pham, K. P. Wong, N. Watson and J. Arrillaga, 'A method of utilising non-source measurements for harmonic state estimation', *Electric Power Systems Research*, **56**, 2000, 231–241.

[7] H. M. Beides and G. T. Heydt, 'Dynamic state estimation of power-system harmonics using kalman filter methodology', *IEEE Transactions on Power Delivery*, **6**, 1991, 1663–1670.

[8] S. S. Matair, N. R. Watson, K. P. Wong, V. L. Pham and J. Arrillaga, 'Harmonic state estimation: a method for remote harmonic assessment in a deregulated utility network', International Conference on Electric Utility Deregulation and Restructuring and Power Technologies, London, UK, 2000.

[9] R. K. Hartana and G. G. Richards, 'Constrained neural network-based identification of harmonic sources', *IEEE Transactions on Industry Applications*, **29**, 1993, 202–208.

[10] Z. P. Du, J. Arrillaga and N. Watson, 'A new symbolic method of observability analysis for harmonic state estimation of power systems', IEEE International Conference on Electrical Engineering, Beijing, China, 1996.

[11] M. Nagpal, W. Xu and J. Sawada, 'Harmonic impedance measurement using three-phase transients', *IEEE Transactions on Power Delivery*, **13**, 1998, 272–277.

[12] M. Sumner, B. Palethorpe, D. W. P. Thomas, P. Zanchetta and M. C. Di Piazza, 'A technique for power supply harmonic impedance estimation using a controlled voltage disturbance', *IEEE Transactions on Power Electronics*, **17**, 2002, 207–215.

[13] J. F. Cardoso, 'Blind signal separation: Statistical principles', *Proceedings of the IEEE*, **86**, 1998, 2009–2025.

[14] A. Hyvarinen and E. Oja, 'Independent component analysis: algorithms and applications', *Neural Networks*, **13**, 2000, 411–430.

[15] A. Hyvarinen, J. Karhunen and E. Oja, *Independent component analysis*. John Wiley & Sons, Inc., New York, USA, 2001.

[16] T. W. Lee, M. Girolami, A. J. Bell and T. J. Sejnowski, 'A unifying information-theoretic framework for independent component analysis', *Computers & Mathematics with Applications*, **39**, 2000, 1–21.

[17] A. Hyvarinen, 'Fast and robust fixed-point algorithms for independent component analysis', *IEEE Transactions on Neural Networks*, **10**, 1999, 626–634.

[18] H. W. Liao and D. Niebur, 'Load profile estimation in electric transmission networks using independent component analysis', *IEEE Transactions on Power Systems*, **18**, 2003, 707–715.

[19] Y. Baghzouz, R. F. Burch, A. Capasso, A. Cavallini, A. E. Emanuel, M. Halpin, A. Imece, A. Ludbrook, G. Montanari, K. J. Olejniczak, P. Ribeiro, S. Rios-Marcuello, L. Tang, R. Thallam and P. Verde, 'Time-varying harmonics: Part I - Characterizing measured data', *IEEE Transactions on Power Delivery*, **13**, 1998, 938–944.

[20] A. Cavallini, G. C. Montanari and M. Cacciari, 'Stochastic Evaluation of Harmonics at Network Buses', *IEEE Transactions on Power Delivery*, **10**, 1995, 1606–1613.

[21] Y. Baghzouz, R. F. Burch, A. Capasso, A. Cavallini, A. E. Emanuel, M. Halpin, R. Langella, G. Montanari, K. J. Olejniczak, P. Ribeiro, S. Rios-Marcuello, F. Ruggiero, R. Thallam, A. Testa and R. Verde, 'Time-varying harmonics: Part II - Harmonic summation and propagation', *IEEE Transactions on Power Delivery*, **17**, 2002, 279–285.

[22] Electric Reliability Council of Texas, "Load Profiling", Feb. 2004 [online]. Available: www.ercot. com.

[23] R. Zimmerman and D. Gan, 'MATPOWER - Matlab routines for solving power flow problems', http://www.pserc.cornell.edu/matpower/matpower.html, 1997.

[24] R. Burch, G. Chang, C. Hatziadoniu, M. Grady, Y. Liu, M. Marz, T. Ortmeyer, S. Ranade, P. Ribeiro and W. Xu, 'Impact of aggregate linear load modeling on harmonic analysis: A comparison of common practice and analytical models', *IEEE Transactions on Power Delivery*, **18**, 625–630.

[25] R. Abu-hashim, R. Burch, G. Chang, M. Grady, E. Gunther, M. Halpin, C. Hatziadoniu, Y. Liu, M. Marz, T. Ortmeyer, V. Rajagopalan, S. Ranade, P. Ribeiro, T. Sims and W. Xu, 'Test systems for harmonics modeling and simulation', *IEEE Transactions on Power Delivery*, **14**, 1999, 579–587.

[26] E. Gursoy and D. Niebur, 'Harmonic load identification using complex independent component analysis', *IEEE Transactions on Power Delivery*, **24**, Issue 1, 2009, 285–292.

20

Enhanced empirical mode decomposition applied to waveform distortions

N. Senroy, S. Suryanarayanan and P. F. Ribeiro

20.1 Introduction

With the pervading application of power electronics and other nonlinear time-varying loads and equipment in modern power systems, distortions in line voltage and current are becoming an increasingly complex issue. In tightly coupled power systems such as the integrated power system (IPS) onboard all-electric ships and islanded microgrids, the estimation and visualization of time-varying waveform distortions present an interesting research avenue [1]. Accurately estimating time-varying distorted voltage and current signals will help to determine innovative power quality indices and thresholds, equipment derating levels and adequate mitigation methods including harmonic filter design. In this context, it is no longer appropriate to use 'harmonics' to describe the higher modes of oscillations present in nonstationary and nonlinear waveform distortions. Harmonics imply stationarity and linearity among the modes of oscillations, while the focus of this chapter is on time-varying waveform distortions. Key issues specific to the problem of estimating time-varying modes in distorted line voltages and currents are: (a) distortion magnitudes are small, and typically range from 1–10% (voltage) and 10–30% (current) of the fundamental, (b) the fundamental frequency may not be constant during the observation, due to load fluctuations and system transients, (c) typical distortion frequencies of interest in electric power quality studies may lie within an octave – posing a challenge of separation.

Time-Varying Waveform Distortions in Power Systems   Edited by Paulo F. Ribeiro
© 2009 John Wiley & Sons, Ltd

Estimating the modes existing in a complex distorted signal may be reposed as an instantaneous amplitude and frequency tracking problem. What is the degree to which the time–frequency–magnitude resolution of the participant modes, in a distorted voltage/current signal, may be realized? Research has already indicated the role of the uncertainty principle in limiting the performance of time–frequency distributions [2]. This chapter approaches the problem as follows:

- adaptively decompose the distorted signal into monocomponent forms existing at different timescales,

- estimate instantaneous frequencies and amplitudes of each of the separated components,

- localize the temporal variations in these amplitudes and frequencies accurately on the timescale.

The Hilbert–Huang (HH) method, including the empirical mode decomposition (EMD), has been developed as an innovative time–frequency–magnitude resolution tool for non-stationary signals [3]. EMD is an adaptive and data driven multiresolution technique, in which a multicomponent waveform is resolved into several components. These components, referred to as intrinsic mode functions (IMF), are expected to be monocomponent in nature. Application of the Hilbert transform yields their analytic forms, from which their instantaneous amplitudes and frequencies can be extracted. However, the original EMD technique fails to separate participating modes whose instantaneous frequencies simultaneously lie within an octave.

The use of a masking signal based EMD was initially proposed to improve the filtering characteristics of the EMD [4]. In this chapter, a hybrid algorithm is presented for building appropriate masking signals for applying EMD to distorted signals [5]. The improved algorithm employs a fast Fourier transform (FFT) to develop masking signals to apply in conjunction with the EMD. Once a masking signal is constructed, the EMD application follows the method of [4]. Since the masking signal based EMD may not always guarantee monocomponent IMFs, further demodulation is recommended, after applying the Hilbert transform. Another alternative to the masking signal based EMD is frequency shifting, suggested in conjunction with EMD to enhance its discriminating capability [6]. The proposed enhancements focus on signals, typically found in power quality applications, with the following characteristics:

- time-varying modes whose instantaneous frequencies may lie in the same octave simultaneously,

- weak higher-frequency signals not 'recognizable' by conventional EMD.

The rest of this chapter is divided into sections as follows. Section 20.2 discusses different approaches to the problem of estimating time-varying waveform distortions. Section 20.3 introduces the original Hilbert–Huang method, along with its limitations. The masking signal based EMD is presented in detail in this section, along with the demodulation techniques suggested for narrowband IMFs. Section 20.4 employs a synthetic signal representative of signals in power systems, to demonstrate the masking signal based EMD. The frequency shifting technique used in conjunction with EMD is presented in Section 20.5 and Section 20.6 presents concluding remarks.

20.2   Estimating nonstationary distortions

The fast Fourier transform (FFT) technique is adopted as a signal processing tool, when linearity and 'stationarity' are satisfied by the participating modes in a distorted waveform [7]. FFT serves as a computationally efficient and accurate technique to decompose such a distorted waveform into sinusoidal components. In the event of nonlinear modal interaction, the FFT spectrum may be spread over a wider range that may lack physical significance. When the distortions are nonstationary, the FFT may be computed over a sliding window [8]. The fixed window width restricts the time–frequency resolution of this method. The sliding window based ESPRIT method, that is, the estimation of signal parameters via rotational invariance techniques, can separate closely spaced harmonics/interharmonics [9]. However, for nonstationary distortions, post-processing techniques are required to ensure continuity in the time–frequency estimation in consecutive windows.

The wavelet transform is a mathematical tool proposed to identify modes of nonperiodic oscillations, as well as those that evolve in time [10]. Kernel functions, referred to as wavelets, are employed to obtain a multiresolution decomposition of the signal into oscillations at different timescales. Thus, different time resolutions are used to adapt to different frequencies. The choice of the mother wavelet depends on the signal being analyzed. The wavelet transform is capable of separating individual frequency components, provided they are sufficiently far apart on the frequency scale. The wavelet technique has been employed to power quality issues with success [11, 12]. An extension of the wavelet transform, the S-transform, is based on a moving, scalable, localizing Gaussian window [13]. The S-transform has superior phase–frequency–time localizing properties and has been applied to analyzing power quality disturbances [14, 15]. In this chapter, the S-transform is used as a benchmark against which the improved HH method is compared. The parameters of the S-transform used in this chapter are: the dilation factor of the wavelet, defined as the straightforward inverse of frequency, and the transform is performed on the analytic form of the real-valued data [13].

The Hilbert–Huang (HH) method was first proposed for geophysics applications [16]. It has, since then, been applied to problems in biomedical engineering [17], image processing [18] and structural safety [19]. In power systems, the HH method has been applied to identify instantaneous attributes of torsional shaft signals [20] and to analyze inter area oscillations [21, 22]. The empirical mode decomposition (EMD) method has also been used to detect and localize transient features of power system events [23]. Research is ongoing to provide an analytical basis of the EMD [24].

20.3   Modified Hilbert–Huang technique

The original HH method has two parts. In the first part, a distorted waveform is decomposed using EMD into multiple intrinsic mode functions (IMFs) that possess well-behaved Hilbert transforms. Each IMF is defined by two principal characteristics: (a) its mean is zero, and (b) the number of local extrema must be equal to, or differ by at most one from, the number of zero crossings within an arbitrary time window. In the second part of the original HH method, the Hilbert transform is applied to obtain the instantaneous frequencies and amplitudes existing in the individual IMFs. The details of the Hilbert–Huang method along with some variations and other improvements are in [3, 19, 25]. This section briefly describes the original EMD method and its performance while separating frequencies lying within one octave of each other (typical in power quality signals). To improve its performance, the need for masking signals has been suggested in [4]. An innovative FFT-based algorithm has been proposed to construct appropriate masking signals to apply with the EMD [5]. A post-processing demodulation technique has also been proposed [5].

20.3.1 Empirical mode decomposition to obtain IMFs

The underlying principle of the Huang's EMD method is the concept of instantaneous frequency defined as the derivative of the phase of an analytic signal [26]. A monocomponent signal, by definition, has a unique well-defined and positive instantaneous frequency represented by the derivative of the phase of the signal. A signal, with multiple modes of oscillation existing simultaneously, will not have a meaningful instantaneous frequency. Accordingly, a distorted signal must be decomposed into its constituent monocomponent signals before the application of a Hilbert transform to calculate the instantaneous frequency.

The essence of EMD is to recognize oscillatory modes existing in timescales defined by the interval between local extrema. A local extremum point is any point on the signal where its derivative is zero, and its second derivative is nonzero. The term local is used to differentiate it from a global extremum point. For instance, within an observation window, there may be several local extrema; however, only one global maximum and global minimum point may be present. Once the timescales are identified, IMFs with zero mean are sifted out of the signal. The steps comprising the EMD method are:

A1. Identify local maxima and minima of the distorted signal, $s(t)$.

A2. Perform cubic spline interpolation between the maxima and the minima to obtain the envelopes $e_M(t)$ and $e_m(t)$, respectively,

$$m(t) = \frac{(e_M(t) + e_m(t))}{2}.$$

A3. Compute the mean of the envelopes.

A4. Extract $c_1(t) = s(t) - m(t)$.

A5. $c_1(t)$ is an IMF if the number of local extrema of $c_1(t)$ is equal to, or differs from, the number of zero crossings by one, AND the average of $c_1(t)$ is reasonably close to zero. If $c_1(t)$ is not an IMF, then repeat steps A1–A4 on $c_1(t)$ instead of $s(t)$, until the new $c_1(t)$ obtained satisfies the conditions of an IMF.

A6. Compute the residue, $r_1(t) = s(t) - c_1(t)$.

A7. If the residue, $r_1(t)$, is above a threshold value of error tolerance, then repeat steps A1–A6 on $r_1(t)$, to obtain the next IMF and a new residue.

In practice, an appropriate stopping criterion, in step A5, avoids 'over-improving' $c_1(t)$ as that can lead to significant loss of information [25]. The first IMF obtained, consists of the highest-frequency components present in the original signal. The subsequent IMFs obtained contain progressively lower-frequency components of the signal. If n orthogonal IMFs are obtained in this iterative manner, the original signal may be reconstructed as

$$s(t) = \sum_n c_i(t) + r(t). \tag{20.1}$$

The final residue exhibits any general trends followed by the original signal.

20.3.2   Separation of modes of oscillation

In experiments conducted on fractional Gaussian noise the EMD technique has been found to behave like a data-driven dyadic filter bank [27]. Further experiments established the ability of the EMD method to decompose white noise into IMFs whose frequency spectrum comprises an octave [28]. The mean frequencies of the extracted IMFs show a doubling phenomenon. It is worthwhile here to note the similarity between an IMF and a real-zero (RZ) signal as defined by [29]. Bandpass signals whose zeros are distinct and real are called RZ signals. Further, Logan [30] examined a subclass of bandpass signals that have no common zeros between the signal and its Hilbert transform, other than real simple zeros. If the bandwidth of such a signal is less than an octave, the signal is completely described by its zero crossings (up to a multiplicative constant). An IMF would fall in this subclass of signals. The implication is that if there are two modes coexisting in a signal, whose frequencies lie within an octave of each other, the straightforward application of the EMD method is unable to separate the two modes.

Consider a synthetic signal [6], of length 0.2 s and sampled at 20 kHz. The signal consists of a 60 Hz fundamental component of magnitude 1.0 p.u.. The signal is distorted by 120 Hz, 300 Hz and 420 Hz frequency components. A range of distorted signals was built by varying the magnitude of distortions from 0.01 p.u. to 0.1 p.u.. All the distorted signals were kept stationary. Masking signals of frequencies 720 Hz and 420 Hz were used along with EMD to extract only the first IMF. In the presence of distortions due to a 420 Hz and a 300 Hz component, it was desired to include only the 420 Hz component in the first IMF using the 720 Hz masking signal. However, traces of the 300 Hz component are bound to be extracted as shown in the demonstrations. Similarly, in the presence of a 300 Hz and a 120 Hz distortion, the 420 Hz masking signal would include most of the 300 Hz component in the second IMF along with some of the 120 Hz component, during EMD. The amplitude of the masking signal determines its effectiveness in excluding/including lower-frequency components in the first IMF during EMD. Large amplitudes of the masking signal would include undesirable lower frequency components in the IMF, while small amplitudes of the masking signal may not adequately include the target frequency component in the IMF.

20.3.3   Separating 420 Hz from 300 Hz

Analyses done on a 60 Hz fundamental distorted by a 300 Hz and 420 Hz component are presented first. The masking signal frequency is chosen to be 720 Hz because it shares an octave with 420 Hz but not the 300 Hz component. The percentages of 300 Hz and 420 Hz components, extracted in the first IMF using EMD with the masking signal of 720 Hz, are shown in Figure 20.1. The amplitude of the masking signal is plotted on the x-axes, while the original magnitudes of the components in the distorted signal are plotted on the y-axes. The objective is to minimize the extraction of the 300 Hz component and maximize the extraction of the 420 Hz component in the first IMF.

As is evident from Figure 20.1, an extremely weak masking signal (<0.05 p.u.) is undesirable as it would not affect the EMD process, and both 300 Hz and 420 Hz components are extracted entirely in the first IMF (regions A and E in the top and bottom plots). In the presence of a relatively stronger masking signal (>0.05 p.u.), some of the 300 Hz component (up to 21%) is included in the first IMF (region B). However, if the masking signal is not strong enough, the 420 Hz component is not sufficiently included in the first IMF, which is

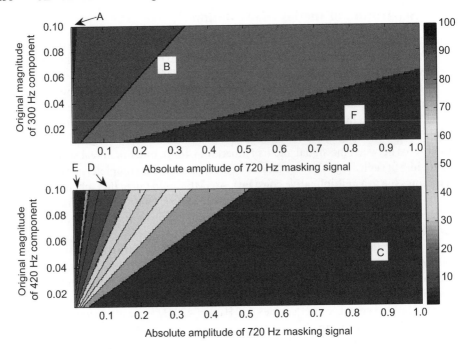

Figure 20.1   Percentage of 300 Hz and 420 Hz component extracted in the first IMF using a 720 Hz masking signal [6]; the absolute amplitude of the masking signal is plotted on the x-axis and the original content of the 300 Hz and 420 Hz components in the signal are plotted on the y-axis of the top and bottom plots respectively

undesirable (region D). If the masking signal is sufficiently strong, the entire 420 Hz signal is included in the first IMF (region C), which is part of the desired objective of using the masking signal. However, an unavoidable 300 Hz component is bound to be included along with the 420 Hz component (region B), which may be minimized. In the presence of a masking signal which is too strong, the 300 Hz will also be entirely included in the first IMF (region F), which is undesirable. In summary, the masking signal magnitude must be selected according to the regions B and C in the top and bottom plots of Figure 20.1.

20.3.4   Separating 300 Hz from 120 Hz

Figure 20.2 shows the scenario when a 60 Hz signal (magnitude 1.0 p.u.) is distorted by a 300 Hz and 120 Hz component. Even though 120 Hz and 300 Hz do not lie in the same octave, a masking signal is nevertheless required to accentuate their distortions to be recognized by the EMD process. A 420 Hz masking frequency was used because it lies in the same octave with 300 Hz and sufficiently far from 120 Hz. The original absolute magnitudes of the two components were varied from 0.01 p.u. to 0.1 p.u. while the masking signal magnitude was varied from 0.01 p.u. to 1.0 p.u.. The objective here is to minimize the extraction of the 120 Hz component and maximize the extraction of the 300 Hz component in the first IMF.

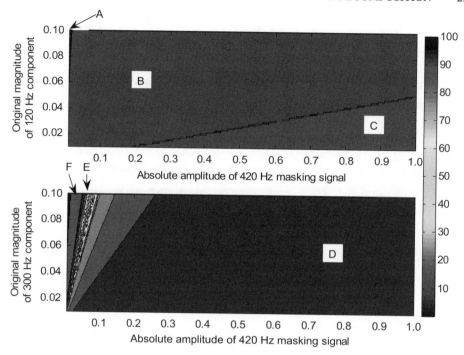

Figure 20.2   Percentage of 120 Hz and 300 Hz component extracted in the first IMF using a 420 Hz masking signal [6]; the absolute amplitude of the masking signal is plotted on the x-axis, and the original content of the 120 Hz and 300 Hz components in the signal are plotted on the y-axis of the top and bottom plots respectively

In Figure 20.2, regions A and F on the top plot and bottom plot respectively, indicate situations where an extremely weak masking signal (less than 0.05 p.u.) has no effect on the EMD process. Hence, both the 120 Hz and the 300 Hz components are extracted in the first IMF, in the presence of such an extremely weak masking signal. When the masking signal is relatively stronger but not strong enough (<0.1 p.u.), it extracts the 300 Hz frequency component insufficiently (region E in lower plot). However, when the masking signal magnitude is sufficiently strong as in region D in the bottom plot, it extracts the entire 300 Hz component, which is the desired objective of using a masking signal. The corresponding region on the top plot is B, which shows an inevitable extraction of 2% of the 120 Hz in the first IMF. If the masking signal is too strong, it extracts the entire 120 Hz component in addition to the 300 Hz component in the first IMF (region C in the upper plot), which is undesirable.

20.4   Algorithm for appropriate masking signal to separate higher frequencies

A typical distorted power quality waveform consists of weak higher-frequency modes whose frequencies may share the same octave. The FFT spectrum of the signal yields its approximate modal content. Masking signals are then required to separate modes of oscillations whose frequencies lie within the same octave, and to accentuate weak higher-frequency signals so that they may be sifted out during the EMD. The appropriate masking signals are constructed as follows [5].

C1. Perform FFT on the distorted signal, $s(t)$, to estimate frequency components $f_1, f_2, \ldots, f_n$, where $f_1 < f_2 < \ldots < f_n$. Note: $f_1, f_2, \ldots, f_n$, are the stationary equivalents of the possibly time-varying frequency components.

C2. Construct masking signals, $mask_2, mask_3 \ldots mask_n$, where $mask_k(t) = M_k \times \sin[2\pi (f_k + f_{k-1}) t]$. The value of M_k is empirical and may be obtained by scaling the magnitude of the appropriate signal (f_k) from the FFT spectrum.

C3. Obtain two signals $s(t) + mask_n$ and $s(t) - mask_n$. Perform EMD (steps A1–A5 from Section 20.3.1) on both signals to obtain their first IMFs only, IMF_+ and IMF_-. Then $c_1(t) = (IMF_+ + IMF_-)/2$.

C4. Obtain the residue, $r_1(t) = s(t) - c_1(t)$.

C5. Perform steps C3–C4 iteratively using the other masking signals and replacing $s(t)$ with the residue obtained, until $n - 1$ IMFs containing frequency components f_2, $f_3, \ldots, f_n$ are extracted. The final residue $r_n(t)$ will contain the remaining component f_1.

From the examples presented above, it is clear that even with masking signals EMD may not produce sufficiently narrow band IMFs. The instantaneous frequency of such an IMF does not hold much physical meaning when related to power system events and phenomena. In such scenarios, some post-processing demodulation is required to extract meaningful instantaneous frequencies and amplitudes.

20.5 Amplitude demodulation of IMF

Each IMF extracted using the masking signal based EMD contains a dominant high-frequency component, along with a remnant lower-frequency component. The amplitude and instantaneous frequency, extracted by Hilbert transform, shows a resultant modulation. This section presents a technique to separate such an IMF into its components. Consider the amplitude modulated (AM) signal,

$$s(t) = A_1 \sin(\omega_1 t) + A_2 \sin(\omega_2 t) \qquad (20.2a)$$

where, $\omega_2 > \omega_1$. The Hilbert transform of $s(t)$ is $s_H(t)$, and the analytical signal, corresponding to $s(t)$, is

$$s_A(t) = s(t) + i.s_H(t) = A_1 e^{i\omega_1 t} + A_2 e^{i\omega_2 t} = A(t).e^{i\phi(t)} \qquad (20.2b)$$

where $A(t)$ is the instantaneous magnitude and $\varphi(t)$ is the instantaneous phase. From Equations (20.2), the instantaneous magnitude is

$$A(t) = \sqrt{A_1^2 + A_2^2 + 2A_1 A_2 \cos[(\omega_1 - \omega_2)t]}. \qquad (20.3)$$

In a modulated signal, the local extrema points may be obtained as

$$|A_1 - A_2| = \text{MIN}\{A(t)\}$$
$$|A_1 + A_2| = \text{MAX}\{A(t)\} \quad . \quad (20.4)$$

Apply cubic spline fitting among the local extrema points to obtain two envelopes corresponding to the maximum envelope, Γ_{max}, and the minimum envelope, Γ_{min}, of the amplitude. The true amplitudes of the two components are then

$$A_1 = (\Gamma_{max} + \Gamma_{min})/2$$
$$A_2 = (\Gamma_{max} - \Gamma_{min})/2 \quad . \quad (20.5)$$

From Equations (20.2), the instantaneous frequency of the signal is defined as $\omega(t) = d\phi/dt$. Also

$$\omega(t) = \text{Im}\left\{\frac{\dot{s}_A(t)}{\dot{s}_B(t)}\right\} = \text{Im}\left\{\frac{A_1 e^{i\omega_1 t}.i\omega_1 + A_2 e^{i\omega_2 t}.i\omega}{A e^{i\omega_1 t} + A e^{i\omega_2 t}}\right\}. \quad (20.6)$$

For the specific case of a modulation between two pure tones, the instantaneous frequency is

$$\omega(t) = \omega_1 + \frac{A_1(t)A_2(t).\cos[(\omega_1-\omega_2)t] + [A_2(t)]^2}{[A_1(t)]^2 + [A_2(t)]^2 + 2A_1(t)A_2(t).\cos[(\omega_1-\omega_2)t]}. \quad (20.7)$$

Substituting $x = A_1/A_2$ in Equation (20.7)

$$\omega(t) = \omega_1 + \frac{\cos[(\omega_1-\omega_2)t] + 1/x}{2\cos[(\omega_1-\omega_2)t] + x + 1/x}.(\omega_1-\omega_2)$$
$$\omega(t) = \omega_1 + \frac{1}{2 + \dfrac{x-1/x}{\cos[(\omega_1-\omega_2)t] + 1/x}}.(\omega_1-\omega_2) \quad . \quad (20.8)$$

From Equation (20.3), the locally maximum magnitude occurs at t_M, when $\cos[(\omega_1-\omega_2)t] = 1$. At this instant, the instantaneous frequency from Equation (20.8) is

$$\omega(t_M) = \omega_1 + \frac{\omega_2-\omega_1}{1+x}. \quad (20.9)$$

Similarly, the locally minimum magnitude occurs at t_M, when $\cos[(\omega_1-\omega_2)t] = -1$. The instantaneous frequency at this instant in time is

$$\omega(t_M) = \omega_1 + \frac{\omega_2-\omega_1}{1-x}. \quad (20.10)$$

Given the instantaneous magnitude and frequency for each IMF, the modulating frequencies may be calculated by solving Equations (20.9) or (20.10). The amplitudes of the modulating components can be calculated from Equation (20.5).

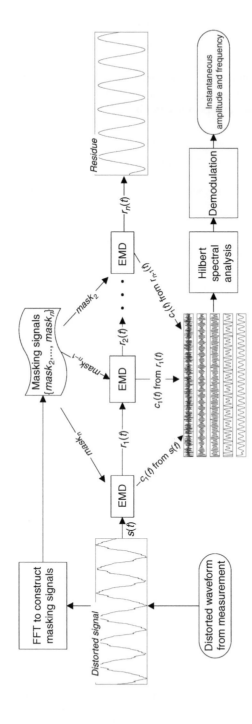

Figure 20.3   Flowchart to compute high-frequency distortion components in a distorted signal, using the improved HH method [5]

A flowchart, shown in Figure 20.3, describes the complete improved HH algorithm for extracting all the frequency components in a distorted signal [5].

20.6 Demonstration

The ability of the masking signal based EMD to separate the components present in a nonstationary distorted signal is demonstrated in this section. Synthetic data are used to validate the improved HH method [5]. A time series was constructed (sampled at 20 kHz) with a fundamental frequency of 60 Hz and magnitude 1.0. Time varying 300 Hz, 420 Hz and constant interharmonic 610 Hz components, with magnitudes shown in Table 20.1, were used to distort the 60 Hz signal. The signal, along with its FFT spectrum, containing only the higher-frequency modes, is plotted in Figure 20.4. As seen in Figure 20.4, the time-varying nature of

Table 20.1   Details of synthetic data

Time (s)	Magnitude of 300 Hz	Magnitude of 420 Hz	Magnitude of 610 Hz
0.0–0.1	0.05	0.03	
0.10005–0.15	0.07	0.06	
0.15005–0.25	0.09	0.05	0.02
0.25005–0.3	0.06	0.07	
0.30005–0.4	0.05	0.04	

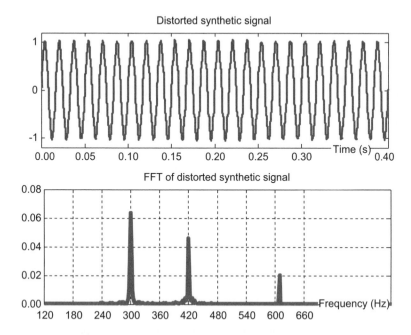

Figure 20.4   Synthetic data of Table 20.1 along with its FFT spectrum from 180 Hz to 840 Hz [5]; the vertical axes are in absolute magnitude

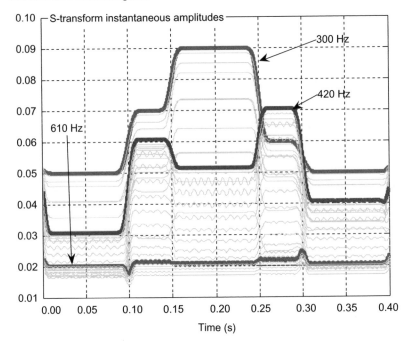

Figure 20.5   Amplitudes of a range of frequencies (247.5–672.5 Hz) extracted using S-transform on the synthetic data of Table 20.1 [5] (note, the significant amplitude of spurious frequencies introduced in the estimation)

the synthetic waveform introduces spurious energy spread around the dominant frequencies in the FFT. The objective is to extract the magnitudes and frequencies of all modes higher than the fundamental mode.

The S-transform was used to analyze the data. A contour plot obtained from the S-transform shows a significant frequency spread. A 120 Hz moving average filter was employed to smooth the results. The spread of frequencies precludes a reliable estimate of the actual modes present in the data. The amplitudes of a range of frequencies from 247.5–672.5 Hz are plotted in Figure 20.5. While the amplitude tracking of the 300 Hz, 420 Hz and 610 Hz is good (artificially highlighted curves), the estimation process clearly introduces fictitious components (artificially greyed).

The improved HH method was employed to study the synthetic data of Table 20.1. The masking signals to extract three IMFs, were constructed using the frequency information from the FFT spectrum. After applying the Hilbert transform, the modulation was removed from the instantaneous magnitude and frequency information using the formulas given in Section 20.3.4. The final magnitude and frequency tracking is shown in Figure 20.6. Apart from some artifacts at the edges of the window, the method successfully tracks both frequency and magnitudes of all the components present in the data.

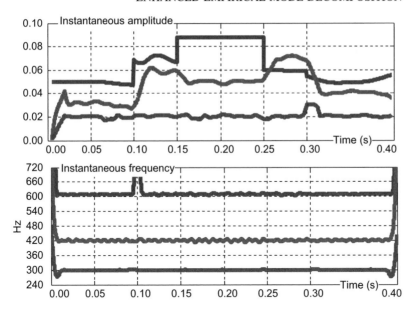

Figure 20.6   Instantaneous magnitude and frequencies of three IMFs extracted using the improved HH method [5]; the data used are synthetic from Table 20.1

20.7   Frequency heterodyne

In this section, frequency shifting is presented as an alternative to using masking signals, as a preprocessing technique before EMD is applied. This technique involves nonlinear mixing of the distorted signal with a pure tone of frequency greater than the highest frequency present in the distorted signal. This principle is similar to the heterodyne detection commonly used in communication theory [31]. Amplitude modulation (AM) involves mixing the signal of interest with a 'carrier signal' to obtain a modulated signal. Such a signal, called double sideband modulated (DSB) with suppressed carrier, consists of two frequency-shifted copies of the original signal on either side of the carrier frequency. To reduce the bandwidth of such an AM signal, one sideband is removed (using an appropriate filter or by a Hilbert transformer), to obtain a single sideband modulation (SSB).

Consider the signal of interest, $s(t)$, with its Hilbert transform $\hat{s}(t)$. The frequency spectrum of $s(t)$ is shifted around a new carrier frequency, F, by employing the analytic representation of the carrier signal, $e^{j2\pi Ft}$. When $s(t)$ is multiplied by $e^{j2\pi Ft}$, all the frequencies present in $s(t)$ are shifted to obtain the DSB signal:

$$f_{\text{DSB}}(t) = s(t).e^{j2\pi F_t}. \qquad (20.11)$$

If the analytic form of $s(t)$ is used instead of just $s(t)$, only the upper sideband is obtained by considering the real part of the product:

$$f_{\text{SSBU}}(t) = \Re e\left\{[s(t) + j\hat{s}(t)]e^{j2\pi F_t}\right\}s(t).e^{j2\pi F_t}. \qquad (20.12)$$

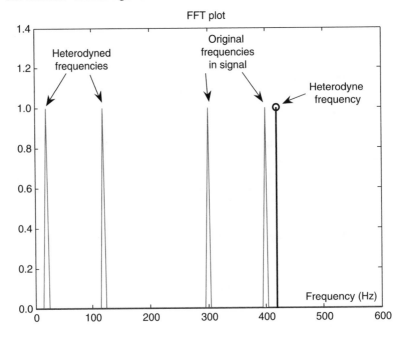

Figure 20.7   Frequency translation using upper sideband modulation [6] (the original frequencies in the signal lie within the same octave); heterodyning with an appropriate frequency 'shifts' the two signals to different octaves

Similarly, the lower sideband signal is obtained as:

$$f_{SSBL}(t) = \Re\{[s(t) + j\hat{s}(t)]e^{j2\pi F_t}\}s(t).e^{-j2\pi F_t}. \tag{20.13}$$

Figure 20.7 shows the frequency translation using the heterodyne technique.

Suppose the distorted signal consists of two frequencies f_1 and f_2 both lying in the same octave. The heterodyne frequency is chosen to be greater than the highest frequency present in the original signal. Using Equation (20.13), the signal is heterodyned by a frequency, F, to locate the lower sideband frequencies, $F - f_1$ and $F - f_2$, in different octaves. Subsequent application of EMD results in separation of the components as individual IMFs of frequencies $F - f_1$ and $F - f_2$. The IMFs are then translated back to the lower sideband frequencies which are the original frequencies using Equation (20.13). The choice of the heterodyning frequency is critical because it must ensure the resultant lower sideband frequencies lie in different octaves.

20.8   Stationary signals

In this section a stationary multicomponent waveform is used to demonstrate the frequency heterodyning before applying the EMD. Consider a synthetic signal consisting of 300 Hz and 420 Hz stationary signals, both of equal absolute magnitude 0.05 p.u. [6]. An ordinary EMD on this signal is unable to separate the two components because 300 Hz and 420 Hz lie within the

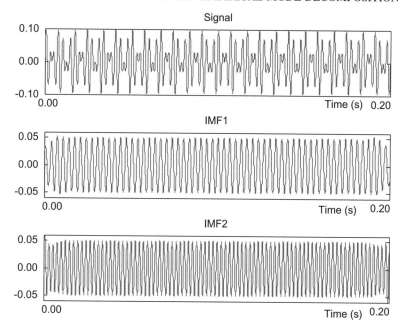

Figure 20.8   Top plot shows a sum of 300 Hz and 420 Hz signals of absolute magnitudes 0.05 each [6]; the two IMFs extracted by EMD after heterodyning with a frequency of 480 Hz are shown in the middle and bottom plots respectively (note the signals obtained are mono-component signals as may be established by an FFT)

same octave. If a masking signal is used along with the EMD as shown in the previous section, the separation of both components is still not perfect as has been demonstrated in the previous section. Therefore, the signal was heterodyned with a frequency of 480 Hz, using Equation (20.13). The lower sideband signal thus obtained, was subjected to EMD to extract two IMFs. The IMFs were translated back to their original frequencies by heterodyning with the same frequency of 480 Hz using Equation (20.13). The two final IMFs obtained contain negligible amounts of other frequencies and are monocomponent for all practical purposes. Figure 20.8 shows the original signal along with two extracted IMFs.

20.9   Nonstationary signals

In this section, a nonstationary synthetic signal, similar to signals in power quality studies, is employed to demonstrate the effectiveness of heterodyning the signal before applying EMD. The signal consists of a 60 Hz fundamental of length 0.2 s, sampled at 20 kHz and of magnitude 1.0 [6]. It is distorted by a time-varying 420 Hz and a constant 300 Hz component. The 420 Hz component decays from 0.09 p.u. to 0.03 p.u. in 0.1 s, and then remains constant at 0.05 p.u. from 0.1 s to 0.2 s. The 300 Hz component was kept constant at 0.05 p.u.. Using Equation (20.13), the signal was heterodyned with a frequency of 480 Hz. The heterodyned signal was used for the EMD process. Two IMFs were extracted and shifted back to their original frequencies using Equation (20.13). The time-varying

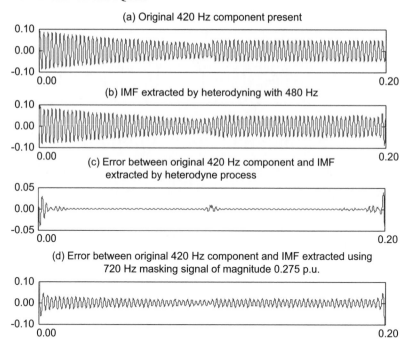

Figure 20.9   The 420 Hz component present in the original signal and its extraction using two different kinds of EMD plotted against time: (a) original 420 Hz component, (b) IMF extracted using a heterodyning frequency of 480 Hz, (c) the error of extraction and (d) the extraction error for an IMF extracted using a 720 Hz masking signal of magnitude 0.275 p.u. [6]

component extraction represents the effectiveness of the heterodyne technique best. The original 420 Hz component (Figure 20.9(a)) present in the signal is compared with the first IMF extracted using the EMD after heterodyning the signal with 480 Hz (Figure 20.9(b)). The error between the two is plotted in Figure 20.9(c)). A masking signal based EMD was also carried out, to obtain the first IMF. A 720 Hz masking signal of magnitude 0.275 p.u. was used. The error of extraction for this case is plotted in Figure 20.9(d). From Figure 20.9, it is clear that heterodyning the signal with an appropriate frequency can enable the EMD to separate closely-spaced frequencies very effectively. The error of extraction when using the frequency heterodyne method, shown in Figure 20.9(c), is smaller than the error of extraction from the masking signal based EMD, shown in Figure 20.9(d).

The error of extraction from the frequency heterodyne method is significant at edges of the window of observation and near switching instants. The heterodyne frequency must be selected higher than the highest frequency present in the original signal. It must be such that all frequencies lying within an octave of the original signal are translated into different octaves during the heterodyning process. Hence, it is imperative to precede the process of choosing a heterodyne frequency with some preliminary spectral analysis. This is significant in the study of time-varying waveform distortions, since such spectral analysis on time-varying waveforms would only yield approximate and possibly inaccurate information.

20.10   Conclusions

The time–frequency and time–magnitude localization abilities of the original HH method have been previously researched. In this chapter, enhancements and modifications to the HH method have been presented for signal processing applications in power systems. The rationale for the modifications is that the original EMD method is unable to separate modes with instantaneous frequencies lying within one octave. The proposed modifications significantly improve the resolution capabilities of the EMD method, in terms of detecting weak higher-frequency modes, as well as separating closely-spaced modal frequencies. The potential applications of this method are in the areas of adaptive harmonic cancellation, empirical transfer function estimation, load signature profiling and power system event diagnosis. Some avenues of application of the hybridized EMD method presented in this chapter occur in [32].

Acknowledgements

The authors acknowledge the financial support of US Office of Naval Research Grant N0014-02-1-0623 and the US Department of Energy Award DE-FG02-05CH11292. The authors also acknowledge the support provided by the Center for Advanced Power Systems at Florida State University to perform the research that led to the development of the method described in this chapter.

References

[1] M. Steurer, 'Real time simulation for advanced time-varying harmonic analysis', IEEE Power Engineering Society General Meeting, Volume 3, San Francisco, California, USA, 2005, pp. 2250–2252.

[2] Y.-J. Shin, A. C. Parsons, E. J. Powers and W. M. Grady, 'Time-frequency analysis of power system disturbance signals for power quality', IEEE Power Engineering Society Summer Meeting, Volume 1, Edmonton, Alberta, Canada, 1999, pp. 402–407.

[3] N. E. Huang, Z. Shen, S. R. Long, M. C. Wu, H. H. Shih, Q. Zheng, N. C. Yen, C. C. Tung and H. H. Liu, 'The empirical mode decomposition and the Hilbert spectrum for nonlinear and non-stationary time series analysis', *Proc. R. Soc. London*, **454**, 1998, 903–995.

[4] R. Deering and J. F. Kaiser, 'The use of masking signal to improve empirical mode decomposition', IEEE International Conference on Acoustics and Speech Signal Processing, Volume 4, 2005, pp. 485–488.

[5] N. Senroy, S. Suryanarayanan and P. F. Ribeiro, 'An improved Hilbert-Huang method for analysis of time-varying waveforms in power quality', *IEEE Trans. Power Sys.*, **22**, 2007, 1843–1850.

[6] N. Senroy and S. Suryanarayanan, 'Two techniques to enhance empirical mode decomposition applied in power quality', IEEE Power Engineering Society General Meeting, Tampa, Florida, USA, 2007, pages 1–6.

[7] A. Oppenheim, R. Schafer and J. Buck, *Discrete-Time Signal Processing*, second edition, Prentice Hall, Englewood Cliffs, New Jersey, USA, 1999.

[8] G. T. Heydt, P. S. Field, C. C. Liu, D. Pierce, L. Tu and G. Hensley, 'Applications of the windowed FFT to electric power quality assessment', *IEEE Trans. Power Delivery*, **14**, 1999, 1411–1416.

[9] M. H. J. Bollen and I. Y.-H. Guo, In *Signal Processing of Power Quality Disturbances*, John Wiley & Sons, Inc., New York, USA, 2006, p. 314.

[10] A. W. Galli, G. T. Heydt and P. F. Ribeiro, 'Exploring the power of wavelet analysis', *IEEE Computer Applications in Power*, **9**, 1996, 37–41.

[11] S. Santoso, E. J. Powers, W. M. Grady and P. Hofmann, 'Power quality assessment via wavelet transform', *IEEE Trans. Power Delivery*, **11**, 1996, 924–930.

[12] J. Driesen, T. V. Craenenbroeck, R. Reekmans and D. V. Dommelen, 'Analysing time-varying power system harmonics using wavelet transform', Proceedings of the IEEE Instrument Measuring Techniques Conference, Brussels, Belgium, 1996.

[13] R. G. Stockwell, L. Mansinha and R. P. Lowe, 'Localization of the Complex Spectrum: The S Transform', *IEEE Trans. Signal Proc.*, **44**, 1996, 998–1001.

[14] M. V. Chilukuri and P. K. Dash, 'Multiresolution S-transform-based fuzzy recognition system for power quality events', *IEEE Trans. Power Sys.*, **19**, 2004, 323–330.

[15] P. K. Dash, B. K. Panigrahi and G. Panda, 'Power quality analysis using S-transform', *IEEE Trans. Power Delivery*, **18**, 2003, 406–411.

[16] N. E. Huang, Z. Shen and S. R. Long, 'A new view of nonlinear water waves: the Hilbert spectrum', *Annual Review of Fluid Mechanics*, **31**, 1999, 417–457.

[17] H. Liang, Q. H. Lin and J. D. Z. Chen, 'Application of the empirical mode decomposition to the analysis of esophageal manometric data in gastroesophageal reflux disease', *IEEE Trans. Biomedical Engineering*, **52**, 2005, 1692–1701.

[18] J. C. Nunes, Y. Bouaoune, E. Delechelle, N. Oumar and P. Bunel, 'Image analysis by bidimensional empirical mode decomposition', *Image and Vision Computing*, **21**, 2003, 1019–1026.

[19] N. E. Huang and S. S. P. Shen, *Hilbert-Huang Transform and Its Applications*, World Scientific, Singapore, 2005.

[20] M. A. Andrade, A. R. Messina, C. A. Rivera and D. Olguin, 'Identification of instantaneous attributes of torsional shaft signals using the Hilbert transform', *IEEE Trans. Power Sys.*, **19**, 2004, 1422–1429.

[21] A. R. Messina and V. Vittal, 'Nonlinear, non-stationary analysis of interarea oscillations via Hilbert spectral analysis', *IEEE Trans. Power Sys.*, **21**, 2006, 1234–1241.

[22] A. R. Messina, V. Vittal, D. Ruiz-Vega and G. Enriquez-Harper, 'Interpretation and visualization of wide area PMU measurements using Hilbert analysis', *IEEE Trans. Power Sys.*, **21**, 2006, 1760–1771.

[23] Z. Lu, J. S. Smith, Q. H. Wu and J. Fitch, 'Empirical mode decomposition for power quality monitoring', in IEEE/PES Transmission and Distribution Exposition: Asia and Pacific, 2005, pp. 1–5.

[24] R. C. Sharpley and V. Vatchev, 'Analysis of intrinsic mode functions', Ph. D. Thesis, Department of Mathematics, University of South Carolina, USA, 2004.

[25] G. Rilling, P. Flandrin and P. Goncalves, 'On empirical decomposition and its algorithms', in IEEE-EURASIP Workshop on Nonlinear Signal Image Processing, Grado, Italy, 2003.

[26] L. Cohen, 'Time frequency distributions – a review', *Proc. IEEE*, **77**, 1989, 941–981.

[27] P. Flandrin, G. Rilling and P. Goncalves, 'Empirical mode decomposition as a filter bank', *IEEE Signal Proc. Lett.*, **11**, 2004, 112–114.

[28] Z. Wu and N. E. Huang, 'A study of the characteristics of white noise using the empirical mode decomposition method', *Proc. R. Soc. London A*, **460**, 2004, 1597–1611.

[29] A. G. Requicha, 'The zeros of entire functions: theory and engineering applications', *Proc. IEEE*, **68**, 1980, 308–328.

[30] B. F. Logan, 'Information in the zero crossings of band-pass signals', *Bell System Tech. J.*, **56**, 1977, 487–510.

[31] S. Haykin, *Communication Systems*, edition? John Wiley & Sons, Inc., New York, USA, 2000.

[32] S. Suryanarayanan and N. Senroy, 'Avenues of Application of an Enhanced Algorithm for Analyzing Time-varying Waveforms', Proceedings of the Natural Power Systems Conference, Powai, India, 2008.

21

Harmonic and interharmonic on adjustable speed drives

R. Langella and A. Testa

21.1 Introduction

Adjustable speed drives (ASDs) based on double-stage conversion systems generate inter-harmonic current components in the supply system side, the DC link and output side, in addition to harmonics typical for single-stage converters [1–4]. Under ideal supply conditions, interharmonics are generated by the interaction between the two conversion systems through the intermodulation of their harmonics [5]. When imbalances, background harmonic and interharmonic distortions are present in supply voltages more complex intermodulation phenomena take place [6].

To characterize the aforementioned process, the interharmonic amplitudes (and phases), frequencies and origins have to be considered. The interharmonic amplitude importance for compatibility problems is evident: a comprehensive simulation or analysis for each working point of the ASD is needed and proper deterministic or probabilistic models must be used in accordance with the aim of the analysis. Various models are available: experimental analog, time-domain and frequency-domain models [7–11]. Each of them is characterized by a different degree of complexity in representing the AC supply system, the converters, the DC link and the AC supplied system. Whichever is the model, particular attention must be devoted to the problem of the frequency resolution and of the computational burden [12].

As for the interharmonic frequencies, whichever is the kind of analysis to be performed, it is very important to forecast them, for a given working point of the ASD. This allows

Time-Varying Waveform Distortions in Power Systems   Edited by Paulo F. Ribeiro
© 2009 John Wiley & Sons, Ltd

evaluating in advance the Fourier fundamental frequency that is the maximum common divisor of all the component frequencies that are present and gives the exact periodicity of the distorted waveform. Other reasons are related to the effects of interharmonics such as light flicker, asynchronous motor aging, dormant resonance excitation, and so on. The frequency forecast requires: (i) under ideal supply conditions, only the study of the interactions between rectifier and inverter, which are internal to the ASD; (ii) when imbalances and/or background harmonic and interharmonic distortion are present in supply voltages, also the study of the further interactions between the supply system and the whole ASD.

As for the origins of a given interharmonic, the availability of a proper symbolism for interharmonics turns out to be very useful. In fact, while the frequency is sufficient information to find out the origin of harmonic components (in a defined scenario or conversion system), the same is not true for interharmonic components. These components have frequencies which vary over a wide range, according to the output frequency, and can assume harmonic frequencies or even become DC components. Moreover, they overlap in situations in which two or more components of different origin assume the same frequency value. In general, the presence of nonideal supply conditions makes it impossible to recognize the origins without complex analyses. Anyway, the knowledge of the normal ASD behavior under ideal conditions may help in recognizing what derives from the ASD internal behavior and what from interactions with nonideal supply systems.

In this chapter, reference is made to ideal supply conditions which allow recognizing the frequencies and the origins of those interharmonics which are generated by the interaction between the rectifier and the inverter inside the ASD. Formulas to forecast the interharmonic component frequencies are developed firstly for line commutated inverter (LCI) drives and then for synchronous sinusoidal pulse width modulation (PWM) drives. Moreover, some thoughts about the amplitude variability of harmonics and interharmonics are given. Afterwards, a proper symbolism is proposed to make it possible to recognize the interharmonic origins. Numerical analyses, performed for both ASDs considered in a wide range of output frequencies, give a comprehensive insight in the complex behavior of interharmonic component frequencies; also some characteristic aspects, such as the degeneration in harmonics or the overlapping of an interharmonic couple of different origins, are described. Finally, some probabilistic modelling aspects are discussed.

21.2   LCI drives

LCI drives are still largely used in high-power applications. In this type of ASD the DC-link is made up of an inductor according to the typical scheme shown in Figure 21.1.

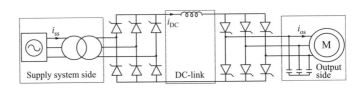

Figure 21.1   LCI drive scheme

21.2.1 Harmonic and interharmonic frequencies

Formulas for the evaluation of harmonic and interharmonic frequencies of the DC-link (DC), supply system (ss) and output side (os) currents are reported. The formulas are obtained applying the principle of modulation theory fully developed in [1, 2] and [6] and summarized in Appendix A (Section 21.8).

21.2.1.1 DC-link

The harmonics due to both the rectifier and the inverter are present in $i_{DC}(t)$. Their frequencies are, respectively:

$$f_{h_{DC}^{ss}}(v) = h_{DC}^{ss}(v) \cdot f_{ss} = vq_{ss} \cdot f_{ss}, \quad v = 1, 2, 3, \ldots \quad (21.1)$$

$$f_{h_{DC}^{os}}(j) = h_{DC}^{os}(j) \cdot f_{os} = jq_{os} \cdot f_{os}, \quad j = 1, 2, 3, \ldots \quad (21.2)$$

h_{DC}^{ss} and h_{DC}^{os} being the order of the DC-link harmonic due to the supply-side rectifier and to the output-side inverter, f_{ss} (f_{os}) the supply system (output) fundamental frequency and q_{ss} (q_{os}) the rectifier (inverter) number of pulses.

The interharmonics due to the intermodulation operated by the two converters are also present [8]. Their frequencies are:

$$f_{ih_{DC}}(v, j) = \left| f_{h_{DC}^{ss}}(v) \pm f_{h_{DC}^{os}}(j) \right|, \quad v = 1, 2, 3, \quad j = 1, 2, 3, \ldots \quad (21.3)$$

where the absolute value, here and in the following formulas, is justified by the symmetry properties of the Fourier transform [16]. Each couple of rectifier and inverter (j) harmonics produces a corresponding interharmonic frequency. In general, two or more couples may even give the same values of interharmonic frequencies.

21.2.1.2 Supply system side

The harmonic frequencies of $i_{ss}(t)$ derive from the modulation of the DC component and of the components (21.1) of $i_{DC}(t)$ operated by the rectifier and are:

$$f_{h_{ss}}(v) = h_{ss}(v) \cdot f_{ss} = |(v-1)q_{ss} \pm 1| \cdot f_{ss} \quad v = 1, 2, 3 \ldots \quad (21.4)$$

h_{ss} being the order of the supply side harmonic. The sign '$+$' in Equation (21.4) determines positive sequences while the sign '$-$' negative sequences.

The interharmonic frequencies of $i_{ss}(t)$ derive from the modulation of the inverter harmonic components in Equation (21.2) of $i_{DC}(t)$ operated by the rectifier and are:

$$f_{ih_{ss}}(v, j) = \left| f_{h_{ss}}(v) \pm f_{h_{DC}^{os}}(j) \right|, \quad v = 1, 2, 3, \ldots \quad j = 1, 2, 3, \ldots \quad (21.5)$$

The sequence of the interharmonics compared to the h_{ss}th harmonic sequence, is:

- the same, when one of the following conditions apply:
  - there is the sign '$+$' in Equation (21.5);
  - there is the sign '$-$' in Equation (21.5) and $f_{h_{ss}} > f_{h_{DC}^{os}}$;

- the opposite, when there is the sign '$-$' in Equation (21.5) and $f_{h_{ss}} < f_{h_{DC}^{os}}$.

- not definable, when $f_{h_{ss}} = f_{h_{DC}^{os}}$, which means that the interharmonics become DC components.

21.2.1.3   Output side

Due to the structural symmetry between the rectifier and the inverter, it is simply necessary to change f_{ss} with f_{os} and q_{ss} with q_{os} in Equations (21.4) and (21.5) respectively, in order to obtain the harmonic and interharmonic frequencies of the output side current, $i_{os}(t)$:

$$f_{h_{os}}(j) = h_{os}(j) \cdot f_{os} = |(j-1)q_{os} \pm 1| \cdot f_{os}, \quad j = 1, 2, 3 \ldots \tag{21.6}$$

$$f_{ih_{os}}(j, \nu) = \left| f_{h_{os}}(j) \pm f_{h_{DC}^{ss}}(\nu) \right|, \quad j = 1, 2, 3, \ldots \quad \nu = 1, 2, 3 \ldots \tag{21.7}$$

h_{os} being the order of the output side harmonic. As for the sequences, the same considerations about Equations (21.4) and (21.5) still apply.

21.2.2   Harmonic and interharmonic amplitudes

Harmonic amplitudes can be predicted by means of analytical formulas but, on the other hand, it does not seem possible to predict the amplitude of interharmonics by means of analytical formulas. It depends on their frequency values and from the system impedance behavior, in particular resonances. To give an idea of the entity of the variability, reference is made to the case study specified in [19] where the results obtained for all the output frequencies from 17.5 to 47.5 Hz of a case study based on an LCI like that of Figure 21.1, for constant flux operation, have been considered. The results are reported in Figure 21.2, where the frequency component amplitudes are referred to the 50 Hz fundamental component.

It is possible to observe that the amplitude of the main interharmonic component, generated by the intermodulation between the fundamental rectifier AC harmonic and the first inverter DC harmonic, is of some percentages (<3 %) for output frequencies around the nominal value, while

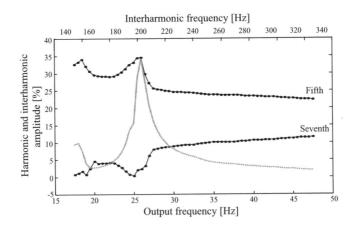

Figure 21.2   LCI harmonic and interharmonic amplitudes referred to the 50 Hz component versus the output frequency: fifth and seventh harmonics (•); main interharmonic (···)

in the output frequency range from 21 to 37.5 Hz it reaches very high amplitudes (up to 35 %). This great variability is a consequence of the resonance phenomenon due to the interaction between the capacitance present on the motor side and the series of the DC link inductance together with the supply side inductance modulated by the rectifier operation (see Figure 21.1).

21.3   PWM drives

Nowadays, from low- to medium high-power applications, voltage source inverters are more and more used in ASDs. The DC-link is made up of a capacitor while, in high-power applications, an inductor is added on the rectifier output side to smooth the current waveform.

Formulas (21.2) and (21.6) become useless because of both the different structure and the inverter operation, and a different analysis has to be considered. The harmonics generated by the inverter depend on the control strategy of the inverter switches, in particular, from the modulation ratio, m_f.

For the sake of brevity, here reference is made to the synchronous sinusoidal PWM, which is the most commonly used for high-power ASDs, and a method to forecast the produced harmonic frequencies is reported in Appendix B (Section 21.9). Other modulation techniques, such as harmonic elimination and random modulations, require different formulas for harmonic (and interharmonic) frequencies evaluation but the phenomenon of intermodulation between rectifier and inverter still takes place.

The rectifier is an uncontrolled diode bridge, being the inverter operated by a PWM technique as reported in the typical scheme shown in Figure 21.3.

21.3.1   Harmonic and interharmonic frequencies

21.3.1.1   DC-link

The harmonics produced by the rectifier, those produced by the inverter and the interharmonics due to the interaction between the two converters are present in both DC-link currents, $i_{DCr}(t)$ and $i_{DCi}(t)$. The harmonic frequencies generated by the rectifier can be calculated using formula (21.1) because the rectifier operation does not change.

The harmonic frequencies generated by the inverter (see Appendix B, Section 21.9) are evaluated as:

$$f_{h_{DC}^{os}}(m_f, j, r) = h_{DC}^{os}(m_f, j, r) \cdot f_{os} = |m_f j \pm r| \cdot f_{os}. \tag{21.8}$$

with m_f the modulation ratio, and j and r integers depending on m_f as reported in Table 21.1; the dependency from m_f is related to the switching strategy adopted as shown in Section 21.6.

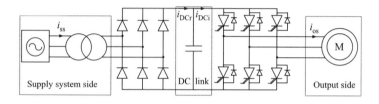

Figure 21.3   PWM drive scheme

Table 21.1   Values of parameters j and r for different m_f choices

m_f	Odd		Even	
Nontriple	j	r	j	r
	Even integers	$=$ even integers	Even integers	$\Rightarrow$ integers
	Odd integers	$=$ odd integers	Odd integers	$\Rightarrow$
Triple	j	r	j	r
	Even integers	$=$ even triple integer	Even integers	$\Rightarrow$ triple integers
	Odd integers	$=$ odd triple integer	Odd integers	$\Rightarrow$

In particular, Table 21.1 shows that:

- both even and odd harmonics are present for even m_f;
- only even harmonics are present for odd m_f;
- only triple harmonics are present for triple m_f.

The interharmonics due to the intermodulation operated by the two converters are also present [9]. Their frequencies are evaluated according to the relationship:

$$f_{ih_{DC}}(\nu, m_f, j, r) = \left| f_{h_{DC}^{ss}}(\nu) \pm f_{h_{DC}^{os}}(m_f, j, r) \right|, \tag{21.9}$$

with $\nu = 1, 2, 3, \ldots, j$ and r as in Table 21.1.

21.3.1.2   Supply system side

For the harmonic frequencies of $i_{ss}(t)$ the same considerations developed in Section 21.3 apply and formula (21.4) is still valid.

The interharmonic frequencies of $i_{ss}(t)$ derive from the modulation of the inverter harmonic components (21.8) operated by the rectifier and can be evaluated according to the following relationship:

$$f_{ih_{ss}}(\nu, m_f, j, r) = \left| f_{h_{ss}}(\nu) \pm f_{h_{DC}^{os}}(m_f, j, r) \right|, \tag{21.10}$$

with $\nu = 1, 2, 3, \ldots, j$ and r as in Table 21.1. As for the sequences, the same considerations about Equations (21.4) and (21.5) still apply.

21.3.1.3   Output side

The harmonic frequencies of $i_{os}(t)$ are evaluated as shown in Appendix B (Section 21.9) giving

$$f_{h_{os}}(m_f, j, k) = h_{os}(m_f, j, k) \cdot f_{os} = \left| m_f j \pm k \right| \cdot f_{os}, \tag{21.11}$$

with j and k as in Table 21.2.

Table 21.2   Values of parameters j and k for different m_f choices

m_f	Odd		Even	
	J	k	j	k
	Even integers	$=$ odd integers	Even integers	$\Rightarrow$ integers
	Odd integers	$=$ even integers	Odd integers	$\Rightarrow$

The interharmonic frequencies of $i_{os}(t)$ derive from the modulation of the rectifier harmonic components (21.1) operated by the inverter and can be evaluated according to the relationship:

$$f_{ih_{os}}(m_f, j, k, v) = \left| f_{h_{os}}(m_f, j, k) \pm f_{h_{DC}^{ss}}(v) \right|, \tag{21.12}$$

with j and k as in Table 21.1, $=1, 2, 3,\ldots$. Also in this case, the sequences of the harmonics in Equation (21.11) determine the sequences of the interharmonics in Equation (21.12) according to the same rules shown in Section 21.3.

21.3.2   Harmonic and interharmonic amplitudes

As for the case of LCI drives, also in the case of PWM drives it does not seem possible to predict by means of analytical formulas the amplitude of interharmonics. To have an idea of the entity of the variability, reference is made to the case study specified in [9] where the results obtained for all the output frequencies from 5 to 50 Hz of a case study based on a PWM like that of Figure 21.3, have been considered. The results are reported in Figure 21.4, where the frequency component amplitudes are referred to the 50 Hz fundamental component.

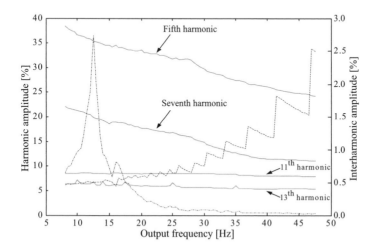

Figure 21.4   *PWM harmonic and interharmonic amplitudes referred to the 50 Hz component versus the output frequency: fifth, seventh, 11th and 13th harmonics (——); main interharmonics (- - -)*

Figure 21.4 confirms that, as is well known, the relative harmonic distortion is greater when the load consumption is lower. On the other hand, the relative interharmonic distortion is not so regular: it may increase with the load and locally be amplified due to resonances.

21.4   Notation proposed

In order to find out the origins of a given interharmonic component produced by the interaction between rectifier and inverter inside the ASD and under ideal supply conditions, it seems useful to introduce a proper symbolism. Reference is made to the interharmonic components in both the supply system and output sides. Though the harmonic order calculation is different in LCI and PWM drives it is, anyway, possible to find a unified and, at the same time, easy to understand symbolism.

Concerning the supply system side interharmonics, it seems suitable to introduce the symbols:

$$I^{u}_{h_{ss},h^{os}_{DC}},\qquad\qquad(21.13)$$

$$I^{d}_{h_{ss},h^{os}_{DC}},\qquad\qquad(21.14)$$

where:

- the subscripts show the common origin of the couple of components (21.13) and (21.14) related to the rectifier harmonic of order h_{ss} (see Equation (21.4)) and to the inverter DC-link harmonic of order h^{os}_{DC} (see Equation (21.2) for LCI and Equation (21.8) for PWM) involved in the intermodulation process;

- the superscript distinguishes the relative position of the components, being 'u' for the component (21.13), which stands for 'up' in frequency due to sign ' + ' in Equation (21.5), for LCI, and Equation (21.10), for PWM, and 'd' for the other component (21.14), which stands for 'down' in frequency due to the sign ' − ' in Equation (21.5), for LCI, and Equation (21.10), for PWM.

An example of the application of these symbols is reported in the following section (Figure 21.6 for LCI and Figure 21.10 for PWM).

Concerning the output side interharmonics, it seems suitable to introduce the symbols:

$$I^{u}_{h_{os},h^{ss}_{DC}},\qquad\qquad(21.15)$$

$$I^{d}_{h_{os},h^{ss}_{DC}},\qquad\qquad(21.16)$$

where:

- the subscripts again show the common origin of the couple of components (21.15) and (21.16) related to the inverter harmonic of order h_{os} (see Equation (21.6) for LCI and Equation (21.11) for PWM) and to the rectifier DC-link harmonic of order h^{ss}_{DC} (see Equation (21.1) for both LCI and PWM) involved in the intermodulation process;

- the use of the superscript is the same as for the supply side components (21.15) and (21.16) but, this time, it is related to the signs ('+' and '−') in Equation (21.7) for LCI and Equation (21.12) for PWM.

An example of the application of these symbols is reported in the following section (Figure 21.7 for LCI and Figure 21.11 for PWM).

21.5   Numerical analyses

Several numerical analyses were carried out on both LCI and PWM typical drives, both characterized by q_{ss} and q_{os} equal to six; their other parameters are fully referenced in [8] and [9] respectively. The models proposed in [8] and [9] were validated by comparing their results with those obtained by means of the well-known software EMTP and Power System Blockset under the Matlab environment.

Some experimental verifications of the formulas developed in this chapter are reported in [3] and [13] for LCI drives and in [14] and [15] for PWM drives. In all the cases the frequency components expected according to the formulas (21.4–21.7) and (21.10–21.12) respectively, were always actually present and of prevalent amplitude. Some additional frequency components were measured and it was demonstrated that their origin was the interaction between the nonideal supply system and ASDs, not covered by the formulas of Sections 21.3 and 21.4.

21.5.1   LCI drive

Figure 21.5 shows the LCI drive interharmonic frequency values versus the output frequency values, for both the supply system and output sides, obtained using Equations (21.5) and (21.7). For the sake of clarity, only some specified interharmonic components are reported.

Figures 21.6(a) and 21.7(a) show the amplitudes of current interharmonics versus their frequencies for $f_{os} = 40$ Hz, and refer to the supply system and output sides, respectively. Figures 21.6(b) and 21.7(b) show how to use the plots of Figures 21.5(a) and 21.5(b), respectively, to forecast the interharmonic component frequencies. The couple of the main interharmonics in Figure 21.6(a), $I_{1,6}^u$ and $I_{1,6}^d$, is generated by the intermodulation between the first supply system harmonic and the sixth DC-link harmonic produced by the inverter. The main interharmonic in Figure 21.7(a) is $I_{11,6}^d$, while $I_{7,6}^d$, of smaller but still remarkable amplitude, seems particularly notable due to its low frequency (20 Hz), which can create problems to asynchronous motor life [17].

21.5.2   PWM drive

The inverter modulation strategy selected is represented in Figure 21.8 in terms of actual, f_{sw}, and mean, f_{sw}^*, switching frequency (see Appendix B, Section 21.9); a constant value of f_{sw}^* for all the output frequencies is assumed equal to 350 Hz. Figure 21.8 also reports the modulation ratio, m_f. It is possible to observe that the modulation ratio varies during the ASD operation so changing the order of the harmonic components injected both into the DC-link and into the output side.

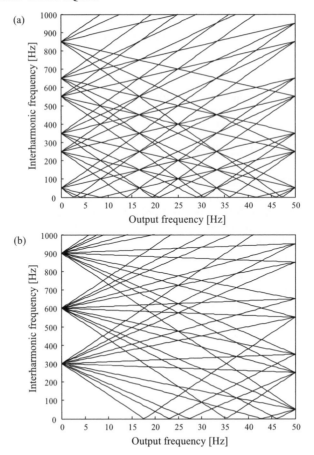

Figure 21.5   *LCI drive interharmonic frequencies versus output frequency: (a) supply system side, all interharmonics due to (ν, j) of (5) = {(1,1), (1,2), (1,3), (2,1), (2,2), (2,3), (3,1), (3,2), (3,3), (4,1), (4,2), (4,3)}; (b) output side, all interharmonics due to (j, ν) of (7) = {(1,1), (1,2), (1,3), (2,1), (2,2), (2,3), (3,1), (3,2), (3,3), (4,1), (4,2), (4,3)}*

The plots in Figure 21.9, obtained by means of Equations (21.10) and (21.12), show the interharmonic frequency values versus the output frequency both in the supply system and output sides, according to the switching frequency trend of Figure 21.8. Only some specified interharmonic components are reported to make it easier to understand the plots: for example in Figure 21.9(a) the interharmonic components due to the intermodulation of the first harmonic of the rectifier with the harmonic components generated by different couples of *j* and *r*, that is to say the intermodulation produced by specified side bands of the DC spectrum produced by the inverter, are represented.

Concerning the supply system side, it is evident that the overlapping of different interharmonic components occurs for some output frequency ranges. The most remarkable overlappings are highlighted by the grey shadows in Figure 21.9(a). Of course, such overlapping

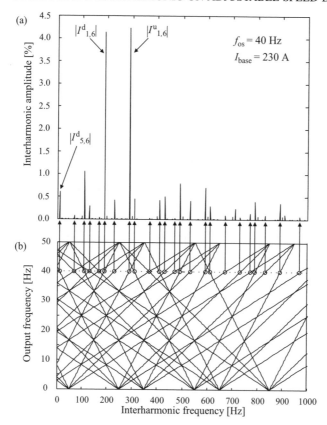

Figure 21.6 LCI drive supply system current interharmonics for $f_{os} = 40\,Hz$: (a) amplitudes in percentage of 230 A versus interharmonic frequency; (b) frequencies (abscissa) versus output frequency (ordinate)

leads to synergic combinations of interharmonic components originated from different inter-modulations. As for the output side, the aforementioned overlapping occurs only for discrete frequency values.

Figures 21.10(a) and 21.11(a) show the amplitudes of the current interharmonics versus their frequencies for $f_{os} = 40\,Hz$ and $m_f = 9$, with reference to the supply system and output sides, respectively.

Figures 21.11(b) and 21.12(b) show how to use the plots of Figures 21.9(a) and 21.9(b) to forecast the interharmonic component frequencies. It is worthwhile noting that overlapping causes the interharmonic components due to the values of the parameters (ν, m_f, j, r) in Equation (21.10) equal to (1,9,0,6) and to (1,9,1,3) have the same frequencies producing $I^u_{1,6}$ and $I^d_{1,6}$ (see Figure 21.8(a)).

In Figure 21.12 the behaviors of the two supply side interharmonic components due to $(1,m_f,0,6)$ and $(1,m_f,1,3)$ in Equation (21.10) in terms of amplitude (Figure 21.12(a) and

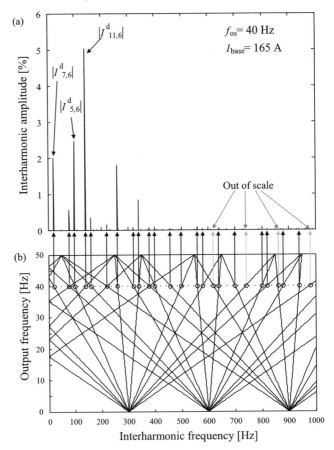

Figure 21.7   LCI drive output current interharmonic for $f_{os} = 40\,Hz$: (a) amplitudes in percentage of 165 A versus interharmonic frequency; (b) frequencies (abscissa) versus output frequency (ordinate)

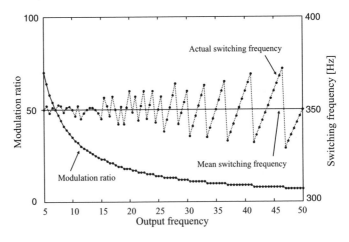

Figure 21.8   PWM drive: switching frequency (f_{sw}) and modulation ratio (m_f)

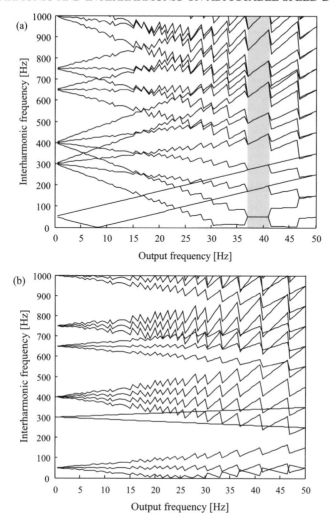

Figure 21.9 PWM drive interharmonic frequencies versus output frequency: (a) supply system side, all interharmonics due to $v=1$, m_f as in Figure 21.8 and (j,r) of Equation $(21.10) = \{(0,6), (1,3), (1,9), (2,0), (2,6), (3,3), (3,9), (4,0), (4,6)\}$; (b) output side, all interharmonics due to m_f as in Figure 21.8, $v=1$ and (j,k) of Equation $(21.12) = \{(0,1), (0,3), (1,0), (1,2), (2,1), (2,3), (3,0), (3,2), (4,1), (4,3)\}$

Figure 21.12(b)) and frequency (Figure 21.12(c)) variations versus the output frequency are shown. The overlapping occurs in the output frequency range (see Figure 21.12(c) in the interval $37 \div 41$ Hz where $m_f = 9$. Finally, it can be observed that the interharmonic due to (1, m_f,1,3) of Equation (21.10) degenerates in the fifth5 harmonic twice ($f_{os} = 15$ Hz and $f_{os} = 20$ Hz).

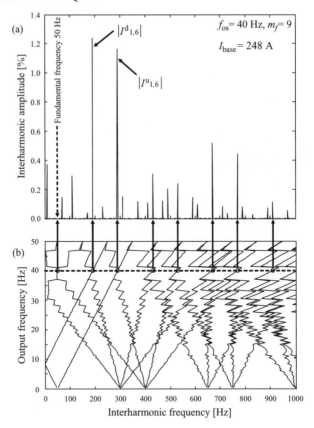

Figure 21.10   PWM drive supply system current interharmonics: (a) amplitudes in percentage of 248 A versus interharmonic frequency; (b) frequencies (abscissa) versus output frequency (ordinate) with switching frequency of Figure 21.8

21.6   Probabilistic modeling aspects [20]

Some considerations on mathematical and computational aspects of the probabilistic modeling of interharmonic time variability are developed here starting from some brief remarks on the harmonic probabilistic modeling aspects and then explicit reference is made to currents but the same considerations apply also for voltages.

As for the harmonic current of order h, it can be represented as a vector $\bar{I}_h$, of amplitude I_h and phase angle φ_h and of Cartesian components X_h and Y_h. The statistical characterization of $\bar{I}_h$ requires the determination of the joint statistics of a pair of real random variables (I_h, φ_h) or (X_h, Y_h). With reference to (X_h, Y_h), and their joint probability density $f_{XY}(x,y)$, also having in mind only numerical approaches that discretize the hyperspace $\Pi_{x,y,f_{XY}}$, assuming $M = Mx$ My discrete values for each coordinate in its definition interval, the following situation arises:

- for each h value of interest the matrix to be utilized for representing the joint pdf (jpdf) assumes dimension $D_h = M^2$ (e.g. if $M = 100$ then $D_h = 10^4$);

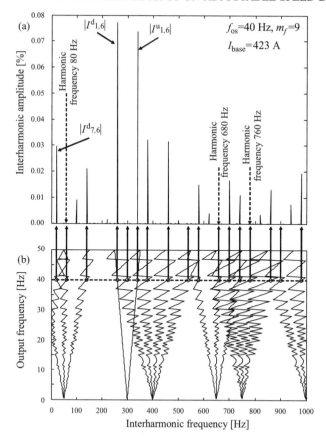

Figure 21.11   *PWM drive output current interharmonics: (a) amplitudes in percentage of 423 A versus interharmonic frequency; (b) frequencies (abscissa) versus output frequency (ordinate) with switching frequency of Figure 21.8*

- for all the N^h harmonic orders of interest the matrices to be utilized assume dimension $D_{Nh} = N^h M^2$ (e.g. if $N^h = 50$, then $D_{Nh} = 5 \times 10^5$).

As for the interharmonic currents, also the frequency, that is the order *ih*, constitutes a real random variable, so the statistical characterization of $\bar{I}_{ih}$ requires the determination of the joint statistics of the three real random variables $(I_{ih}, \varphi_{ih}, F_{ih})$ or (X_{ih}, Y_{ih}, F_{ih}). With reference to (X_{ih}, Y_{ih}, F_{ih}) and to their joint probability density function $f_{XYF}(x,y,f)$, also having in mind only numerical approaches that discretize the hyperspace $\Pi_{x,y,f,f_{XYF}}$, assuming discrete values for x and y as above for harmonics, and M_f values for frequencies from 0 to f_{max}, that is the maximum frequency of interest, the following situation arises:

- for each interharmonic of order *ih*, the matroid to be utilized for representing the jpdf assumes dimension $D_{ih} = M_f M^2$ (e.g. if $M = 100$, and $M_f = 2500$ that corresponds to a frequency resolution of 1 Hz, then $D_{ih} = 2.5 \times 10^7$);

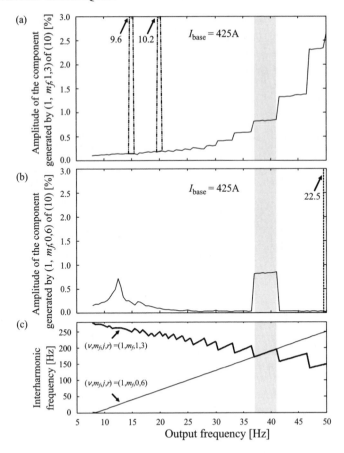

Figure 21.12   PWM drive supply system interharmonics versus output frequency, with switching frequency of Figure 21.8: amplitude of the component generated by the values of the parameters (v,m_f,j,r) in Equation (21.10) equal to (a) (1,m_f,1,3) and (b) (1,m_f,0,6); (c) corresponding interharmonic frequency versus output frequency

- for all the N^{ih} interharmonic orders of interest, the matroids to be utilized assume dimension $D_{Nih} = N^{ih}M_fM^2$ (e.g. if $N^{ih} = 10N^h$, then $D_{Nih} = 2.5 \times 10^8$).

Further on the discussed aspects, also the decomposition in positive, negative and zero sequences should be considered (see Section 22.2.1), but this has not been done here for the sake of simplicity.

As for modelling harmonics and interharmonics, the matroids to be utilized assume a total dimension: $D_{Nh + Nih} = D_{Nh} + D_{Nih} = M^2(N^h + N^{ih}M_f)$.

21.6.1   Fixed frequency resolution

If there is no interest in distinguishing interharmonics for their origin, the problem formulation results are simplified. Once introduced a fixed resolution frequency f_r, assuming that both the harmonic and interharmonic frequencies are (or appear[1]) integer multiples of it, all the components can be treated as harmonics h' of f_r. This means that the number of components of interest $N^{h'}$ is equal to the ratio between the maximum frequency of interest and f_r, that is to say $N^{h'} = N^h \cdot (f_{ss}/f_r)$, f_{ss} being the system fundamental frequency. The following situation arises for x', y', $f'_{x'y'}$:

- for each h' value of interest the matrix to be utilized for representing the jpdf assumes dimension $D_{h'} = M^2$ (e.g. if $M = 100$ then $D_{h'} = 10^4$);

- for all the $N^{h'}$ orders of interest, the matrices utilized assume dimension $D_{Nh'} = N^{h'} M^2$ (e.g. if $N^{h'} = f_{max}/f_r = 2500/5 = 500$, then $D_{Nh'} = 5 \times 10^6$ or if $N^{h'} = f_{max}/f_r = 2500/1$ 2500, then $D_{Nh'} = 2.5 \times 10^7$).

21.6.2 IEC subgrouping

Moreover, if there is no interest in distinguishing interharmonics for their exact frequency, and there is only interest in their amplitudes and frequency position between the harmonics, that is to say accepting the IEC grouping technique, the problem formulation results are further simplified.

In this case, the amplitude of the interharmonic subgroup of order $n + 0.5$, $C_{n+0.5-200\text{-ms}}$, is defined as the rms value of all the interharmonic components between adjacent harmonic subgroups.

Once introduced as harmonic and interharmonic subgroups, all the subgroups can be treated as harmonics h'' of a resolution frequency which is half the fundamental. This means that $N^{h''}$ is equal to the ratio between the maximum frequency of interest and $f_{ss}/2$, that is to say $N^{h''} = 2N^h$. The phase angles become meaningless because of the group definition and the subgroup of order h'' can be represented by its amplitude I_G and its probability density function $f_{IG}(I_G)$

The following situation arises:

- for each h'' value of interest, the vector to be utilized for representing the pdf assumes dimension $D = M$ (e.g. if $M = 100$, that is a very low value, then $D_{h''} = 10^2$);

- for all the $N^{h''} = 2N^h$ subgroups, the vectors to be utilized assume dimension $D_{Nh''} = N^{h''} M = 2 N^h M$ (e.g. if $N^{h''} = 2 N^h = 100$ then $D_{Nh''} = 10^6$).

21.6.3 Considerations

Whichever practical solution is adopted to include interharmonics in distortion probabilistic modeling, the computational burden and storage memory requirements are increased. In particular, the solution of fixed frequency resolution increases of about two order of quantities

---

[1] Interharmonic components that are in between the bins spaced at f_r Hz would spill over primarily into adjacent interharmonic bins with a minimum of spill into harmonic bins. So, all components of the spectral analysis appear as harmonics of f_r even when they are originated by interharmonics at frequencies which are not an integer multiple of f_r.

the dimension of the problem, while the solution based on IEC subgrouping doubles the dimension of the problem but looses information which can be very important in some frequency ranges or for the classical problem of current summation.

21.7   Conclusions

The interharmonic generation process has been addressed with reference to high-power adjustable speed drives based on double stage conversion systems using line commutated or pulse width modulated inverters. Reference has been made to ideal supply conditions. Formulas to forecast the interharmonic frequencies due to the interaction between the rectifier and inverter have been developed and a proper symbolism has been proposed to recognize the interharmonic origins. Numerical analyses have been performed for both the considered ASDs in a wide range of output frequencies, giving a comprehensive insight into the complex behavior of interharmonic component frequencies. Some characteristic aspects, such as the degeneration of interharmonic components in harmonics or the overlapping of a couple of interharmonics of different origins, have been highlighted. Finally, some probabilistic modeling aspects have been described.

21.8   Appendix A

Reference is made to a q-pulse bridge converter operating synchronously with the AC side; the current on the DC side is affected by a ripple superimposed on the direct component, I_{DC}. The results apply to both rectifier and inverter operation and, for this reason, the subscripts 'ss' and 'os' are omitted.

As is well known, the bridge can be seen (Figure 21.13) as having two inputs: the 'switching function', $S_w(t)$, and the 'modulating function', that is the current on the DC side, $i_{DC}(t)$.

In other words, the bridge acts as a modulator and its output, the current on the AC side, $i_{AC}(t)$, is said to be 'modulated'.

The switching function $S_w(t)$ can be expressed using the Fourier series:

$$S_w(t) = \left| \bar{S}_w^0 \right| \cdot \cos(\omega_{AC} t + \varphi_{(\bar{S}_w^0)}) + \sum_{k=1}^{+\infty} \left| \bar{S}_w^k \right| \cdot \cos((kq \pm 1)\omega_{AC} t + \varphi_{(\bar{S}_w^k)}) \qquad (21.17)$$

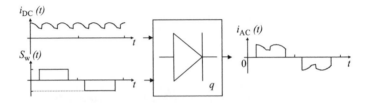

Figure 21.13   q-pulse bridge converter

where $|\bar{S}_w^k|$ is the peak magnitude of sinusoidal components and $\varphi_{()}$ its phase angle, ω_{AC} is the fundamental angular frequency of the AC side, and q is the number of pulses of the bridge converter.

The 'modulating' function $i_{DC}(t)$ can be expressed in a general form as the sum of the direct current and of superimposed sine waves:

$$i_{DC}(t) = I_{DC} + \sum_{h=1}^{+\infty} |\bar{I}_{DC}^h| \cos(h\omega_{DC}t + \varphi_{(\bar{I}_{DC}^h)}), \qquad (21.18)$$

where $|\bar{I}_{DC}^h|$ is the peak magnitude of sinusoidal components on the DC side and $\varphi_{()}$ its phase angle; ω_{DC} can assume any value and it is not necessarily an integral multiple of ω_{AC}.

The 'output' of the bridge, $i_{AC}(t)$, can be found by multiplying the switching function $S_w(t)$ by the 'modulating' function $i_{DC}(t)$:

$$
\begin{aligned}
i_{ac}(t) = & |\bar{S}_w^0| I_{DC} \cdot \cos(\omega_{ac}t + \varphi_{(\bar{S}_w^0)}) \\
& + \sum_{k=1}^{+\infty} |\bar{S}_w^k| I_{DC} \cdot \cos((kq \pm 1)\omega_{ac}t + \varphi_{(\bar{S}_w^k)}) \\
& + \sum_{k=1}^{+\infty} |\bar{S}_w^0| |\bar{I}_{dc}^h| \cdot \cos(\omega_{ac}t + \varphi_{(\bar{S}_w^k)}) \cdot \cos(h\omega_{dc}t + \varphi_{\bar{I}_{dc}^h}) \\
& + \sum_{k=1}^{+\infty} \sum_{h=1}^{+\infty} |\bar{S}_{wa}^k| |\bar{I}_{dc}^h| \cdot \cos((kq \pm 1)\omega_{ac}t + \varphi_{(\bar{S}_w^k)}) \cdot \cos(h\omega_{dc}t + \varphi_{\bar{I}_{dc}^h})
\end{aligned} \qquad (21.19)
$$

which, by applying the Werner formula, can be written as:

$$
\begin{aligned}
i_{ac}(t) = & |\bar{S}_w^0| I_{DC} \cdot \cos(\omega_{ac}t + \varphi_{(S_w^{-0})}) + \\
& + \sum_{k=1}^{+\infty} |S_w^{-k}| I_{DC} \cdot \cos((kq \pm 1)\omega_{ac}t + \varphi_{(S_w^{-k})}) + \\
& + \sum_{h=1}^{+\infty} |S_w^{-0}| |I_{dc}^{-h}| \cdot \cos(\omega_{ac}t + \varphi_{(S_w^{-0})}) \cdot \cos(h\omega_{dc}t + \varphi_{(I_{dc}^{-h})}) + \\
& + \sum_{k=1}^{+\infty} \sum_{h=1}^{+\infty} |S_{wa}^{-k}| |I_{dc}^{-h}| \cdot \cos((kq \pm 1)\omega_{ac}t + \varphi_{(S_w^{-k})}) \cdot \cos(h\omega_{dc}t + \varphi_{(I_{dc}^{-h})})
\end{aligned} \qquad (21.20)
$$

The last two series of the formula above show that, under the considered conditions, the three-phase bridge converters generate interharmonics, which are passed into the AC side, and which are in addition to the normal theoretical harmonics of ω_{AC} (of first, fifth, seventh, 11th, 13th, etc., orders).

21.9 Appendix B

Reference is made to the circuit of Figure 21.14 in which, for the sake of simplicity, a constant DC voltage, V_{DC}, is considered.

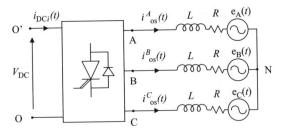

Figure 21.14   PWM inverter supplying steady-state motor model

Figure 21.15 shows the sinusoidal PWM: if synchronous PWM is adopted, m_f must be an integral number.

The voltage $v_{AO}(t)$, defined with reference to Figure 21.14, is obtained by comparing the carrier signal, $\Delta(t)$, with the modulation signal, $c_{AO}(t)$ (see Figure 21.15). Similarly, $v_{BO}(t)$ and $v_{CO}(t)$ are obtained by comparing $\Delta(t)$ with two modulation signals, $c_{BO}(t)$ and $c_{CO}(t)$, with a phase shifting from $c_{AO}(t)$ of $-120°$ and $+120°$, respectively.

Referring to the symbolism of Figure 21.14 and supposing that the inverter does not dissipate or generate power and that the load is balanced, it results:

$$i_{DCi}(t) = \frac{v_{AO}(t)}{V_{DC}} \cdot i_{os}^A(t) + \frac{v_{BO}(t)}{V_{DC}} \cdot i_{os}^B(t) + \frac{v_{CO}(t)}{V_{DC}} \cdot i_{os}^C(t). \tag{21.21}$$

The harmonic components of $i_{DCi}(t)$ are obtained by modulating harmonics of $v_{AO}(t)$, $v_{BO}(t)$ and $v_{CO}(t)$ with those of $i_{os}^A(t)$, $i_{os}^B(t)$ and $i_{os}^C(t)$, respectively. It is worthwhile noting that $v_{XO}(t)/V_{DC}$, with $X = A$, B, C, represents the well-known inverter switching function.

If m_f is an integer, $v_{AO}(t)$, $v_{BO}(t)$ and $v_{CO}(t)$ have f_{os} as fundamental frequency and only harmonic components are present. Moreover, if m_f is an odd integer, $v_{AO}(t)$, $v_{BO}(t)$ and $v_{CO}(t)$ contain only odd harmonics and their DC components, V^0, are exactly $V_{DC}/2$; instead, if m_f is an even integer, they contain both odd and even harmonics and the DC components may be different from $V_{DC}/2$.

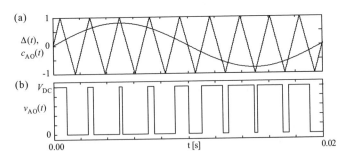

Figure 21.15   Sinusoidal PWM: (a) $c_{AO}(t)$ is the sinusoidal modulation signal for the phase A, $\Delta(t)$ is the carrier signal; (b) $v_{AO}(t)$ is the voltage defined with reference to Figure 21.14

A general expression can be written:

$$v_{XO}(t) = V^0 + v_{XO}^1(t) + \tilde{v}_{XO}(t), \qquad (21.22)$$

where $X = A, B, C$ and

$$\tilde{v}_{XO}(t) = \sum_{n=2}^{+\infty} A_n \sin(2\pi n f_{os} t + \varphi_n). \qquad (21.23)$$

The amplitudes A_n are calculated as:

$$A_n = \frac{m_a V_{DC}}{2} \left\{ \sum_{p=1}^{+\infty} \left[1 - (-1)^{n+p \cdot (m_f-1)} \right] \cdot R(n, p, m_f, m_a, \varphi_X) \right\}, \qquad (21.24)$$

where φ_X is the phase of $c_{XO}(t)$ and:

$$R(\cdot) = \frac{J_{pm_f+n}(p\varepsilon) \cdot e^{ipm_f \varphi_X} - J_{pm_f-n}(p\varepsilon) \cdot e^{-ipm_f \varphi_X}}{p\varepsilon}, \qquad (21.25)$$

J_p being the p-order first kind Bessel function, m_a the amplitude modulation ratio and $\varepsilon = (m_f \cdot m_a \cdot \pi)/2$.

Some simplifications can be obtained if m_f is odd; assuming that:

$$n = jm_f + k, \qquad (21.26)$$

Equations (21.23, 21.24) and (21.25) can be rewritten as:

$$\tilde{v}_{XO}(t) = \frac{2V_{DC}}{\pi} \sum_{j=1}^{+\infty} \frac{1}{j} \sum_{k=-\infty}^{+\infty} R(j, k, m_a) \sin\left[2\pi(jm_f + k)f_{os}t + k\varphi_X\right], \qquad (21.27)$$

with

$$R(j, k, m_f) = \begin{cases} 0 & j+k, \quad \text{even} \\ J_k\left(\dfrac{j\pi m_a}{2}\right) & j+k, \quad \text{odd}. \end{cases} \qquad (21.28)$$

However, if m_f is even, j and k can assume all integer values.

So, the harmonic frequency components of $v_{XO}(t)$ can be evaluated according to the following relationship:

$$f_{h_{os}} = |jm_f \pm k| \cdot f_{os}, \qquad (21.29)$$

where j and k vary as shown in Table 21.2.

Due to the inverter load model adopted, harmonic components of $i_{os}^X(t)$ are located at the same frequencies as those of the corresponding voltages. The harmonic frequencies of $i_{DCi}(t)$

can be deduced by means of the modulation theory: the harmonic components of $i_{os}^X(t)$ are modulated according to Equation (21.21) and its harmonic frequencies are evaluated according to:

$$f_{h_{DC}^{os}} = \left| jm_f \pm (k \pm 1) \right| \cdot f_{os} \tag{21.30}$$

So, formula (21.8) is obtained by (21.30), writing $(k \pm 1)$ as r.

Finally, it is worthwhile observing that the modulation strategy depends on the relation between the actual switching frequency, f_{sw}, and f_{os} according to:

$$f_{sw}(f_{os}) = m_f(f_{os}) \cdot f_{os}, \tag{21.31}$$

$m_f(f_{os})$ being, in general, a proper function representing the modulation ratio, and assuming integer values for synchronous PWM:

$$m_f(f_{os}) = \text{round}\left(\frac{f_{sw}^*}{f_{os}}\right), \tag{21.32}$$

$round(.)$ being the function which gives the integer nearest to the argument and f_{sw}^* is the asynchronous switching frequency which depends on the control strategy [18].

References

[1] R. Yacamini, 'Power system harmonics. IV. Interharmonics', *Power Engineering Journal*, **10**, 1996, 185–193.

[2] IEEE Interharmonic Task Force, Interharmonic in Power System, Cirgré 36.05/CIRED 2 CC02 Voltage Quality Working Group, http://grouper.ieee.org/groups/harmonic/iharm/docs/ihfinal.pdf.

[3] R. Carbone, D. Menniti, R. E. Morrison, E. Delaney and A. Testa, 'Harmonic and Intherarmonic Distortion in Current Source Type Inverter Drives', *IEEE Trans. on Power Delivery*, **10**, 1995, 1576–1583.

[4] M. B. Rifai, T. H. Ortmeryer and W. J. McQuillan, 'Evaluation of Current Interharmonics from AC Drives', *IEEE Trans. on Power Delivery*, **15**, 2000, 1094–1098.

[5] R. Carbone, D. Menniti, R. E. Morrison and A. Testa, 'Harmonic and Intheraharmonic Distortion Modelling in Multiconverter System', *IEEE Trans. on Power Delivery*, **10**, 1995, 1685–1692.

[6] R. Carbone, A. Lo Schiavo, P. Marino and A. Testa, 'Frequency coupling matrixes for multi stage conversion system analysis', *European Transactions on Electric Power Systems*, **12**, 2002, 17–24.

[7] W. Xu, H. W. Dommel, M. Brent Hughes, G. W. K. Chang and L. Tan, 'Modelling of Adjustable Speed Drives for Power System Harmonic Analysis', *IEEE Trans. on Power Delivery*, **14**, 1999, 595–601.

[8] R. Carbone, F. De Rosa, R. Langella and A. Testa, 'A New Approach to Model AC/DC/AC Conversion Systems', IEEE/PES Summer Meeting, Vancouver, Canada, July 2001.

[9] R. Carbone, F. De Rosa, R. Langella, A. Sollazzo and A. Testa, 'Modelling of AC/DC/AC Conversion Systems with PWM Inverter', IEEE/PES Summer Meeting, Chicago, USA, July 2002.

[10] M. Sakui and H. Fujita, 'Calculation of harmonic currents in a three-phase converter with unbalanced power supply conditions', *IEE Proc.B*, **139**, 1992, 478–484.

[11] L. Hu and R. Morrison, 'The Use of the Modulation Theory to Calculate the Harmonic Distortion in HVDC Systems Operating on an Unbalanced Supply', *IEEE Trans. on Power Systems*, **12**, 1997, 973–980.

[12] D. Castaldo, F. De Rosa, R. Langella, A. Sollazzo and A. Testa, 'Waveform Distortion Caused by High Power Adjustable Speed Drives Part II: Probabilistic Analysis', *ETEP*, **13**, 2003, 355–363.

[13] R. Carbone, V. Mangoni, C. Mirone, P. Paternò and A. Testa, 'Interharmonic distortion in the static strarting-up system of a hydro-electric pumped-storage power station', Seventh IEEE ICHQP, Las Vegas, Nevada, USA, October 1996.

[14] P. Caramia, G. Carpinelli, P. Varilone, D. Gallo, R. Langella, A. Testa and P. Verde, 'High Speed Ac Locomotives: Harmonic And Interharmonic Analysis At A Vehicle Test Room', Ninth IEEE International Conference on Harmonics and Quality of Power, Orlando, Florida, USA, October 2000.

[15] R. Carbone, F. De Rosa, R. Langella, A. Sollazzo and A. Testa, 'Modeling Waveform Distortion Produced by High Speed AC Locomotive Converters', IEEE PowerTech., Bologna, Italy, June 2003.

[16] A. V. Oppenheim and R. W. Schaffer, *Discrete time signal processing*, Prentice-Hall International Inc., 1989.

[17] J. Policarpo G. de Abreu and A. E. Emanuel, 'Induction motors loss of life due to voltage imbalance and harmonic: a preliminary study', Ninth ICHQP, Orlando, Florida, USA, October 2000.

[18] J. M. D. Murphy and F. G. Turnbull, *Power electronic control of AC motors*, Pergamon Press, New York, USA, 2005.

[19] R. Carbone, F. De Rosa, R. Langella, A. Sollazzo and A. Testa, 'Modelling of AC/DC/AC Conversion Systems with PWM Inverter', Proceedings of the IEEE Summer Power Meeting, Chicago, Illinois, USA, July 2002.

[20] R. Langella and A. Testa, 'Interharmonics from a Probabilistic Perpective', IEEE PES Annual Meeting, San Francisco, California, (USA), June 2005.

22

Tracking time-varying power harmonic distortions

C. A. Duque, P. M. Silveira, T. Baldwin and P. F. Ribeiro

22.1 Introduction

Although it is well known that Fourier analysis is in reality only accurately applicable to steady-state waveforms, it is a tool which is widely used to study and monitor time-varying signals, which are commonplace in electrical power systems. The disadvantages of Fourier analysis, such as frequency spillover or problems due to sampling (data window) truncation can often be minimized by various windowing techniques, but they nevertheless exist. This chapter demonstrates that it is possible to track and visualize amplitude and time-varying power systems harmonics, without frequency spillover caused by time–frequency techniques. This new tool allows for a clear visualization of time-varying harmonics which can lead to better ways of tracking harmonic distortion and understanding time-dependent power quality parameters. It also has the potential to assist with control and protection applications.

While estimation technique is concerned with the process used to extract useful information from the signal, such as amplitude, phase and frequency, signal decomposition is concerned with the way that the original signal can be split into other components, such as harmonics, interharmonics, subharmonics, and so on.

Harmonic decomposition of a power system signal (voltage or current) is an important issue in power quality analysis. There are at least two reasons to focus on harmonic

Time-Varying Waveform Distortions in Power Systems   Edited by Paulo F. Ribeiro
© 2009 John Wiley & Sons, Ltd

decomposition instead of harmonic estimation: (a) if separation of the individual harmonic component from the input signal can be achieved, the estimation problem becomes easier; (b) the decomposition is carried out in the time-domain, such that the time-varying behavior of each harmonic component can be observed.

Some existing techniques can be used to separate frequency components. For example, short time Fourier transform (STFT) and wavelet transforms [1], are two well-known decomposition techniques. Both can be seen as a particular case of filter bank theory [2]. The STFT coefficient filters are complex numbers which generate a complex output signal whose magnitude corresponds to the amplitude of the harmonic component being estimated [2]. The main disadvantage of this method is the difficulty with the construction of an efficient bandpass filter with lower frequency spillover. On the other hand, although wavelet transform utilizes filters with real coefficients, the common wavelet mothers do not have good magnitude response in order to prevent frequency spillover. Besides, the traditional binary tree structure is not able to divide the spectrum conveniently for harmonic decomposition.

Adaptive notch filter and phase-locked loop (PLL) [3–4] have been used for extracting time-varying harmonics components. However, these methods work well only if a few harmonics components are present at the input signal. In the other case the energies of adjacent harmonics spill over each other and the decomposed signal becomes contaminated. In [5] the authors presented a technique based on multistage implementation of narrow low-pass digital filters to extract stationary harmonic components. Thus, no digital analytical technique has been proposed or used to track time-varying power systems harmonics without frequency spillover. Consequently, the method proposed in this chapter adequately utilizes filter banks to avoid the frequency mixture, particularly associated with time-varying signals.

Attempts to visualize time-varying harmonics using wavelet transforms have been proposed in [6] and [7]. However, the structures were not able to decouple the frequencies completely.

The methodology proposed in this chapter is constructed to separate the odd and the even harmonic components, until the 15th. It uses selected digital filters and down-sampling to obtain the equivalent bandpass filters centered at each harmonic. After the signal is decomposed by the analysis bank, each harmonic is reconstructed using a nonconventional synthesis bank structure. This structure is composed of filters and up-sampling that reconstructs each harmonic to its original sampling rate.

The immediate use for this method is the monitoring of time-varying individual power systems harmonics. Future use may include control and protection applications, as well as interharmonic measurements.

The chapter is divided into method description, odd/even harmonic extraction and simulation results.

22.2   Method description

Multirate systems employ a bank of filters with either a common input or summed output. The first structure is known as an analysis filter bank [8] as it divides the input signal in different subbands in order to facilitate the analysis or the processing of the signal. The second structure is known as a synthesis filter bank and is used if the signal needs to be reconstructed. Together with the filters the multirate systems must include the sampling

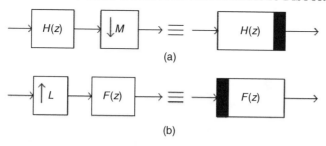

Figure 22.1   Basic structures used in a multirate filter bank and its equivalent representation for $L = M = 2$: (a) decimator; (b) interpolator

rate alteration operator (up- and down-sampling). Figure 22.1 shows two basic structures used in a multirate system, where the up- and down-sampling factor, M and L, equal 2. Figure 22.1(a) shows a decimator structure composed of a filter followed by the down-sampler and Figure 22.1(b) the interpolator structure composed by a up-sampler followed by a filter. The decimator structure is responsible for reducing the sampling rate while the interpolator structure increases it. The filters $H(z)$ and $F(z)$ are typically bandpass filters.

The direct way to build an analysis filter bank, in order to divide the input signal in its odd harmonic component, is represented in Figure 22.2. In this structure the filter $H_k(z)$ is a bandpass filter centered in the kth harmonic and must be projected to have 3 dB bandwidth lower than $2f_0$, where f_0 is the fundamental frequency. If only odd harmonics are supposed to be present in the input signal, the 3 dB bandwidth can be relaxed to be lower than $5f_0$. Note that Figure 22.2 is not a multirate system, because the structure does not include sampling rate alternation, which means that there is only one sampling rate in the whole system.

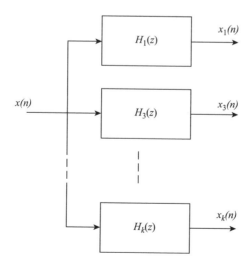

Figure 22.2   Analysis filter bank to decompose the input signal in its harmonics components

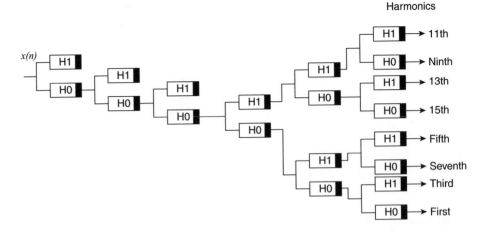

Figure 22.3   Multirate equivalent structure to the filter bank in Figure 22.2

The practical problem concerning the structure shown in Figure 22.2 is the difficulty of designing each individual bandpass filter. This problem becomes more challenging when a high sampling rate must be used to handle the signal and the consequent abrupt transition band.

In this situation the best way to construct an equivalent filter bank is to utilize the multirate technique. Figure 22.3 shows how an equivalent structure to that shown in Figure 22.2 can be obtained using the multirate approach. The filters $H_0(z)$ and $H_1(z)$ are orthogonal filters [6], with the first one as a low-pass filter and the second one as a high-pass filter.

Figure 22.4 shows the amplitude response for the analysis bank. This figure was obtained using a sampling rate equal to 256 samples by cycle and FIR filters of 69th order. These filters correspond to so-called orthogonal filter which are also known as power-symmetric filter [8].

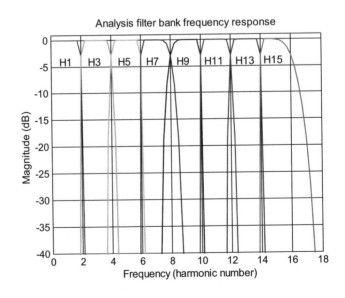

Figure 22.4   Amplitude response

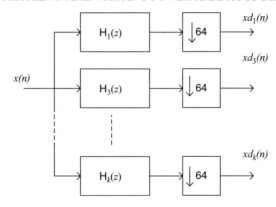

Figure 22.5   Equivalent multirate analysis filter bank

The main difference between the structure shown in Figure 22.2, and the other one obtained using the multirate technique, is the decimator at the output of the bank, as shown in Figure 22.5. In fact, Figure 22.5 is equivalent to Figure 22.3. It was obtained using the multirate noble identities [2] and moving the down-sampler factor inside Figure 22.2 to the right-hand side. The decimated signal at the output of each filter has a sampling rate 64 times lower then the input signal. To reconstruct each harmonic into its original sampling rate it is necessary to use the synthesis filter bank structure.

22.3   Synthesis filter banks

The synthesis filter bank used here is a different implementation of the conventional one. As the aim is to reconstruct each harmonic instead of the original signal, the filter bank must be divided in order to obtain the corresponding harmonic. Figure 22.6 shows the filter bank utilized to reconstruct the harmonic components.

It is important to note that the amplitude response of the synthesis filter bank is similar to Figure 22.4.

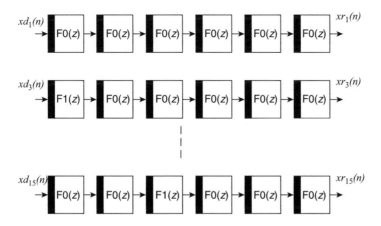

Figure 22.6   Modified synthesis filter bank

The analysis of Figure 22.4 reveals that the filter bank in this configuration is unable to filter even harmonics appropriately, which means that these components will appear with the adjacent odd harmonics. To overcome this problem an additional filter stage is included. This filter is composed of a second-order infinite impulse response (IIR) band-pass filter.

22.4 Extracting even harmonics

To extract the even harmonics the same bank can be used together with a preprocessing of the input signal. The Hilbert transform is then used to implement a technique known as single side band (SSB) modulation [8]. The SSB modulation moves all frequencies in the input signal to the right by f_0 Hz. In this way, even harmonics are changed to odd harmonics and vice versa.

Figure 22.7 shows the whole system for extracting odd and even harmonics.

22.5 Simulation results

This section presents two examples: the first is a synthetic signal, which has been generated in Matlab using a mathematical model, and the second is a signal obtained from the Electromagnetic Transient Program including DC systems (EMTDC) with its graphical interface Power Systems Computer Aided Design (PSCAD™). This program can simulate any kind of power system with high fidelity and the resulting signals of interest are very close to physical reality.

22.5.1 Synthetic signal

The synthetic signal utilized can be represented by:

$$x(t) = \sum_{h=1}^{N} A_h \sin(h\omega_0 t).f(t) + g(t) \tag{22.1}$$

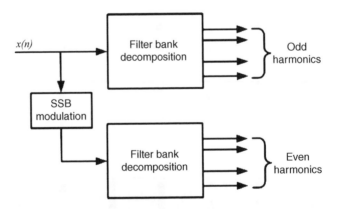

Figure 22.7 Whole structure for harmonic extraction

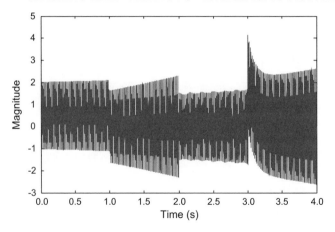

Figure 22.8   Synthetic signal used

where h is the order (1 up to 15) and A is the magnitude of the component, ω_0 is the fundamental frequency, and finally, $f(t)$ and $g(t)$ are exponential functions (crescent, decrescent or alternated one) or simply a constant value $\in \Re$. Besides, $x(t)$ is partitioned into four different segments in such a way that the generated signal is a distorted one with some harmonics in steady state and others varying in time, including abrupt and modulated change of magnitude and phase, as well as a DC component. Figure 22.8 illustrates the synthetic signal.

The structure shown in Figure 22.7 has been used to decompose the signal into 16 different harmonic orders, including the fundamental (60 Hz) and the DC component.

Figure 22.9 shows the decomposed signal with its corresponding components from DC up to the 11th harmonic. The left-hand column represents the original components and the right-hand column the corresponding components obtained through the filter bank.

For simplicity and space limitation the higher components are not shown. However, it is important to remark that all waveforms of the time-varying harmonics are extracted with efficiency along the time.

Naturally, intrinsic delays will be present during the transitions from the previous to the new state. Figure 22.10 shows some components of both the original and decomposed signal in a short timescale interval.

22.5.2   Simulated signal

It is well known that during energization a transformer can draw a large current from the supply system, normally called an inrush current, whose harmonic content is high. Although today's power transformers have lower harmonic content Table 22.1 shows the typical harmonic components present in the inrush currents [9]. These values are normally used as references for protection reasons, but they do not take into account the time-varying nature of this phenomenon.

In recent years, improvements in materials and transformer design have lead to inrush currents with lower distortion content [10]. The magnitude of the second harmonic, for

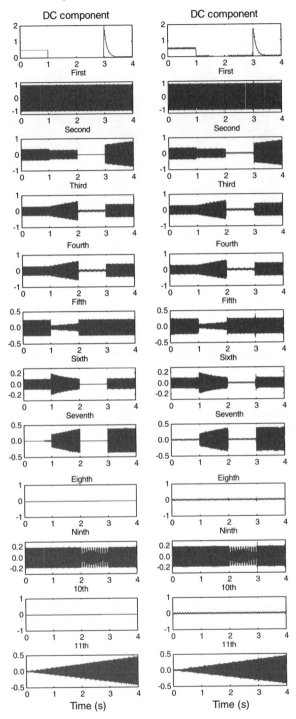

Figure 22.9   First column – original components; second column – decomposed signals

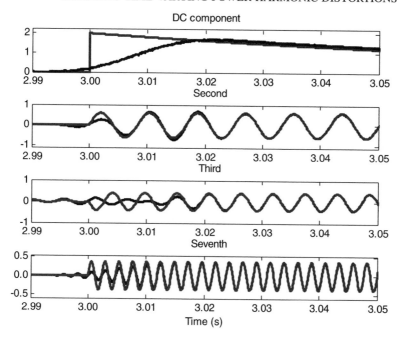

Figure 22.10   Comparing original and decomposed component

example, has dropped to approximately 7% depending on the design [11]. However, independent of these new improvements, it is always important to emphasize the time-varying nature of the inrush currents. While doing so, a transformer energization case was simulated using EMTDC/PSCAD and the result is shown in Figure 22.11.

Using the methodology proposed to visualize the harmonic content of the inrush current, Figure 22.12 reveals the rarely seen time-varying behavior of the waveform of each harmonic component, where the left-hand column shows the DC and even components and the right-hand column the odd ones. This could be used to understand other physical aspects not observed previously.

Table 22.1   Typical harmonic content of the inrush current

Order	Content (%)
DC	55
2	63
3	26.8
4	5.1
5	4.1
6	3.7
7	2.4

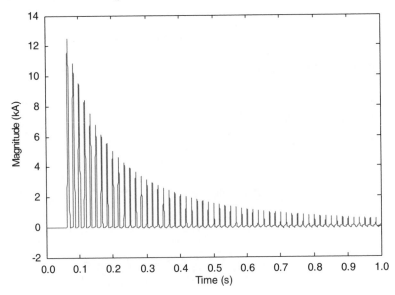

Figure 22.11   Inrush current in phase A

22.6   Final consideration and future work

The methodology proposed has some intrinsic limitations associated with the analysis of interharmonics as well as with real-time applications. The interharmonics (if they exist) are not filtered by the bank, so it would corrupt the nearby harmonic components. The limitation with real-time applications is related to the transient time response produced by the filter bank. For example, in the presence of abrupt change in the input signal, such as in inrush currents, the transient response can last more than 5 cycles, which is not appropriate for applications whose time delay must be as short as possible. The computational effort for real-time implementation is another challenge that the authors are investigating. In fact the high-order filter (69th) used in the bank structure demands high computational effort. By using multirate techniques it is possible to show that the number of multiplications (per second) to implement one branch of the analysis filter bank is around one million. Some low price digital signal processors (DSPs) available on the market are able to execute 300 million float point operations per second. This shows that, despite the higher computational effort of the structure, it is feasible to be implemented in hardware. However, new opportunities exist for overcoming the limitations and the development of improved and alternative algorithms. For example, the authors are investigating the possibility of extracting interharmonic components and developing a similar methodology based on the discrete Fourier transform (DFT). The first results show similar visualization capabilities with the promise of providing a reduced transient time response and lower computational burden. This process has the potential of addressing specific protective relaying needs such as detecting a high-impedance fault during transformer energization and detection of ferro-resonance.

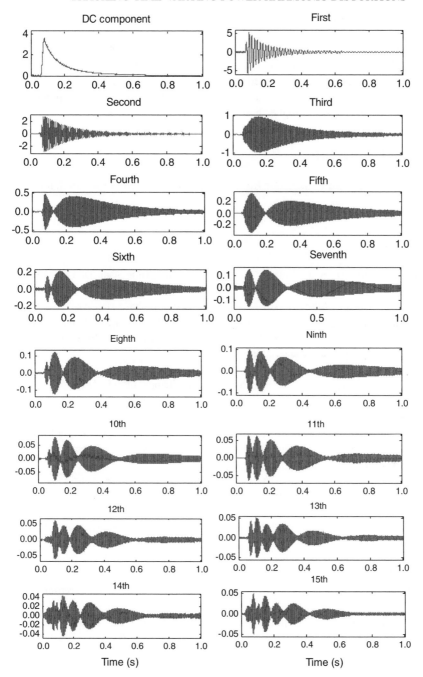

Figure 22.12   Decomposition of the simulated transformer inrush current

22.7   Conclusions

This chapter has presented a new method for time-varying harmonic decomposition based on multirate filter banks theory. The technique is able to extract each harmonic in the time domain. The composed structure was developed to work with 256 samples per cycle and to track up to the 15th harmonic. The methodology can be adapted through convenient preprocessing for different sampling rates and higher harmonic orders.

Acknowledgments

The authors would like to thank Dr Roger Bergeron and Dr Mathieu van den Berh for their useful and constructive suggestions.

References

[1] Gu Yuhua and M. H. J. Bollen, 'Time-Frequency and Time-Scale Domain Analysis', *IEEE Transaction on Power Delivery*, **15**, 2000, 1279–1284.

[2] P.P. Vaidyanathan, *Multirate Systems and Filter Banks*, Prentice Hall, 1993.

[3] M. Karimi-Ghartemani, M. Mojiri and A. R. Bakhsahai, 'A Technique for Extracting Time-Varying Harmonic based on an Adaptive Notch Filter', Proceedings of the IEEE Conference on Control Applications, Toronto, Canada, August 2005.

[4] J. R. Carvalho, P. H. Gomes, C. A. Duque, M. V. Ribeiro, A. S. Cerqueira and J. Szczupak, 'PLL based harmonic estimation', IEEE PES Conference, Tampa, Florida, USA, 2007.

[5] C.-L. Lu, 'Application of DFT filter bank to power frequency harmonic measurement', *IEE Proc. of Generation Transmission and Distribution*, **152**, 2005, 132–136.

[6] P. M. Silveira, M. Steurer and P F. Ribeiro, 'Using Wavelet decomposition for Visualization and Understanding of Time-Varying Waveform Distortion in Power System', Seventh CBQEE, Brazil, August 2007.

[7] V. L. Pham and K. P. Wong, 'Antidistortion method for wavelet transform filter banks and nonstationay power system waveform harmonic analysis', *IEE Proc. of Generation, Transmission and Distribution*, **148**, 2001, 117–122.

[8] Sanjit K. Mitra, *Digital Signal Processing – A computer-based approach*, Third edition, Mc-Graw Hill, 2006.

[9] C.R. Mason, *The Art and Science of Protective Relaying*, John Wiley & Sons, Inc., New York, USA, 1956.

[10] B. Gladstone, 'Magnetic Solutions, Solving Inrush at the Source', *Power Electronics Technology*, **30**, 2004, 14–27.

[11] F. Mekic, R. Girgis, Z. Gajic and E. teNyenhuis, 'Power Transformer Characteristics and Their Effect on Protective Relays', PROC 33rd Western Protective Relay Conference, Place, October 2006.

23

Enhanced DFT for time-varying harmonic decomposition

P. M. Silveira, C. A. Duque, T. Baldwin and P. F. Ribeiro

23.1 Introduction

Signals decomposition techniques are concerned with the way that the original signal can be split into individual components including harmonics, interharmonics, subharmonics and so on. Normally, signal decomposition is carried out in the time domain, such that the time-varying behavior of each harmonic component is observable. This subject is an important issue in power quality analysis for different reasons, including the analysis of loads behavior, failure detection, pattern recognition of events and so on.

There are different techniques that can be used to separate frequency components; among them the most used have been short time Fourier transform (STFT) and wavelets transforms [1–3]. Unfortunately, the structures using wavelets are not able to decouple the frequencies completely [2].

Other techniques have been proposed when the fundamental frequency is time varying and the sampling frequency is not synchronous, such as adaptive notch filter [4], phase-locked loop (PLL) [5, 6], resonator-in-a-loop filter bank [7] and a multistage implementation of narrow low-pass digital filters valid to extract stationary harmonic components [8].

For most of the applications in power quality one can work with a synchronous sampling frequency, as well as consider the fundamental frequency practically constant with no interharmonic. In these cases, other approaches can be used, such as [9] and [10].

Time-Varying Waveform Distortions in Power Systems   Edited by Paulo F. Ribeiro
© 2009 John Wiley & Sons, Ltd

In [9] the authors presented a new methodology to separate the harmonic components up to the 15th harmonic using the multirate and filter bank approach. The method is able to track time-varying power harmonic frequencies without frequency spillover. An alternative to this approach is to use a sliding window recursive DFT (SWR-DFT) [10], which has a low computation burden, no phase delay and a short transient time.

This chapter presents an improvement to be added to the SWR-DFT method presented in [10], including a dyadic down-sampling before each group of harmonics to be extracted and tracked. The advantage of this new strategy compared with previous ones is the reduced processing time and reduction of computational effort without loss of information.

23.2   Sliding window recursive DFT

Just as with continuous-time signals, a periodic discrete-time signal x with period p can be described as a sum of sinusoids like in Equations (23.1) and (23.2). The second one (rectangular form) is, of course, related to the first one through Equations (23.3) and (23.4). $H = (p-1)/2$ for p odd and $H = p/2$ for p even.

$$x(k) = a_0 + 2 \sum_{h=1}^{H} A_h.\cos(h\Omega_0 k + \theta_h) \tag{23.1}$$

$$x(k) = a_0 + 2 \sum_{h=1}^{H} Y_{C_h}(k).\cos(h\Omega_0 k) - Y_{S_h}(k).\sin(h\Omega_0 k) \tag{23.2}$$

$$A_h = \sqrt{(Y_{C_h})^2 + (Y_{S_h})^2} \tag{23.3}$$

$$\theta_h = \arctan\left(-\frac{Y_{S_h}}{Y_{C_h}}\right). \tag{23.4}$$

Thus, the rectangular (quadrature) terms $Y_{C_h}^k$ and $Y_{S_h}^k$ can be obtained by using the expressions in Equations (23.5) and (23.6),

$$Y_{C_h}^k = \frac{2}{N} \sum_{l=0}^{N-1} x_{(k-N+l)}.\cos\left(\frac{2\pi.h.l}{N}\right) \tag{23.5}$$

$$Y_{S_h}^k = \frac{2}{N} \sum_{l=0}^{N-1} x_{(k-N+l)}.\sin\left(\frac{2\pi.h.l}{N}\right) \tag{23.6}$$

where N is the number of samples per cycle and k is the actual sample.

These expressions are very common in algorithms of protection numerical relay and normally are performed just to extract the fundamental component phasor ($h = 1$). The moving or sliding window concept is then applied, that is, as a new sample becomes available, the oldest is discarded and the new one is included in the calculation, in such a way that N is always the same during the processing task. The sine and cosine coefficients are defined as functions of N for each component h. For $h = 1$ the algorithm using Equation (23.5) is known as a full-cycle

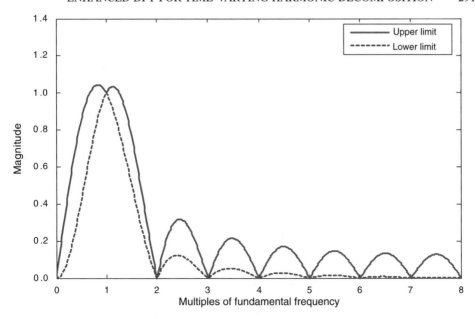

Figure 23.1   Response frequency full-cycle DFT filter

DFT [11], which can also be represented as a difference equation in its general form
(Equation (23.7)):

$$y[k] = \sum_{m-o}^{M} \frac{b_m}{a_o}.x[k-m] - \sum_{n=1}^{N} \frac{a_n}{a_o}.y[k-n].$$

(23.7)

Adopting $a_n = 0$ and $a_0 = 1$, Equation (23.6) becomes a nonrecursive numerical filter,
whose frequency response can easily be found from a difference equation, in the Z-domain,
designed for 16 samples/cycle, as illustrated in Figure 23.1.

DFT calculations using Equations (23.5) to (23.7) represent more calculations than are
actually necessary in practice [11] and by simple adjustment the full-cycle window can
become a recursive form of a full-cycle algorithm to compute the rectangular terms Y_C^k and Y_S^k,
as the structure shown in Figure 23.2. If the same structure is applied for each integer $h \neq 1$, the
phasors of each harmonic are then obtained, according to the recursive Equations (23.8)
and (23.9).

$$Y_{C_h}^k = Y_{C_h}^{k+1} + (x_k - x_{k-N})\cos\left(\frac{2\pi h}{N}k\right)$$

(23.8)

$$Y_{S_h}^k = Y_{S_h}^{k+1} + (x_k - x_{k-N})\sin\left(\frac{-2\pi h}{N}k\right)$$

(23.9)

where x_k is the newest sample corresponding to N and x_{k-N} is the oldest sample corresponding
to a fundamental full cycle earlier.

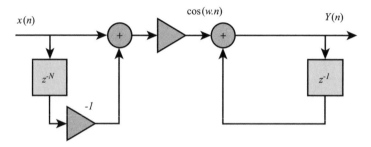

Figure 23.2   Recursive filter to compute the quadrature term Y(n) [9]

23.3   The decomposition structure

Normally, the DFT recursive algorithm has been used to extract and compute the amplitude and phase of the fundamental for protection purposes [11], but not the waveform. Nevertheless, the main objective of this work is, in fact, to obtain the fundamental waveform, as well as the waveform of each individual harmonic. This task can be performed by considering and using the rectangular form (Equation (23.2)) that has become possible from all the methodology based on Fourier theory. The implementation of this approach can be accomplished in two ways:

a. using the sine and cosine coefficients previously calculated and stored – in this case, the algorithm must perform an internal product using a vector of coefficients in each observable window;

b. using a digital sine–cosine generator – this second way is more effective and has been adopted to decompose and analyze some signals from power system events, as will be demonstrated.

A digital sine–cosine generator is presented in [12], but it can be implemented with some minor modifications according to the following matrix equation:

$$\begin{bmatrix} s_1(n) \\ s_2(n) \end{bmatrix} = \begin{bmatrix} \cos(w_h) & \sin(w_h) \\ -\sin(w_h) & \cos(w_h) \end{bmatrix} \cdot \begin{bmatrix} s_1(n-1) \\ s_2(n-1) \end{bmatrix} \tag{23.10}$$

where $s_1(n)$ is a sine function and $s_2(n)$ is a cosine function. In adopting this sine–cosine generator, both the decomposition and the reconstruction tasks can run parallel to each other, according to Figure 23.3.

For extracting N harmonics it is necessary to employ an N structure as shown in this Figure 23.3, but there are some advantages when using it, such as:

• low computational effort, suitable for real-time decomposition implementation;

• no phase delay;

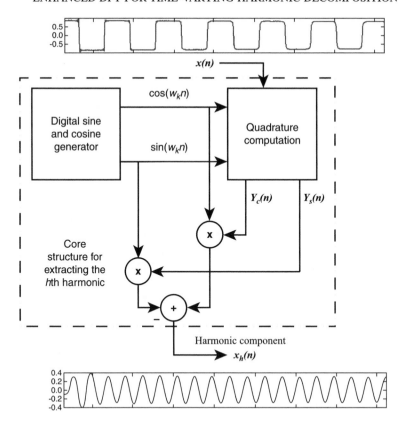

Figure 23.3   The core structure for extracting the h*th harmonic [9]*

- transient time equal to the sliding window width – window of one cycle, the convergence is reached after one cycle.

On the other hand, the disadvantages of the method are related to the limitations of the DFT:

- a synchronous sampling is needed;
- interharmonics are a source of error to the process.

23.4 Dyadic down-sampling

The number of mathematical operations necessary to track time-varying harmonics can be substantially reduced if a reduced sampling rate is used. Therefore, the sliding window recursive DFT has been implemented for different groups of harmonics and, for each group, a different number of samples is used.

Consider a signal whose sampling frequency is 15.360 Hz or 256 samples/cycle of 60 Hz. Of course, this sampling rate, according to Shannon theory (Nyquist criteria), is more than

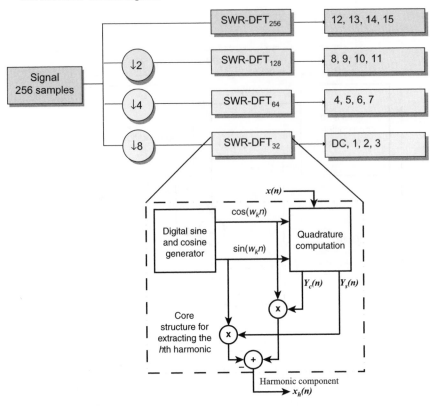

Figure 23.4   SWR-DFT using dyadic down-sampling

enough to compute and visualize up to the 15th harmonic. So, why not reduce the sampling rate according to the desired harmonic with a desired resolution? To answer this question an experimental algorithm has been implemented using dyadic down-sampling [13], according to Figure 23.4.

In the SWR-DFT algorithm each time-step needs just one addition, one subtraction and one multiplication to perform a complete cycle for each rectangular term, according to Equations (23.8) and (23.9). This is repeated for all harmonic components. If there are N samples per cycle, all the operation must be multiplied by N in order to accomplish a complete time period of 60 Hz. Considering, for example, a signal with $N = 256$, the number of operations to accomplish a complete cycle can be calculated as: 3 (operators $+ - {}^{*}$) $\times$ 256 (samples) $\times$ 16 (components) $\times$ 2 (rectangular terms Y_s and Y_c) resulting in 24 576 operations. However, adopting the down-sampling strategy, the operation number is reduced to a half in each subsequent level.

In Figure 23.4, four groups of harmonics are represented, including the fundamental and the DC component. The dyadic (2^n) down-sampling up to eight is adopted to reduce the computational effort. In this case, the number of operations per cycle is 11 520, representing a reduction of 53% in operation numbers.

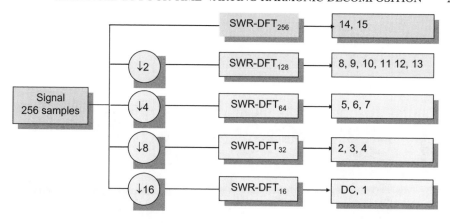

Figure 23.5   SWR-DFT using dyadic down-sampling up to 16

Several other down-sampling strategies can be adopted, depending on the desired resolution for each harmonic. Figure 23.5 is an example. Also, no dyadic down-sampling may be performed. Nevertheless, it is important to take care with the aliasing error.

By adopting the scheme in Figure 23.5, for example, the spectrum of the 15th harmonic will superimpose onto the fundamental component and, consequently, the amplitude and phase of the fundamental will be affected. One strategy to avoid this error is to implement an anti-aliasing filter before SWR-DFT$_{16}$ or to subtract the 15th harmonic from the original signal as soon it has been extracted.

23.5   Simulation results

A structure similar to Figure 23.4, but including one more level of down-sampling, has been used to track time-varying harmonic signals generated by simulations. Two examples are shown below.

23.5.1   Synthetic signal

These kinds of hypothetical signals, generated using a mathematical model, are important to test these classes of algorithms because the content of the signals is known. Thus, the process results can be compared and analyzed to observe the errors. Thus, several synthetic signals have been generated in Matlab to test the structure presented in this chapter. For example, the signal shown in Figure 23.6 has 16 components (DC up to 15th). The signal is divided into four different segments in such a way that the result is distorted with some harmonics in steady state and others time-varying modulated by a constant or by a exponential functions (crescent or decrescent) or simply with abrupt changes of magnitude and phase, as well as a DC component.

Figure 23.7 shows some components that are present in the signal from DC up to the 15th harmonic. The left-hand column represents the original components and the right-hand

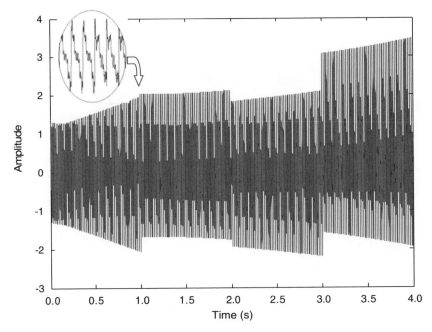

Figure 23.6 Synthetic signal used

column the same components obtained through the SWR-DFT with a dyadic down-sampling in five levels, according to the following scheme: 256 samples/cycle for 12th to 15th harmonic; 128 for eighth to 11th; 64 for fourth to seventh; 32 for second to thirdand, finally, 16 for DC and first component.

For reasons of simplicity and space limitation not all components are shown in Figure 23.7. For example, the fifth harmonic was generated with a DC component (a small DC step). Although it is not shown in the figure, it will appear exactly as it is, when the DC component is extracted. It is important to note that all waveforms of the time-varying harmonics that are contained in the signal have been extracted efficiently and accurately. Naturally, the effect of the down-sampling can be observed in the different outputs: they have different numbers of samples (in the right-hand column).

There is an important observation to make with regard to the 15th harmonic in the adopted scheme. As has been said, this component will interfere with the fundamental value. Figure 23.8 illustrates the result of another simulation, in which the 15th harmonic jumps up from 0 to 0.2 causing an error to the fundamental, with the same value, when this component is extracted.

23.5.2 Simulated signal

Simulated signals can be obtained from different 'electromagnetic transient programs', such as ATP, SimPower-Matlab, PSCAD, and so on. These signals, depending on the precision of the models, will represent the real world with great fidelity. Therefore, it is very important to use them to test any kind of algorithm to be implemented in intelligent electronic devices (IEDs).

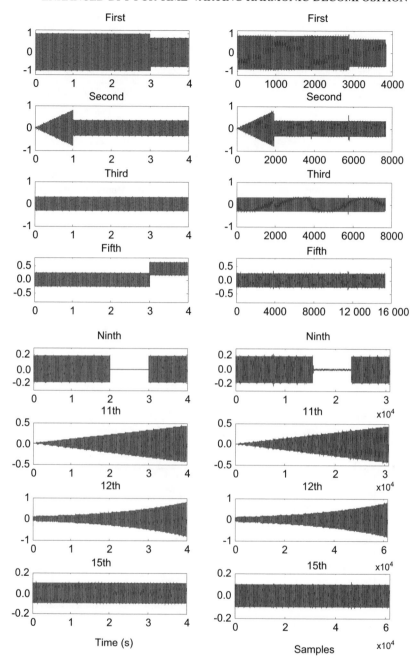

Figure 23.7   Left-hand column – original components; right-hand column – decomposed signals

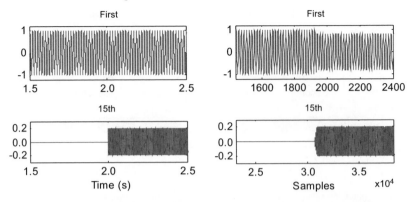

Figure 23.8   Alias error due to 15th harmonic

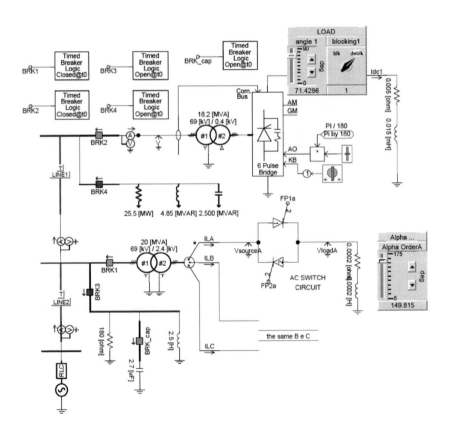

Figure 23.9   Modeled system in PSCAD

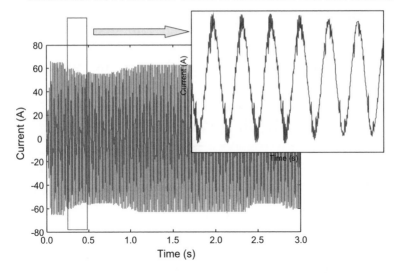

Figure 23.10   Variation of current caused by variable load

Figure 23.9 shows a piece of a system that has been modeled in PSCAD, which contains a source and two sections of a transmission line feeding transformers and linear and nonlinear loads, such as a six-pulse bridge. Some disturbances are provoked in this system, such as load imbalance, load rejection and failures in the converters pulse system through the control interface. The signals captured during the simulations have served to analyze and understand some time-varying harmonics that appears during these events.

Figure 23.10 is an example of a current signal that has been decomposed and whose results can be seen in Figure 23.11. The odd harmonics will vary during a load-imbalance disturbance associated with an angle shooting variation. These harmonics can be tracked and observed using the SWR-DFT proposed above.

23.6   Conclusions

This chapter has presented a method for time-varying harmonic decomposition based on sliding window recursive DFT using a dyadic down-sampling strategy.

From the results presented in this chapter, as well as several analyses with other signals, it is possible to conclude that the combined techniques of recursive DFT and down-sampling strategy bring some advantages to the decomposition of nonstationary signals, when compared with other methodologies previously cited. A lower computational burden, no phase delay and a short transient time are important aspects to be taken into account when implementing this tool in future IEDs for real-time applications. On the other hand, the disadvantages are inherent to all DTF based algorithms, that is, the need for synchronous sampling and the influence by the presence of interharmonics. However, other strategies to solve the problems of interharmonics and synchronized time-steps have been studied and will be presented opportunely.

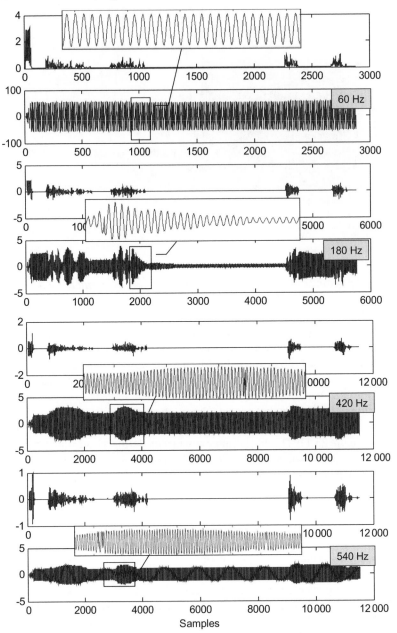

Figure 23.11 Behavior of the time-varying harmonics during system events

References

[1] Y. Gu and M. H. J. Bollen, 'Time-Frequency and Time-Scale Domain Analysis', *IEEE Trans. on Power Delivery*, **15**, 2000, 1279–1284.

[2] P. M. Silveira, M. Steurer and P. F. Ribeiro, 'Using Wavelet decomposition for Visualization and Understanding of Time-Varying Waveform Distortion in Power Systems', VII CBQEE, Brazil, August 2007.

[3] V. L. Pham and K. P. Wong, 'Antidistortion method for wavelet transform filter banks and nonstationay power system waveform harmonic analysis', *IEE Proc. Gener., Transm., Distrib.*, **148**, 2001, 117–122.

[4] M. Karimi-Ghartemani, M. Mojiri and A. R. Bakhsahai, 'A Technique for Extracting Time-Varying Harmonic based on an Adaptive Notch Filter', Proceedings of the IEEE Conference on Control Applications, Toronto, Canada, August 2005.

[5] J. R. Carvalho, P. H. Gomes, C. A. Duque, M. V. Ribeiro, A. S. Cerqueira and J. Szczupak, 'PLL based harmonic estimation', IEEE PES Conference, Tampa, Florida, USA, 2007.

[6] J. R. Carvalho, C. A. Duque, M. V. Ribeiro, A. S. Cerqueira and P. F. Ribeiro, 'Time-Varying Harmonic Distortion Estimation using PLL Based Filter Bank and Multirate Processing', Seventh Conferência Brasileira sobre Qualidade de Energia Elétrica, Santos-SP, Brazil, 2007. (Available: http://www.labsel.ufjf.br/.)

[7] H. Sun, G. H. Allen and G. D. Cain, 'A new filter-bank configuration for harmonic measurement', *IEEE Trans. on Instrumentation and Measurement*, **45**, 1996, 739–744.

[8] C.-L. Lu, 'Application of DFT filter bank to power frequency harmonic measurement', *IEE Proc. Gener., Transm,. Distrib.*, **152**, 2005, 132–136.

[9] C. A. Duque, P. M. Silveira, T. Baldwin and P. F. Ribeiro, 'Novel method for tracking time-varying power power harmonic distortion without frequency spillover', IEEE PES, Pittsburgh, Philadelphia, USA, July 2008.

[10] P. M. Silveira, C.A. Duque, T. Baldwin and P. F. Ribeiro, 'Time-Varying Power Harmonic Decomposition using Sliding-Window DFT', IEEE International Conference on Harmonics and Quality of Power, Wollongong, Australia, 2008.

[11] A. G. Phadke and J. S. Thorp, *Computer Relaying for Power Systems*, Research Studies Press Ltd, New Orleans, Louisiana, USA, 1988.

[12] R. Hartley and K. Welles, 'Recursive Computation of the Fourier Transform', IEEE International Symposium on Circuits and Systems, Volume 3, 1990, pp. 1792–1795.

[13] C. S. Burrus, R. A. Gopinath and H. Guo, *Introduction to Wavelets and Wavelet Transforms - A Primer*, Prentice-Hall Inc., New Jersey, USA, 1998.

24

Enhanced PLL based filter for time-varying harmonic decomposition

J. R. Carvalho, C. A. Duque, M. V. Ribeiro, A. S. Cerqueira and P. F. Ribeiro

24.1   Introduction

With the increased application of power electronics, controllers, motor drives, inverters and FACTS devices in modern power systems, distortions in line voltage and current have been increasing significantly. These distortions have affected the power quality of the power systems and to keep it under control the monitoring of harmonic and interharmonic distortion is an important issue [1–3].

The fast Fourier transform (FFT) is a suitable approach for estimating the spectral content of a stationary signal, but it loses accuracy under time-varying conditions [4] and, as a result, other algorithms must be used. The short-time Fourier transform (STFT) can partly deal with time-varying conditions but it has the limitation of fixed window width, chosen a priori, and this imposes a limitation on the analysis of low-frequency and high-frequency nonstationary signals at the same time [5].

The IEC Standard drafts [6] have specified signal processing recommendations and definitions for harmonic and interharmonic measurements. These recommendations utilize DFT over a rectangular window of exactly 12 cycles for 60 Hz (10 cycles for 50 Hz) and frequency resolution of 5 Hz. However, different authors [7, 8] have shown that the detection

Time-Varying Waveform Distortions in Power Systems   Edited by Paulo F. Ribeiro
© 2009 John Wiley & Sons, Ltd

and measurement of interharmonics, with acceptable accuracy, is difficult to achieve using the IEC specification.

Unlike the previous methods, that follow the IEC Standard, other techniques based on the Kalman filter, the adaptive notch filter or PLL approaches have been applied to harmonic and interharmonic estimation. The main disadvantage of Kalman filter based harmonics estimation is the higher-order model required to estimate several components.

In [3], the enhanced phase-locked loop (EPLL) [1] is used as the basic structure for harmonic and interharmonic estimation, and several of such structures are arranged together. Each one is adjusted to estimate a single sinusoidal waveform. The convergence takes about 18 cycles, but it can take more than 100 cycles for higher harmonic frequencies, mostly if there is a fundamental frequency deviation.

In [9], a new multirate filter bank structure for harmonic and interharmonic extraction is presented. The method uses EPLL as an estimation tool in combination with sharp bandpass filters and down-sampler devices. As a result, an enhanced and low computational complexity method for parameters estimation of time-varying frequency signals is attained.

This chapter describes a phase-locked loop (PLL) based harmonic estimation system which makes use of an analysis filter bank and multirate processing. The filter bank is composed of a bandpass adaptive filter. The initial center frequency of each filter is purposely chosen to be equal to harmonic frequencies. However, the adaptation makes it possible to track time-varying frequencies as well as interharmonic components. A down-sampler device follows the filtering stage which reduces the computational burden, because an under-sampling operation is realized. Finally, the last stage is composed of a PLL estimator which provides estimates for amplitude, phase and the apparent frequency of input signal. The true values of frequencies are obtained from the apparent frequency using simple algebraic equations. Simulations show that this approach is precise and faster than other PLL structures.

This work presents a new version of the proposed method in [9]. This new version uses the concept of apparent frequency and the under-sampling principle. These concepts are correlated with each other. In addition to reducing the computational effort of the overall estimator, they guarantee that robust structures can be implemented in fixed point processors.

This chapter is organized as follows: Section 24.2 presents some concepts about digital filter banks. Section 24.3 describes the multirate processing, the concepts of under-sampling and apparent frequency. Section 24.4 describes the proposed structure and the recursive equations of the PLL method [1]. Section 24.5 presents numerical results. Finally, in Section 24.6, some concluding remarks are stated.

24.2   Digital filter banks

A digital filter bank [10] is a collection of digital bandpass filters with either a common input (the analysis bank) or a summed output (the synthesis bank). The object of discussion in this section is the analysis bank.

The analysis filter bank decomposes the input signal $x[n]$ into a set of M subband signals $y_1[n], y_2[n], \ldots, y_M[n]$, each one occupying a portion of the original frequency band. Figure 24.1 shows a typical analysis filter bank.

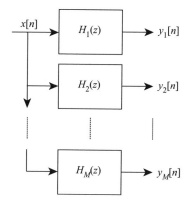

Figure 24.1   A typical analysis filter bank

In this work, the filter bank differs from traditional filter banks found in the literature [10]. In particular, bandpass filters split the input signal into the spectrum. These filters are conventional parametric bandpass filters [11] given by

$$H(z) = \frac{1-\alpha}{2} \cdot \frac{1-z^{-2}}{1-\beta(1+\alpha)z^{-1}+\alpha z^{-2}}.$$ (24.1)

The above transfer function has a narrow bandwidth, when the poles are a pair of complex conjugates near the unit circle. The parameter α controls this proximity defining the 3 dB bandwith of the filter. The maximum magnitude value of Equation (24.1) occurs at discrete frequency ω_0 which is related to β, by the expression $\beta = \cos(\omega_0)$. The magnitude response of Equation (24.1) is plotted in Figure 24.2 for the bandpass centered at the fundamental

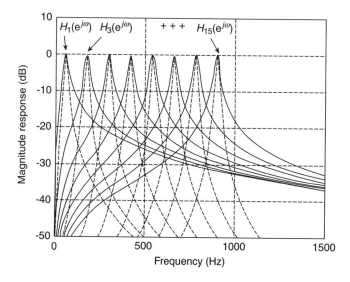

Figure 24.2   Magnitude responses of the bandpass filters of the analysis bank

frequency (60 Hz) and some of its harmonics. Solid lines are for the second-order filter of Equation (24.1) while dashed lines show the response for a cascade structure of two second-order filters.

Although the parameter α near one produces a sharper magnitude response, it increases the transient response time. This fact is very important because the convergence time of the estimator is proportional to the duration of the transient period. For example, using Equation (24.1) with $\alpha = 0.98$ and an input signal of 60 Hz with 128 points per cycle, the transient virtually decays at about four cycles.

24.3 Multirate processing

The two basic components in sampling rate modification are: the down-sampler, to reduce the sampling rate, and the up-sampler, to increase the sampling rate [11]. The block diagram representation of these two components is shown in Figure 24.3. The down-sampler will be described in this section.

A down-sampler with a down-sampling factor M, where M is a positive integer, creates an output sequence $y[n]$ with a sampling rate M times smaller than the sampling rate of the input sequence $x[n]$. In other words, this device keeps every Mth sample of the input signal, removing the other $(M-1)$.

The input–output relationship can be written as

$$y[n] = x[nM].\tag{24.2}$$

In the frequency domain it can be shown that

$$Y(e^{j\omega}) = \frac{1}{M}\sum_{k=0}^{M-1} X(e^{j(\omega-2\pi k)/M}).\tag{24.3}$$

Equation (24.3) implies that the DTFT of the down-sampled output signal $y[n]$ is a sum of M uniformly shifted and stretched versions of the DTFT of input $x[n]$, scaled by a factor $1/M$. It can be shown that aliasing due to down-sample operation is absent if and only if the input signal is band-limited to $\pm\pi/M$.

Figure 24.4 illustrates an example of the down-sample effect in the frequency domain by direct application of Equation (24.3). The solid curve is the spectrum of the input signal. With $M = 4$, the dashed line is the shifted and stretched spectrum of the output signal. It can be seen that the peak value has been moved from 0.1563 radians to 0.6250 radians, while the frequency remains equal to 300 Hz.

$$\begin{array}{ccc} x[n] \quad \boxed{\downarrow M} \quad y[n] & \quad & x[n] \quad \boxed{\uparrow L} \quad y[n] \\ \text{(a)} & & \text{(b)} \end{array}$$

Figure 24.3   (a) Block diagram representation of a down-sampler; (b) block diagram representation of an up-sampler

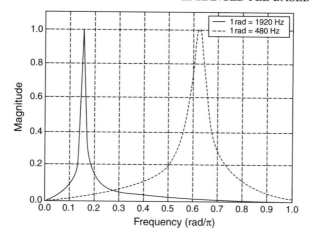

Figure 24.4   Effect of the down-sample operation in the frequency domain

An alternative and simple interpretation of Equation (24.3) can be carried out assuming a single sinusoidal signal at the input. Figure 24.5(a) shows a circle where the sinusoidal component of frequency f is correctly placed with an angle $\theta = (f/f_N) \cdot \pi$ radians, where f_N is the Nyquist frequency. After down-sampling with a factor M, the input exhibits a new angle position: $\theta_M = M\theta$, as shown in Figure 24.5(b).

There are some singularities to be considered here. First, if $\theta_M < \pi$ the output has the same frequency in Hz as its input, as in Figure 24.4 Secondly, if $\pi < \theta_M < 2\pi$ the output was obtained by under-sampling [11] the input, which means that the sampling theorem was not applied. In this case it is necessary to find the value of the output frequency f', called the apparent frequency, by analyzing the angle $2\pi - M\theta$ and the new Nyquist frequency. This is illustrated in Figure 24.5(b). Finally, if $\theta_M > 2\pi$, an under-sampling was performed again and the full revolutions must be discounted. In this case it is necessary to find the value of the apparent frequency f' by analyzing the angle $M\theta - 2\pi$ and the new Nyquist frequency.

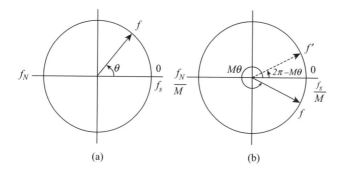

Figure 24.5   Alternative interpretation of the down-sample effect in a single sinusoidal signal: (a) original position of the component with frequency f; (b) position of the component after the down-sample operation

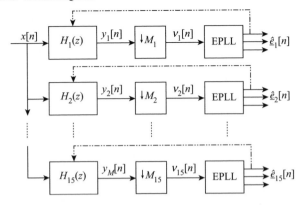

Figure 24.6   Proposed structure for harmonic estimation

24.4   Proposed structure

Figure 24.6 shows the proposed structure for harmonic estimation. Fifteen bandpass filters compose the filter bank. Each bandpass filter has been designed previously with initial central frequencies at the fundamental power frequency or one of the harmonic frequencies. The input signal $x[n]$ ideally has a frequency of $f_0 = 60\,\text{Hz}$ and the sampling rate used is $f_S = 128 \cdot f_0$.

After the filtering stage, the down-sampler device reduces the sampling rate, performing the under-sampling of $y_k[n]$ for $k = 5, 6, \ldots, 15$, according to Table 24.1. Here we point out another difference between this filter bank structure and the one commonly found in the literature – the down-sampling factors are not equal. There are many possible values of M_k. Table 24.1 shows typical values used for simulation and their effects on changing the frequency of the input signal $y_k[n]$. We can apply this reasoning to a row of frequencies given in Table 24.1. For example, the ninth harmonic sequence $y_9[n]$ passed through a down-sampler device with factor $M_9 = 16$ results in an output sequence $v_9[n]$ with an apparent frequency equal to 60 Hz.

Table 24.1   Typical values for down-sampling factors and their effects on changing the frequency of the input signal

k	M_k	Frequency (Hz) $y_k[n]$	Frequency (Hz) $v_k[n]$	k	M_k	Frequency (Hz) $y_k[n]$	Frequency (Hz) $v_k[n]$
1	16	60	60	9	16	540	60
2	8	120	120	10	11	600	98.18
3	16	180	180	11	16	660	180
4	12	240	240	12	12	720	80
5	16	300	180	13	16	780	180
6	14	360	188.57	14	15	840	184
7	16	420	60	15	16	900	60
8	14	480	68.57	—	—	—	—

Finally, the last stage is the estimator stage. This is composed of the enhanced PLL (EPLL) system [1], which is responsible for extracting three parameters from its input signal $v_k[n]$. These parameters are the magnitude, the frequency (apparent frequency) and the total phase, that is

$$\hat{\underline{e}}_k[n] = \begin{bmatrix} \hat{A}_k[n] & \hat{\omega}_k[n] & \hat{\phi}_k[n] \end{bmatrix}^T. \tag{24.4}$$

The EPLL discrete-time recursive equations are:

$$\begin{aligned}
\hat{A}_k[n+1] &= \hat{A}_k[n] + \mu_1 e_k[n] \sin(\hat{\phi}_k[n]) \\
\hat{\omega}_k[n+1] &= \hat{\omega}_k[n] + \mu_2 e_k[n] \cos(\hat{\phi}_k[n]) \\
\hat{\phi}_k[n+1] &= \hat{\phi}_k[n] + T_S \hat{\omega}_k[n] + \mu_3 e[n] \cos(\hat{\phi}_k[n])
\end{aligned} \tag{24.5}$$

where μ_1, μ_2 and μ_3 are constants that determine the speed of convergence, T_S is the sampling period and $e_k[n]$ is the error signal given by

$$e_k[n] = v_k[n] - \hat{A}_k[n] \sin(\hat{\phi}_k[n]). \tag{24.6}$$

From Figure 24.6 it can be seen that the estimated frequency of each EPLL block is used to update its respective bandpass filter. There are many strategies for doing this and here we always chose to update when the index time n was a multiple of a constant J.

24.5 Simulation results

This section presents the performance of the proposed method under a variety of conditions applied to the input signal. Simulation results are also compared with those presented in references [1–3]. A cascade structure of two filters (Equation (24.1)) was used with $\alpha = 0.98$. The EPLL constants μ_1, μ_2 and μ_3 were $300T_S$, $500T_S$ and $6T_S$, respectively.

24.5.1 Presence of harmonics

This case shows the behavior of the system when the input is composed of the fundamental with odd harmonics:

$$\begin{aligned}
x(t) =\ & V_M \sin(\omega_0 t) + \frac{1}{3} V_M \sin(3\omega_0 t) + \frac{1}{5} V_M \sin(5\omega_0 t) \\
& + \frac{1}{7} V_M \sin(7\omega_0 t) + \frac{1}{9} V_M \sin(9\omega_0 t) + \frac{1}{11} V_M \sin(11\omega_0 t) \\
& + \frac{1}{13} V_M \sin(13\omega_0 t) + \frac{1}{15} V_M \sin(15\omega_0 t)
\end{aligned} \tag{24.7}$$

Figure 24.7 shows results of the amplitude estimation. Unlike the DFT methods, in which steady-state occurs in one cycle, Figure 24.7(a) shows that the steady state is reached in approximately five cycles. Although slower than a DFT, this result is faster than the one described in [3]. This is mostly due to the use of the bandpass filters, which permits gains μ_1, μ_2

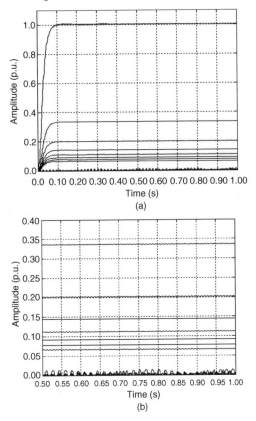

Figure 24.7   Amplitude estimation for a signal corrupted with harmonics: (a) total time simulation; (b) zoom at the steady-state harmonic estimation

and μ_3 to increase without increasing the steady-state error. Figure 24.7(b) shows a zoom in the estimation of harmonics components.

24.5.2   Presence of additive white Gaussian noise

In this case the behavior of the system is evaluated when the input signal is corrupted by a zero-mean white Gaussian additive noise, $n(t)$, as

$$x(t) = V_M \sin(\omega_0 t) + n(t) \tag{24.8}$$

Figure 24.8 shows the results of the amplitude estimation for the input signal of Equation (24.8) and a signal-to-noise ratio (SNR) equal to 15 dB. The solid line is the output of the EPLL block and the dashed line is the output of a moving-average filter (MAF) used to smooth the estimate. It can be noted that the transient time does not increase significantly when using the MAF.

Since the noise is a random signal, every simulation will result in a particular estimation curve that has the average amplitude tending towards V_M (or, equivalently, 1 p.u.). An

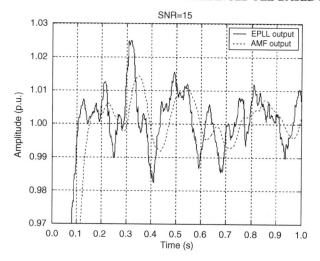

Figure 24.8 Example of amplitude estimation for a signal corrupted with noise; a moving average filter is used to improve estimation reducing the error

important analysis is performed by obtaining the relationship between the error of the estimated amplitude and the SNR. This is achieved by making several simulations for each value of SNR. The result is shown in Figure 24.9.

Comparing this with results obtained in [2], it can be seen that the proposed method presents practically the same performance. While the error in [2] for SNR = 10 dB is 4%, here the error for this value of SNR is close to 5% for the EPLL output and 3.5% for the AMF output. However, the convergence time in our simulations remains at about five cycles.

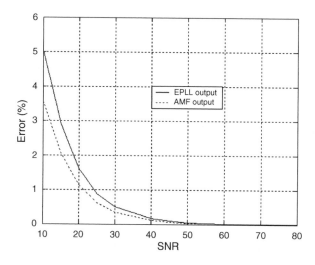

Figure 24.9 Relationship between amplitude error and SNR

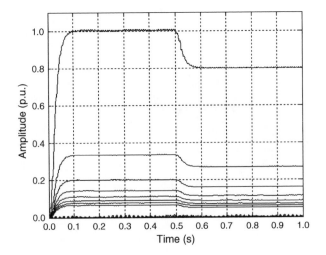

Figure 24.10   Performance of the system for a step change in the amplitude of the input signal

24.5.3   Disturbance in amplitude

An important situation for which the system must be tested occurs when the amplitude of the signal varies. Here the signal is assumed to be composed of the fundamental, odd harmonics and white Gaussian noise (SNR $= 40\,$dB):

$$x(t) = V_M \sin (\omega_0 t) + \sum_{\substack{k=3 \\ \text{odd}}}^{15} \frac{1}{k} V_M \sin (k\omega_0 t) + n(t). \tag{24.9}$$

The total time simulated is 1.0 s. At 0.5 s the amplitude of the signal is reduced to 4/5. Figure 24.10 shows the amplitude estimation curves. The transitory response due to a decrease in the amplitude extinguishes in about five cycles as well as the transitory response at the beginning of the simulation. The error observed in the simulations was less than 1% for the fundamental component and it increased for the high-order components reaching nearly 2%.

24.5.4   Presence of interharmonics

This case discusses the system behavior when an interharmonic is added to the fundamental component:

$$x(t) = V_M \sin (\omega_0 t) + \frac{1}{k} V_M \sin (k\delta\omega_0 t). \tag{24.10}$$

The parameter δ is the frequency deviation of the kth component. Figure 24.11 shows the estimated frequency for $\delta = 1.15$ and $k = 3$, which corresponds to an interharmonic added to the fundamental of 207 Hz.

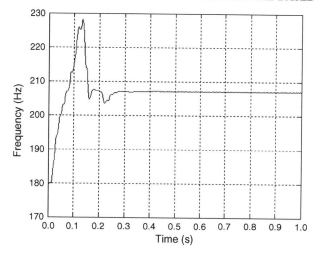

Figure 24.11   Frequency tracking of an interharmonic component

Basically, there was a slight increase in the transitory response, but with nine cycles the frequency estimated value is within the error band of 2%. It was observed that there was a slight increase in the transitory response of the estimated amplitude as well. This is due to the adaptive characteristic of the filter.

24.5.5   Disturbance in frequency

This case deals with the change in the frequency of the input signal, that is:

$$x(t) = \sum_{\substack{k=1 \\ \text{odd}}}^{15} \frac{1}{k} V_M \sin(k\omega_0(t)) + n(t). \tag{24.11}$$

The total duration of the simulation is 2.0 s. At 1.0 s the fundamental frequency jumps to 61 Hz. This implies that the kth harmonic frequency is shifted to $f_k = k \cdot 60 + k$. Figure 24.12 presents results for this environment. Figure 24.12(a) shows the frequency deviation (k) for each harmonic while Figure 24.12(b) shows the effect of a step change in frequency on the amplitude estimation. From Figure 24.12(a) it can be seen that the transitory response of the frequency estimation remains at about five or six cycles, considering a 2% band error. This response is faster than that related in [3].

24.5.6   Flicker

Finally, this case deals with voltage fluctuations, that is, the disturbance referred to as flicker:

$$x(t) = V_M[1 + 0.05\sin(2\pi t)]\sin(\omega_0 t) + n(t). \tag{24.12}$$

Equation (24.12) shows that the amplitude of the input signal varies with a sinusoidal frequency of 1 Hz and the maximum amplitude is 5% of the fundamental amplitude. This

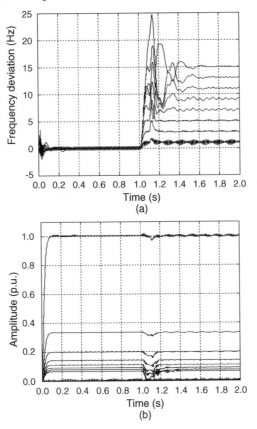

*Figure 24.12   Frequency tracking for a step change of 1 Hz in the fundamental frequency:
(a) frequency deviation, (b) amplitude estimation*

signal is plotted in Figure 24.13(a). Figure 24.13(b) presents the behavior of the estimator. It
can be seen that variations in amplitude are detected, with a very small delay. Moreover, the
estimated frequency is not significantly affected.

24.5.7   Comparative computational effort

The computational effort between the new approach and the approach presented in [3] (single
rate) is compared in terms of the number of multiplications, additions and table searches
(functions sin and cos). The total operation number for processing one cycle, that is, 128
samples, is presented in Table 24.2. To obtain these numbers it is necessary take into account the
down-sampling factor used in Table 24.1 and the number of operations in the bandpass filter and
EPLL estimator. Note that the computational effort for the multirate approach is greater than
that for the approach of [3] only for the number of additions. However, the multirate approach
needs only 82% of multiplications and 7% of the table searches required for the approach of [3].
It is important to highlight that the EPLL analyzed in this work did not include the extras filters
as proposed in [3] to smooth the estimates. If these filters were included the computed data in
Table 24.2 would be even more favorable for the proposed method.

Table 24.2 The number of operations realized within one cycle of the fundamental to estimate up to the 15th harmonic

	Additions	Multiplications	Table searches
Single rate	9600	15360	3840
Multirate	14125	12616	274

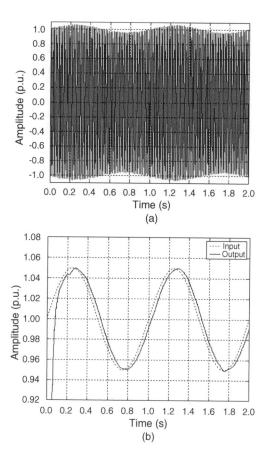

Figure 24.13 *Response of the method for a voltage fluctuation (flicker) environment: (a) input signal, (b) estimated amplitude*

24.6 Conclusions

This chapter presented an improved structure of PLL-based harmonic estimation [9]. It has been shown that parameters of a high-order harmonic can be extracted performing an undersampling of this high-frequency signal. The frequency estimated is an apparent frequency that can be converted to its actual value using algebraic relations. The simulations results have shown that the increase in transitory response was not significant. The computational effort is reduced compared with [3] and the previous structure [9]. Finally, the fact that PLL estimates an apparent frequency, lower than 240 Hz, makes the implementation in a fixed point DSP-based system more robust.

References

[1] M. Karimi-Ghartemani and M. R. Iravani, 'A Nonlinear Adaptive Filter for Online Signal Analysis in Power Systems: Applications', *IEEE Tranactions. on Power Delivery*, **17**, 2002, 617–622.

[2] M. Karimi-Ghartemani and M. R. Iravani, 'Robust and frequency-adaptive measurement of peak value', *IEEE Transactions on Power Delivery*, **19**, 2004, 481–489.

[3] M. Karimi-Ghartemani and M. R. Iravani, 'Measurement of harmonics/inter-harmonics of time-varying frequencies', *IEEE Transactions on Power Delivery*, **20**, 2005, 23–31.

[4] Y. Baghzouz, R. F. Burch, A. Capasso, A. Cavallini, A. E. Emanuel, M. Halpin, A. Imece, A. Ludbrook, G. Montanari, K. J. Olejniczak, P. Ribeiro, S. Rios-Marcuello, L. Tang, R. Thaliam and P. Verde, 'Time-varying harmonics: Part I—Characterizing measured data', *IEEE Transactions on Power Delivery*, **13**, 1998, 938–944.

[5] M. V. Chilukuri and P. K. Dash, 'Multiresolution S-transform-based fuzzy recognition system for power quality events', *IEEE Transactions on Power Delivery*, **19**, 2004, 323–329.

[6] IEC Standard Draft 61000-4-7, *General guide on harmonics and inter-harmonics measurements, for power supply systems and equipment connected thereto*, 2000.

[7] D. Gallo, R. Langella and A. Testa, 'Desynchronized Processing technique for Harmonic and interharmonic analysis", *IEEE Transactions On Power Delivery*, **19**, 2004, 993–1001.

[8] L. L. Lai, C. T. Tse, W. L. Chan and A. T. P. So, 'Real-time frequency and harmonic evaluation using artificial neural networks', *IEEE Transactions on Power Delivery*, **14**, 1999, 52–59.

[9] J. R. Carvalho, P. H. Gomes, C. A. Duque, M. V. Ribeiro, A. S. Cerqueira and J. Szczupak, 'PLL based harmonic estimation', IEEE PES Conference, Tampa, Florida, USA, 2007.

[10] P. P. Vaidyanathan, *Multirate Systems and Filter Banks*, Prentice Hall, Englewood Cliffs, New Jersey, USA, 1993, pp. 113, 151–157.

[11] S. K. Mitra, *Digital Signal Processing: A Computer-based Approach*, Third edition, McGraw-Hill, New York, USA, 2006, pp. 383–385, 177, 740.

25

Prony analysis for time-varying harmonics

L. Qi, S. Woodruff, L. Qian and D. Cartes

25.1 Introduction

The proliferation of nonlinear loads in power systems has increased harmonic pollution and led to a deterioration in power quality. Not requiring a prior knowledge of existing harmonics, Prony analysis detects frequencies, magnitudes, phases and especially damping factors of exponential decaying or growing transient harmonics. Prony analysis is implemented to supervise power system transient harmonics, or time-varying harmonics. Further, to improve power quality when transient harmonics appear, the dominant harmonics identified from Prony analysis are used as the harmonic references for harmonic selective active filters. Simulation results of two test systems during transformer energizing and induction motor starting confirm the effectiveness of the Prony analysis in supervising and canceling power system transient harmonics.

In today's power systems, the proliferation of nonlinear loads has increased harmonic pollution. Harmonics cause many problems in connected power systems, such as reactive power burden and low system efficiency. Harmonic supervision is highly valuable in relieving these problems in power transmission systems. Further, harmonic selective active filters can be connected in power distribution systems to improve power quality. Normally, Fourier transform-based approaches are used for supervising power system harmonics. However, the accuracy of the Fourier transform is affected when time-varying harmonics exist. Without prior knowledge of frequency components, some harmonic filters require phase locked loops (PLL) or frequency estimators for identifying the specific harmonic frequency before the corresponding reference is generated [1–4].

Prony analysis is applied as an analysis method for harmonic supervisors and as a harmonic reference generation method for harmonic selective active filters. Prony analysis, as an

Time-Varying Waveform Distortions in Power Systems   Edited by Paulo F. Ribeiro
© 2009 John Wiley & Sons, Ltd

autoregressive spectrum analysis method, does not require frequency information prior to filtering. Therefore, additional PLL or frequency estimators described earlier are not necessary. Due to the ability to identify the damping factors, transient harmonics can be correctly identified for the Prony-based harmonic supervision and harmonic cancellation. Two important operations in power systems – the energizing of a transformer and the starting of an induction motor [5–7], are studied for harmonic supervision and cancellation.

25.2   Prony theorem

Since Prony analysis was first introduced into power system applications in 1990, it has been widely used for power system transient studies [8], but rarely used for power quality studies. Prony analysis is a method of fitting a linear combination of exponential terms to a signal as shown in Equation (25.1) [9]. Each term in Equation (25.1) has four elements: the magnitude A_n, the damping factor σ_n, the frequency f_n and the phase angle θ_n. Each exponential component with a different frequency is viewed as a unique mode of the original signal $y(t)$. The four elements of each mode can be identified from the state space representation of an equally sampled data record. The time interval between each sample is T.

$$y(t) = \sum_{n=1}^{N} A_n e^{\sigma_n t} \cos(2\pi f_n t + \theta_n) \quad n = 1, 2, 3, \cdots, N. \tag{25.1}$$

Using Euler's theorem and letting $t = MT$, the samples of $y(t)$ are rewritten as Equation (25.2):

$$y_M = \sum_{n=1}^{N} B_n \lambda_n^M \tag{25.2}$$

$$B_n = \frac{A_n}{2} e^{j\theta_n} \tag{25.3}$$

$$\lambda_n = e^{(\sigma_n + j2\pi f_n)T}. \tag{25.4}$$

Prony analysis consists of three steps. In the first step, the coefficients of a linear predication model are calculated. The linear predication model (LPM) of order N, shown in Equation (25.5), is built to fit the equally sampled data record $y(t)$ with length M. Normally, the length M should be at least three times larger than the order N:

$$y_M = a_1 y_{M-1} + a_2 y_{M-2} + \cdots + a_N y_{M-N}. \tag{25.5}$$

Estimation of the LPM coefficients a_n is crucial for the derivation of the frequency, damping, magnitude and phase angle. To estimate these coefficients accurately, many algorithms can be used. A matrix representation of the signal at various sample times can be formed by sequentially writing the linear prediction of y_M repetitively. By inverting the matrix representation, the linear coefficients a_n can be derived from Equation (25.6). An algorithm, which uses singular value decomposition for the matrix inversion to derive the LPM coefficients, is called a SVD algorithm:

$$
\begin{bmatrix} y_N \\ y_{N+1} \\ \vdots \\ y_{M-1} \end{bmatrix} = \begin{bmatrix} y_{N-1} & y_{N-2} & \cdots & y_0 \\ y_N & y_{N-1} & \cdots & y_1 \\ \vdots & \vdots & \vdots & \vdots \\ y_{M-2} & y_{M-3} & \cdots & y_{M-N-1} \end{bmatrix} \begin{bmatrix} a_1 \\ a_2 \\ \vdots \\ a_N \end{bmatrix}.
\tag{25.6}
$$

In the second step, the roots λ_n of the characteristic polynomial shown as Equation (25.7) associated with the LPM from the first step, are derived. The damping factor σ_n and frequency f_n are calculated from the root λ_n according to Equation (25.4):

$$
\lambda^N - a_1 \lambda^{N-1} - \cdots - a_{N-1}\lambda - a_N = (\lambda - \lambda_1)(\lambda - \lambda_2) \cdots (\lambda - \lambda_n) \cdots (\lambda - \lambda_N)
\tag{25.7}
$$

In the last step, the magnitudes and the phase angles are solved in the least square sense. According to Equation (25.2), Equation (25.8) is built using the solved roots λ_n:

$$
\mathbf{Y} = \varphi \mathbf{B}
\tag{25.8}
$$

$$
\mathbf{Y} = \begin{bmatrix} y_0 & y_1 & \cdots & y_{M-1} \end{bmatrix}^T
\tag{25.9}
$$

$$
\varphi = \begin{bmatrix} 1 & 1 & \cdots & 1 \\ \lambda_1 & \lambda_2 & \cdots & \lambda_N \\ \vdots & \vdots & \vdots & \vdots \\ \lambda_1^{M-1} & \lambda_2^{M-1} & \cdots & \lambda_N^{M-1} \end{bmatrix}
\tag{25.10}
$$

$$
\mathbf{B} = \begin{bmatrix} B_1 & B_2 & \cdots & B_N \end{bmatrix}^T.
\tag{25.11}
$$

The magnitude A_n and phase angle θ_n are thus calculated from the variables B_n according to Equation (25.3). The greatest advantage of Prony analysis is its ability to identify the damping factor of each mode in the signal. Due to this advantage, transient harmonics can be identified accurately.

25.3   Selection of prony analysis algorithm

Three algorithms normally used to derive the LPM coefficients are the Burg algorithm, the Marple algorithm and the singular value decomposition (SVD) algorithm [11–13]. They are compared for implementing Prony analysis in transient harmonic studies. The non-recursive SVD algorithm utilized the Matlab pseudo-inverse (pinv) function. This pinv function uses LAPACK routines to compute the singular value decomposition for the matrix

Table 25.1   Estimated dominant harmonics (EDH) on a nonstationary signal

	EDH	Ideal	Burg	Marple	SVD
Frequencies (Hz)	#1	60	60.1690	59.9986	59.9987
	#2	300	298.2309	279.3917	299.9951
	#3	420	419.3031	420.0081	420.0138
	#4	660	657.8118	659.9380	659.9578
	#5	780	779.1504	779.9914	780.0137
Damping factors (s^{-1})	#1	0	−0.0037	−0.0027	−0.0012
	#2	−6	−1.3173	0.2127	−6.0403
	#3	−4	−0.0940	0.1245	−4.0638
	#4	0	−0.5625	−0.1881	−0.1097
	#5	0	−3.4003	−0.6494	−0.1752
Magnitudes (A)	#1	1	1.0001	0.9997	1.0002
	#2	0.2	0.1478	0.1441	0.2002
	#3	0.1	0.0819	0.0809	0.1003
	#4	0.02	0.0184	0.0203	0.0204
	#5	0.01	0.0104	0.0107	0.0103
Phase angles (°)	#1	0	3.1693	0.0320	3.1693
	#2	45	79.0299	44.9906	45.0567
	#3	30	41.4376	30.2150	29.9158
	#4	0	36.8397	−0.8574	0.7913
	#5	0	12.7567	0.1850	0.3631

inversion [14]. To choose the appropriate algorithm, the three algorithms are applied on the same signals with the same Prony analysis parameters. The signals are synthesized in the form of Equation (25.1) plus a relatively low level of noise to approximate real transient signals. The sampling frequency is selected as being equal to four times the highest harmonic and the length of data is six times of one cycle of the lowest harmonic [15]. Table 25.1 lists the estimation results from the three algorithms on one transient signal. More estimation results on synthesized power system signals were derived by the authors for different studies [16]. From the comparison of the estimation results of various signals to approximate power system transient harmonics, the SVD algorithm has the best overall performance on all estimation results and thus is selected as the appropriate algorithm for Prony analysis.

25.4   Tuning of prony parameters

Since the estimation of data is an ill-conditioned problem [12], one algorithm could perform completely differently on different signals. Therefore, Prony analysis parameters should be adjusted by trial and error to achieve the most accurate results in different situations. Although the parameter tuning is a trial and error process, there are still some rules to follow. A general guidance on parameter adjustment is given in the rest of this section.

A technique of shifting time windows by Hauer [5] is adopted for continuously detecting dominant harmonics in a Prony analysis based harmonic supervisor. The shifting time window for Prony analysis has to be filled with sampled data before correct estimation results are derived. The selection of the equal sampling intervals between samples and the data length in an analysis window depends on the simulation time step and the estimated frequency range.

The equal sampling frequency follows the Nyquist sampling theorem and should be at least twice the highest frequency in a signal. Since the Prony analysis results are not accurate for a very high sampling frequency [15], two or three times the highest frequency is considered to produce accurate Prony analysis results. Similarly, the length of the Prony analysis window should be at least one and half times one cycle of the lowest frequency of a signal.

Besides the sampling frequency and the length of Prony analysis window, the LPM order is another important Prony analysis parameter. A common principle is that the LPM order should be no more than one third of the data length [8,15]. The data length and LPM order could be increased together in order to accommodate more modes in simulated signals. It is quite difficult to make the first selection of the LPM order since the exact number of modes of a real system is hard to determine. In our case studies, a guess of 14 is a good start. If the order is found to be not high enough, the data length of the Prony analysis window should be increased in order to increase the LPM order.

The general guidance for tuning Prony analysis parameters is applicable to other applications of Prony analysis. Not requiring the specific frequency of a signal for Prony analysis, the tuning method is not sensitive to fine details of the signal and thus extensive retuning for different types of transients in the same system is unlikely to be necessary for Prony analysis.

25.5   Prony analysis and Fourier transform

As described earlier, the Fourier transform is widely used for spectrum analysis in power systems. However, signals must be stationary and periodic for the finite Fourier transform to be valid. The following analysis explains why results from the Fourier transform are inaccurate for exponential signals.

The general form of a nonstationary signal can be found in Equation (25.1). If the phase angle of the signal is equal to zero, and the magnitude is equal to unity, then the general form can be simplified into Equation (25.12). The initial time of the Fourier analysis is taken to be t_0 and the duration of the Fourier analysis window is T, which is equal to the period of the analysis signal for accurate spectrum analysis:

$$x(t) = e^{\sigma t}\cos(2\pi f t). \tag{25.12}$$

The Fourier transform during t_0 to $t_0 + T$ is calculated using Equation (25.13). The first term on the right-hand side of Equation (25.13) is equal to zero according to Equation (25.14). Therefore, the magnitude of the signal in terms of the Fourier transform is given in Equation (25.15). The ratio k between the magnitude of the Fourier transform in Equation (25.15) and the actual magnitude $e^{\delta t_0}$ is shown as Equation (25.16), which indicates the average effect of the Fourier analysis window:

$$a_n = \frac{2A}{T} \int_{t_0}^{t_0+T} e^{\delta t}\cos(2\pi f t)\cos(2\pi f t)dt = \frac{A}{T} \int_{t_0}^{t_0+T} e^{\delta t}[\cos(4\pi f t) + 1]dt$$

$$= \frac{A}{T} \int_{t_0}^{t_0+T} e^{\delta t}\cos(4\pi f t)dt + \frac{A}{T} \int_{t_0}^{t_0+T} e^{\delta t}dt \tag{25.13}$$

$$\frac{1}{T}\int_{t_0}^{t_0+T} e^{\delta t}\cos(4\pi ft)dt = \frac{1}{T}\frac{e^{\delta t}}{\delta^2 + \left(\frac{4\pi}{T}\right)^2}\left[-\delta\cos(4\pi ft) + 4\pi f\sin(4\pi ft)\right]\Bigg|_{t_0}^{t_0+T} = 0 \qquad (25.14)$$

$$a_n = \frac{1}{T}\int_{t_0}^{t_0+T} e^{\delta t}dt = \frac{1}{\delta T}e^{\delta t}\Bigg|_{t_0}^{t_0+T} = \frac{1}{\delta T}\left[e^{\delta(t_0+T)} - e^{\delta(t_0)}\right] = e^{\delta t_0}\frac{(e^{\delta T}-1)}{\delta T} \qquad (25.15)$$

$$k = \frac{e^{\delta T}-1}{\delta T}. \qquad (25.16)$$

Let us consider a fast damping signal and a slow damping signal with damping factors δ equal to -100 and -0.01, respectively. If the frequency f is equal to 60 Hz, then the duration T is equal to 0.0167 s. According to Equation (25.16), the ratio k between the Fourier magnitude and the real magnitude are derived as 0.4861 and 0.9999. Therefore, with rapid decaying factors, the magnitude derived from the Fourier transform is averaged by its analysis window and not even close to its actual magnitude. If the damping factor is equal to zero, the ratio k becomes one and the Fourier magnitude exactly reflects the real signal magnitude. With rapidly decaying signals, Fourier analysis results depend greatly on the length of the analysis window. For example, if the time duration T of the analysis window is two cycles long and the damping ratio is -100, then the ratio k decreases to 0.2888. Therefore, prior knowledge of the involved frequency is quite important for selecting the proper length of the Fourier analysis window and obtaining accurate results.

A conflict exists in selecting the length of the Fourier analysis window. In order to reduce the error due to the average effect, the length of the Fourier analysis window should decrease. However, the fewer periods there are in the record, the less random noise is averaged out and the less accurate the result will be. Some compromise must be made between reducing noise effects and increasing Fourier analysis accuracy. The length of the Prony analysis window is not as sensitive as the Fourier analysis window. If the frequency of an analyzed signal is within a certain range, it is not necessary to change the length of the analysis window. A long window can be used to deal with noise and still detect decaying modes accurately.

25.6   Case studies

With the SVD algorithm two cases were studied to implement Prony analysis for power quality study. The first case studies the harmonic supervision during transformer energizing. The second case studies the harmonic cancellation during motor starting. The test systems are realized in the simulation environment of PSIM and Matlab. The parameters of the test systems can be found in the Appendix (Section 25.8).

25.6.1   Case 1: harmonic supervision

Figure 25.1 shows the configuration of Test System 1, which models a part of a transmission system at the voltage level of 500 kV including a voltage source, a local LC load bank, a

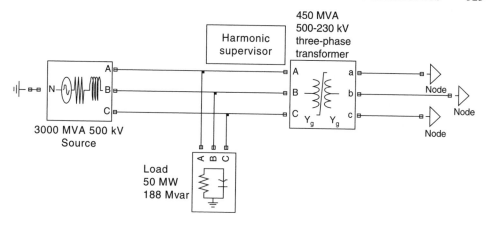

Figure 25.1   Configuration of Test System 1

three-phase transformer and a harmonic supervisor. The voltages and currents at the transformer primary side are inputs of the harmonic supervisor; the outputs are the harmonic description of the voltages and currents, which can be harmonic magnitudes and phase angles from Fourier analysis or harmonic waveforms from Prony analysis.

In our study, the Fourier transform analysis utilizes the FFT function provided in the SimPowerSystems Toolbox in Matlab. One cycle of simulation has to be completed before the outputs give the correct magnitudes and angles since the FFT function uses a running average window [17]. As described earlier, shifting time windows are used in Prony analysis for continuously detecting dominant harmonics. In this Prony-based harmonic supervisor for transformer energizing, since the fundamental frequency is considered as the lowest frequency, the time duration of the Prony analysis window is 0.036 s, which is longer than two cycles of the fundamental frequency. The time interval between any two windows is 0.6 ms. The sampling frequency is 833 Hz, which is sufficient for identifying up to the 13th harmonic in the system. The data length within a time window is 60. The order of the linear prediction model is 20, which is equal to one third of the data length.

The simulated system is designed to be resonant at the fourth harmonic [17]. Both even and odd harmonic components are produced during the transformer energizing. Among them, the second harmonic is the dominant harmonic. The magnitudes of the harmonics during energizing vary with time [5]. No abrupt changes are found in time-varying harmonic magnitudes. An exponential function is thus able to approximate to these time-varying harmonics.

Figure 25.2 shows the harmonic supervision results of Test System 1 using the Prony analysis and Fourier transform based harmonic supervisors. Figure 25.2(a) and (b) show the fundamental and fourth harmonic voltage waveforms derived from Prony and the magnitudes from FFT. Because there is no decaying or growing in the fundamental component, the magnitude obtained from FFT agrees with the magnitude of the fundamental voltage waveform from Prony. However, for the fourth harmonic, the magnitude from FFT is much smaller than that from Prony since the fourth harmonic decays quickly. Figure 25.2(c) and (d)

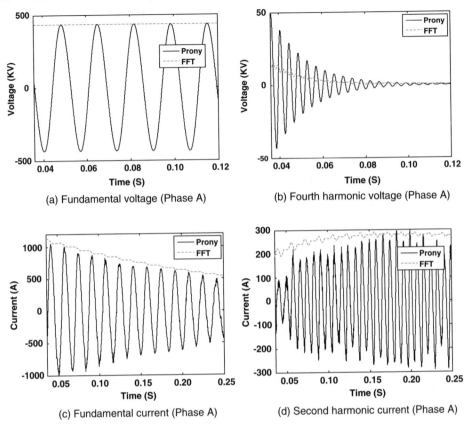

Figure 25.2   Prony and FFT results for Case 1

show the fundamental and second harmonic currents from Prony analysis and the magnitudes from FFT. Due to the slow decaying and growing speeds, the magnitudes from Prony analysis agree well with the magnitudes from FFT.

The effectiveness of the Prony-based harmonic supervisor is verified from the results shown earlier. From the comparison shown above, without damping or with small damping, the harmonic supervision results from Prony analysis and FFT are almost equal in the sense of harmonic magnitudes. With fast damping, Prony analysis derives more accurate harmonic supervision results than FFT does.

25.6.2   Case 2: harmonic cancellation

Figure 25.3 shows the configuration of Test System 2, is models a part of a distribution system at the voltage level of 480 V. The test system includes a voltage source, a nonlinear load including an induction motor and a diode rectifier load, a harmonic selective active filter using a three-phase active voltage source insulated gate bipolar transistor (IGBT) converter and a

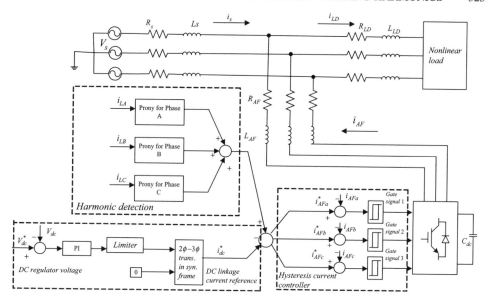

Figure 25.3   Configuration of Test System 2

controller systems associated with the active filter. The nonlinear load represents a type of load combination, induction motors plus power electronic loads, typical in power distribution systems. To describe the detailed motor starting dynamics, the induction motor is modeled by a set of nonlinear equations [17], which are different from the commonly used linear equivalent circuit to model induction motors in power quality studies [5]. The motor has a constant load torque of 0.2 Nm.

The control system of the active filter can be separated into two parts, one for controlling harmonic reference generation using Prony analysis or Fourier transform and one for controlling the DC link bus voltage. The dominant harmonics of the three-phase load currents are derived from Prony analysis or Fourier transform and used as the reference for the harmonic selective active filter. To control DC bus voltage, a proportional integral (PI) controller is used to generate the DC link current reference. Three hysteresis controllers then generate gate signals for the IGBT converter.

As described earlier, some Prony analysis parameters used earlier in transformer energizing are adjusted by trial and error to achieve the most accurate results in motor starting. In this Prony-based active filter to cancel transient harmonics during motor starting, since there could be subharmonics during motor starting, the time duration of the Prony analysis window is reduced to 0.024 s. The sampling frequency is 2500 Hz, which is sufficient to identify up to the 20th harmonic in the system. With the careful selection of these Prony analysis parameters, the spurious harmonics besides the dominant harmonics are small enough that their effects can be neglected.

During motor starting, the magnitude of the starting current can become as high as several times the rated current. This high motor starting current contains mainly an exponentially decaying fundamental current and a small portion of decaying harmonic or

Table 25.2   Estimated dominant harmonics (EDH) for Case 2

	EDH	$T=0.1$	$T=0.1786$	$T=0.2403$
Frequencies (Hz)	#1	60.0880	60.9739	60.0395
	#2	300.2141	121.2281	299.8692
	#3	420.4627	299.9159	420.9291
Magnitudes (A)	#1	44.0179	82.5488	58.9011
	#2	8.8723	12.0627	9.2252
	#3	3.0207	8.8103	3.4445
Damping factors (s^{-1})	#1	0.0450	-14.6057	0.1782
	#2	0.2094	-116.4742	-0.1920
	#3	0.4268	-2.8884	-0.5291

subharmonic currents. Due to the exponentially decaying fundamental current, the load voltage changes exponentially. With this exponentially changing voltage supplied to the rectifier load, the harmonic currents drawn by the rectifier load appear as exponentially changing harmonic currents. Detecting and canceling harmonics in the total load currents during motor starting is thus critical to the overall power system reliability and power quality.

Table 25.2 shows the three most dominant frequencies identified from Prony analysis at three time instants selected arbitrarily from the time duration before, during and after motor starting transient. Because the second harmonic current is induced by starting an induction motor [5], it is found that the second dominant harmonic current changes from the fifth harmonic current before motor starting to the second harmonic current after motor starting. This second harmonic current is damped out during motor starting with a high damping factor. The magnitudes of the fifth and seventh harmonic currents also change exponentially during motor starting, but are almost kept constant before and after motor starting transients. The magnitude of the fundamental current increases after motor starting since the total load current increases with the motor load added as a part of the system load.

Figure 25.4 shows the simulation results of the transient harmonic cancellation by the Prony-based and Fourier based active filter during motor starting. At first, the system runs at steady state with only an ideal diode rectifier load connected. After the active filter takes action at 0.06 s, the induction motor is switched into the test system at 0.12 s. The motor starting transient lasts for several cycles and dies down at approximately 0.2 s. The test system finally settles down at steady state with both the motor and the diode rectifier load connected. The cancellation of stationary harmonics before and after motor starting has been verified by the authors [15]. The harmonic cancellation by the Prony filter and the Fourier filter during motor starting is compared in Figure (c) and (d). It is seen that the improvement of the source current by the Fourier based active filter is much less than the improvement by the Prony based filter.

The effectiveness of the Prony-based harmonic selective active filter to cancel transient harmonics during motor starting is verified. Using the active filter, the dominant second, fifth and seventh harmonics in the load currents are cancelled and the power quality during motor starting is improved. Due to the length of Prony analysis window, it is observed in

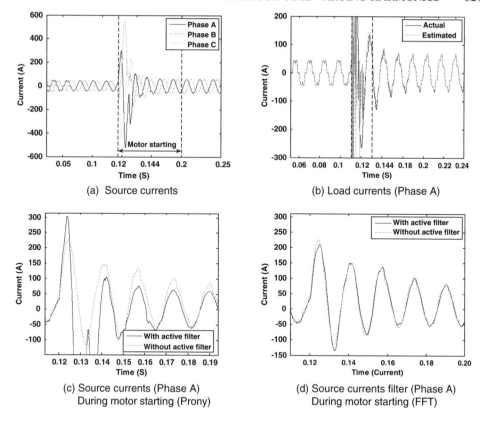

Figure 25.4   Simulation results for Case 2

Figure 25.4(b) that there are oscillations in the estimated load current at the beginning of motor starting.

25.7   Conclusions and future work

The effectiveness of the Prony analysis is verified for the harmonic supervision and cancellation when transient or time-varying harmonics occur in power systems. The unique features of Prony analysis, such as frequency identification without prior knowledge of frequency and the ability to identify damping factors, are useful to power system quality study. Further studies can be carried out for power quality study. Just as Prony analysis was used with harmonic selective active filters, Prony analysis may also be applied to other measures to improve power quality. In this study, the computational speed of the nonrecursive SVD algorithm for Prony is slow. With more efficient algorithms developed, Prony analysis can be applied for online harmonic monitoring and harmonics cancellation using real-time hardware-in-loop (RT-HIL) technology. Due to the insufficient information input into Prony analysis, inaccurate harmonic estimation from Prony analysis may adversely affect harmonic controllers at the beginning of transient periods. These disadvantages of applying Prony analysis are being considered and will be improved in future studies.

25.8   Appendix

Table 25.3   Parameters of source of Test System 1

V_{RMSLL} (KV)	f (Hz)	R_S (Ω)	L_S (H)
500	60	5.55	0.221

Table 25.4   Parameters of a local capacitive load of Test System 1

R (Ω)	C (F)
0.6606	0.0011

Table 25.5   Parameters of a three-phase transformer of Test System 1

R_1 (Ω)	L_1 (H)	R_2 (Ω)	L_2 (H)	R_M (Ω)	Ratio (KV/KV)
1.1111	44.4444	0.2351	9.4044	0.041	500/230

Table 25.6   Transformer saturation characteristics Test System 1

λ (p.u.)	I (p.u.)
0	0
1.20	0.0024
1.52	1.0

Table 25.7   Parameters of source of Test System 2

V_{RMSLL} (V)	f (Hz)	R_S (Ω)	L_S (H)
480	60	10^{-5}	10^{-6}

Table 25.8   Parameters of a six-pole induction motor of Test System 2

R_s (Ω)	L_s (H)	R_r (Ω)	L_r (H)	L_M (H)	J (kgm^2)
0.294	0.00139	0.156	0.00074	0.041	0.4

Table 25.9   Parameters of a rectifier load of Test System 2

R_L (Ω)	C_L (F)	R_{LD} (Ω)	L_{LD} (H)
15	0.004	0.001	0.003

Table 25.10   Parameters of an active filter of Test System 2

V_{Ref} (V)	C_{DC} (F)	R_{AF} (Ω)	L_{AF} (H)
800	0.002	0.0001	0.002

Table 25.11   Parameters of a PI controller of Test System 2

K_P	K_I
1.0	1.0

Table 25.12   Parameters of a hysteresis controller of Test System 2

T_{on} (S)	T_{off} (S)	G_{on}	G_{off}
0.001	−0.001	1	0

References

[1] P.-T. Cheng, S. Bhattacharya and D. Divan, 'Operations of the dominant harmonic active filter (DHAF) under realistic utility conditions', *IEEE Transactions on Industry Applications*, **37**, 2001, 1037–1044.

[2] P.-T. Cheng, S. S Bhattacharya and D. M. Divan, 'Control of Square-Wave Inverters in High-Power Hybrid Active Filter Systems', *IEEE Transactions on Industry Applications*, **34**, 1998, 458–472.

[3] P.-T. Cheng, S. S Bhattacharya and D. D. Divan, 'Line Harmonics Reduction in High-Power Systems Using Square-Wave Inverters-Based Dominant Harmonic Active Filter', *IEEE Transactions on Power Electronics*, **14**, 1999, 265–272.

[4] P.-T. Chen, S. Bhattacharya and D. Divan, 'Experimental Verification of Dominant Harmonic Active Filter for High-Power Applications', *IEEE Transactions on Industry Applications*, **36**, 2000, 567–577.

[5] J. Arrillaga, B. C. Smith, N. R. Watson and A. R. Wood, *Power System Harmonic Analysis*, John Wiley & Sons, Ltd, Chichester, UK, 1997.

[6] V. S. Moo, Y. N. Chang and P. P. Mok, 'A Digital Measurement Scheme for Time-Varying Transient Harmonics', *IEEE Trans. Power Delivery*, **10**, 1995, 588–594.

[7] S. J. Huang, C. L. Huang and C. T. Hsieh, 'Application of Gabor Transform Technique to Supervise Power System Transient Harmonics', *IEE Proceedings Generation, Transmission and Distribution*, **143**, 1996, 461–466.

[8] J. F. Hauer, C. J. Demeure and L. L. Scharf, 'Initial Results in Prony Analysis of Power System Response Signals', *IEEE Trans. Power Systems*, **5**, 1990, 80–89.

[9] O. Chaari, P. Bastard and M. Meunier, 'Prony's method: An Efficient Tool for the Analysis of Earth Fault Currents in Petersen-coil-protected Networks', *IEEE Trans. On Power Delivery*, **10**, 1995, 1234–1242.

[10] F. B. Hildebrand, *Introduction to Numerical Analysis*, McGraw-Hill Book Company, Inc., New York, USA, 1956.

[11] L. Marple, 'A New Autoregressive Spectrum Analysis Algorithm', *IEEE Trans. Acoustics, Speech, and Signal Processing*, **ASSP-28**, 1980, 441–454.

[12] P. Barone, 'Some Practical Remarks on the Extended Prony's Method of Spectrum Analysis', *Proc. IEEE*, **76**, 1988, 284–285.

[13] S. M. Kay and S. L. Marple, 'Spectrum Analysis – A Modern Perspective', *Proc., IEEE*, **69**, 1981, 1380–1419.

[14] E. Anderson, Z. Bai, C. Bischof *et al.*, LAPACK User's Guide [http://www.netlib.org/lapack/lug/lapack_lug.html], Third edition, SIAM, Philadelphia, 1999.

[15] M. A. Johnson, I. P. Zarafonitis and M. Calligaris, 'Prony Analysis and Power System Stability Some Recent Theoretical and Application Research', Proceedings of IEEE PES General Meeting, Volume 3, July 2000, pp. 1918–1923.

[16] L. Qi, L. Qian, D. Cartes and S. Woodruff, 'Initial Results in Prony Analysis for Harmonic Selective Active Filters', Proceedings of IEEE PES General Meeting, Montreal, Quebec, Canada, June 2006.

[17] Hydro-Quebec, TransEnergie, SymPowerSystems For Use With Simulink, Online only, The Mathworks Inc., July 2002. http://www.mathworks.com/products/simpower/

[18] Powersim Inc., *PSIM User's Guide Version 6.0*, Powersim Inc., June 2003.

Appendix A: Time-varying harmonic currents from large penetration electronic equipment

A. Capasso, R. Lamedica and A. Prudenzi

A1 Introduction

The international literature of this sector has not yet covered time-varying harmonic absorption of electronic equipment in detail, even though the ever increasing penetration of such equipment makes the prediction of harmonic impact an urgent problem.

In order to investigate time-varying behavior of harmonic spectra adequately specific harmonic monitoring systems are required which allow a continuous recording of harmonic quantities [1–4]. Specific and quite expensive commercial products or custom-made equipment are required.

The continuous monitoring can be of a certain utility but also for low-demand single-phase nonlinear loads, since this practice allows a better characterization of harmonic spectra and an improved understanding of the impact due to the various stages of typical operation as well. To this end, some selected results which were obtained from a wide monitoring activity performed in the laboratory over the last few years are reported [5–9].

The measurement activity can be performed by using customized monitoring equipment with the simplified scheme illustrated in Figure A1.

The equipment permits the simultaneous and synchronous sampling of multiple single-phase voltage and current signals by using two different acquisition boards dedicated to voltage and current channels, respectively. Details of the equipment characteristics are reported in [5–9].

This appendix deals with the characterization of time-varying harmonic currents from large penetration electronic equipment. The time-varying characterization of single-phase

Time-Varying Waveform Distortions in Power Systems   Edited by Paulo F. Ribeiro
© 2009 John Wiley & Sons, Ltd

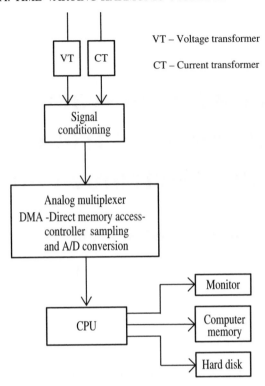

Figure A1   Simplified schematic of the monitoring equipment

nonlinear appliances of large penetration into the end-user market such as a desktop PC, printer, photocopier and cell phone battery charger is presented first. Then, an experimental procedure is presented which helps predict the mutual influence between several single-phase items of electronic equipment and the low-voltage distribution network.

A2   Experimental results for single loads

The results are relevant for the following nonlinear appliance samples, with power ratings ranging from less than 10 W to several hundred W:

- desktop PC,
- printer (both laser and ink-jet),
- photocopier,
- cell phone battery charger.

Most of the data obtained has been processed in order to determine probability density functions (pdf) and distribution functions which can well predict the investigated time-varying behavior.

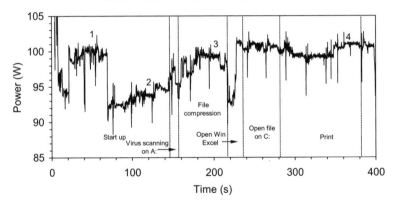

Figure A2   Demand diagram recorded for some common operations of PC

A2.1   Desktop PC

A significant variation in the time of a harmonic current can be identified during the various typical operation phases of a desktop PC. The monitoring activity has covered several PC types. Some selected results are reported here regarding a compatible 133 MHz Pentium processor.

The investigated PC included a 15-inch color monitor. The nameplates provide respectively:

- PC: AC 100–240V/180 mA/50–60 Hz,

- monitor: AC 100–120 V/200–240V, 1/2 A, 50–60 Hz.

The PC's absorption (including the monitor) during typical operation phases has been continuously monitored. Recordings have used a time interval step of 4 cycles (80 ms) for average calculations. The electrical quantities have been monitored with the PC supplied from an outlet of the laboratory. Figures A2 and A3 show the variations in power demand during some more common operations. Figure A4 shows the current harmonic spectrum (main lower-order characteristic harmonics) for the points marked in Figures A2 and A3.

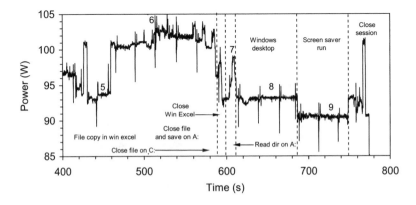

Figure A3   Demand diagram recorded for some common operations of PC

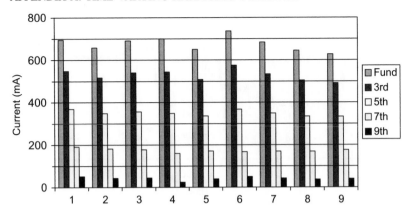

Figure A4   Harmonic current spectrum (main harmonics) recorded in operations evidenced in Figure A3

In Figures A5 and A6 the pdfs for each harmonic current order are reported which illustrates that a certain harmonic spectrum variability with time does exist even though harmonic spectrum patterns do not significantly change. Furthermore, some more detailed investigations have shown that most part of the variability detected must be attributed to the monitor's demand modulation.

Similar variability has also been found for the phase angle of harmonic currents.

A2.2   Laser printer

An analogous monitoring activity has been applied to laser printers. In particular, some selected results obtained for a very diffuse older type of printer are reported below.

The laser printer's main characteristics were:

- electrical ratings: AC 220–240V/50 Hz/850 W,

- printing speed: about four pages/minute.

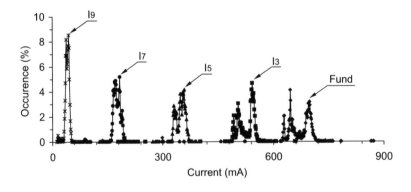

Figure A5   Main harmonic currents magnitude distribution frequency histogram

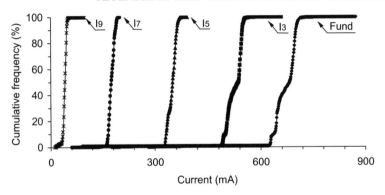

Figure A6   Harmonic currents magnitude distribution frequency diagram for PC

The typical operation phases have firstly been isolated and then correlated to harmonic spectra. Thus, the two main phases of printing and stand-by have been differentiated.

Figure A7 reports voltage and current snapshots as monitored for a laser printer during its typical operation main stages. Figure A8 reports the power demand profile recorded during some selected duty-cycles.

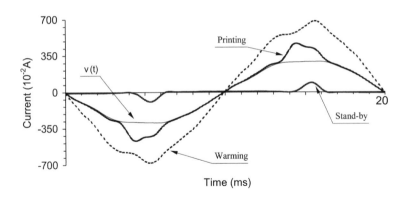

Figure A7   Voltage and current wave shapes monitored for a laser printer

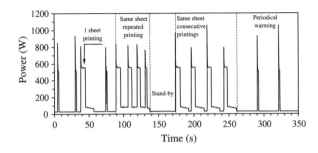

Figure A8   Power demand profile recorded for selected operations of a laser printer

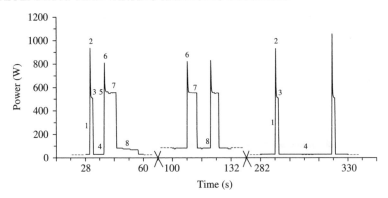

Figure A9   Selected intervals of the power demand profile of Figure A8

In Figure A9, the expanded view of three different operation stages, of the diagram in Figure A8, is reported showing few operation points. For every numbered point Figure A10 illustrates the relevant main characteristic current harmonic spectrum.

Figures A9 and A10 show that the laser printer harmonic current spectrum also includes even harmonics. These harmonics are recorded during printing starting transient conditions (points 1, 2, 5 and 6 of Figure A9) with a magnitude which is not negligible.

In Figures A11 and A12 both the distribution frequency histograms and the cumulative frequency distributions for each harmonic current magnitude are reported for the whole operation time interval of Figure A8. The cumulative frequency histograms, in particular, show the various harmonic current magnitude levels that can be recorded during printer operation better. Among them the operation stages, demonstrated by point 7 and point 4 in Figure A9, result more frequently and correspond, to printing and stand-by operations of the equipment, respectively. Such terms are therefore used for the following considerations.

The points shown by 1, 2, 5 and 6 correspond to the stage defined as 'warming' in Figure A7. From Figure A11 it is also evident that the magnitude of the even harmonics is statistically small although a significant dispersion is also arguable from Figure A12.

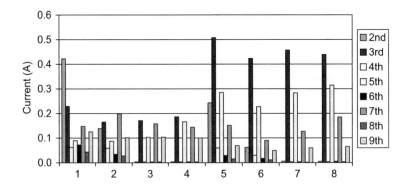

Figure A10   Harmonic current spectrum (main harmonics) recorded at instants shown in Figure A9

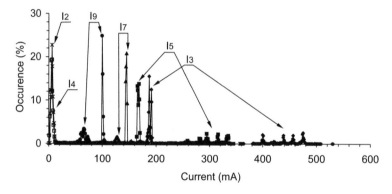

Figure A11   Main harmonic currents magnitude distribution frequency histograms

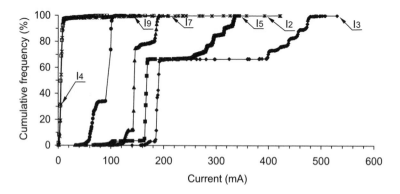

Figure A12   Harmonic currents magnitude cumulative frequency diagram

The phase-angle distribution frequency histograms for only odd harmonic currents are reported in Figure A13, where according to the points mentioned above the two operation stages of printing and stand-by are clearly demonstrated.

A2.3   Ink-jet printers

The monitoring activity has also covered ink-jet printers. In particular a very diffuse low-cost color type has been investigated with the following main characteristics:

- electrical ratings: AC 220–240V/50–60 Hz/0.2 A,

- printing speed: about one page/minute.

The measurement results have been processed in the same way as with the laser printer.

Figure A14 shows voltage and current wave shapes as recorded for stand-by and printing operation stages. Figures A15, A16 and A17 are reported for illustrating the harmonic currents magnitude spectrum variability for the different operation points shown in Figure A16. In particular, the harmonic spectrum relevant to the different points even reveals significant differences both in magnitude and in pattern terms. Figure A15 shows the differences between

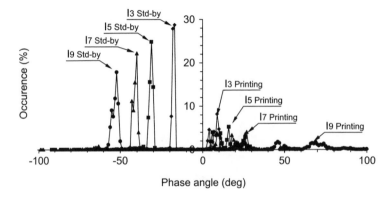

Figure A13   Main harmonic currents phase-angle frequency histograms

ink-jet and laser printing processes. The repeated peaks shown in Figure A15 are in fact due to the reiterated movements of the printing head along the paper sheet during printing.

Figures A18 and A19 show the recorded frequency distribution and cumulative frequency histograms of harmonic currents magnitude. The two operation stages of stand-by and printing are evident. These stages correspond to the points 1 and 5 as shown in Figure A16. Figure A20

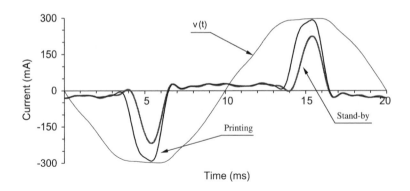

Figure A14   Voltage and current wave shapes monitored for an ink-jet printer

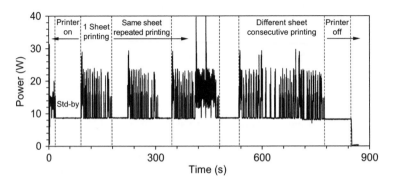

Figure A15   Power demand profile recorded for typical operations of an ink-jet printer

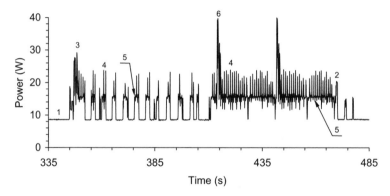

Figure A16   Selected intervals of the power demand profile of Figure A15

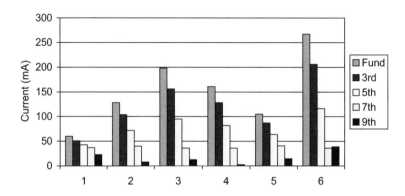

Figure A17   Harmonic current spectrum (main harmonics) recorded at instants shown in Figure A16

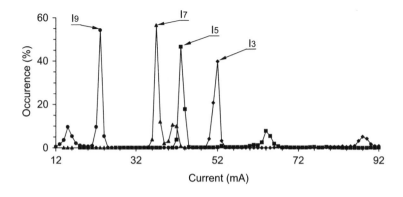

Figure A18   Main harmonic currents magnitude frequency distributions

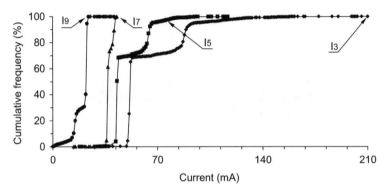

Figure A19   Harmonic currents magnitude cumulative frequency distributions

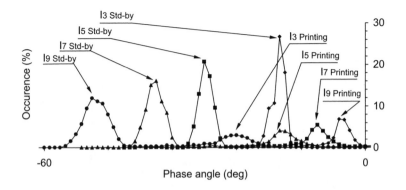

Figure A20   Main harmonic currents phase angle frequency distributions

shows the phase angle frequency distribution histogram obtained for harmonic currents. The two stages mentioned above are opportunely demonstrated.

Finally, for such equipment the continuous monitoring activity also demonstrated a range of variability of recordings which were not negligible.

A2.4   Photocopiers

The harmonic currents absorbed by photocopiers (PH) have also been monitored. Among several samples monitored, some selected results were obtained for a sample which copied up to 25 pages per minute, as reported in Figures A21 and A23. The monitoring results allow a correct differentiation between the two main operation phases of stand-by and copying.

A2.5   Cellular phone battery chargers

The cellular phones require battery chargers (BC) that are typically equipped with single-phase switching mode power suppliers. Several types of diverse battery chargers have been investigated in the laboratory. A nearly continuous harmonic current monitoring activity has been performed in order to capture time-varying characteristics of harmonic absorption due to the battery chargers' typical duty-cycle.

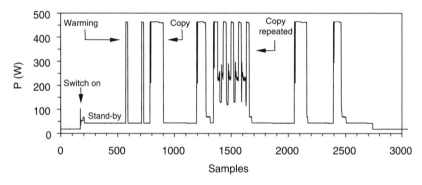

Figure A21   Demand diagram recorded for some typical PH operation phases

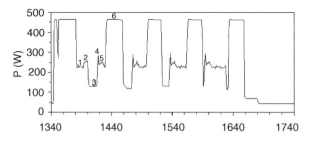

Figure A22   Selected intervals of the demand profile of Figure A9

The experimental activity concerned two battery charger types from the same manufacturer, equipping different phone models: type BC1, an earlier model still widely used by customers, and type BC2, a more modern model.

The monitoring activity has been conducted on:

- BC1, rapid BC with only a Li-Ion battery,

- BC2, rapid BC with (a) a Li-Ion battery and (b) a Ni-MH battery.

The experimental sessions conducted on the above mentioned equipment have been identified as BC1, BC2a and BC2b.

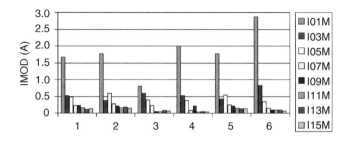

Figure A23   Harmonic current spectrum (main harmonics) recorded at instants shown in Figure A10

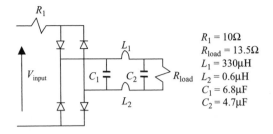

Figure A24   Schematic of BC1's power supplier

A2.5.1   BC Equivalent circuit and input wave shapes

Figure A24 shows the simplified schematic of the BC1 power supplier input stage. Figure A25 shows voltage and current snapshots taken for BC1 during charging and stand-by operations and for BC2 charging Li-Ion and Ni-MH batteries, respectively.

A2.5.2   Session BC1

The BC nameplate provides:

- input: AC 100–240V/180 mA/50–60 Hz,

- output: DC 10V/740 mA.

A complete charging duty-cycle has been continuously monitored. Recordings have used a time interval step of 80 ms (4 cycles) for average calculations. The electrical quantities have been monitored with the BC supplied from an outlet of the laboratory.

Figure A26 shows the BC power demand for a whole charging cycle. The demand level results decrease with an increase in charging level. In Figure A27, the expanded view of three different charging intervals, of the diagram in Figure A26, is shown in order to illustrate the demand modulation operated during charge and controlled by battery charging status better. For every point shown in Figures A27, A28 illustrates the relevant current harmonic spectrum (limited to the main lower-order characteristic harmonics).

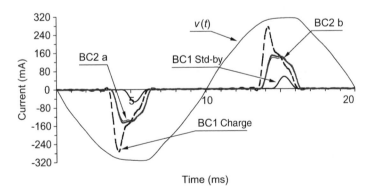

Figure A25   Voltage and current wave shapes monitored for BC1 and BC2

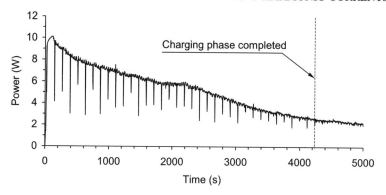

Figure A26   Demand diagram recorded for a whole charging cycle of BC1

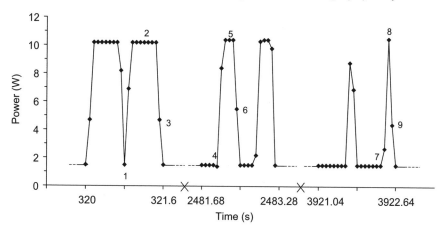

Figure A27   Charging subcycles for BC1 showing demand modulation

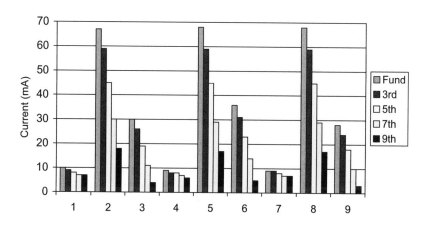

*Figure A28   Harmonic current spectrum (main harmonics) recorded at instants shown in
Figure A27*

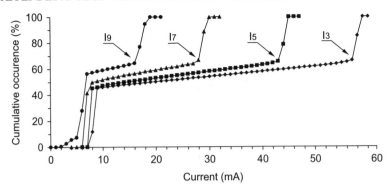

Figure A29   Harmonic currents magnitude cumulative frequency diagram for BC1

The charging cycle of BC1 is performed with partial duty-cycles lasting about 150 s each with average demand progressively decreasing as a function of the battery charge status. Each partial duty-cycle provides in its turn subcycles, such as those illustrated in Figure A27, in which the current absorption varies mainly between a maximum (charge) and a minimum (stand-by) level.

As illustrated in Figure A28 the harmonic spectrum of the two main states of operation are very different. Some differences also exist between points 3, 6 and 9 and points 2, 5 and 8 of Figure A27 harmonic spectra. The differences are even larger for higher harmonic orders. In Figure A29 the cumulative frequency distributions for each harmonic current magnitude are also reported which illustrate that up to 20% of measurements are distributed between the two main demand levels (points similar to 3, 6 and 9 in Figure A27).

It can be concluded that the BC1 harmonic spectrum presents a huge variability during a whole charging cycle and that such a variability cannot be strictly correlated to the actual current absorption. For the whole charging cycle, lasting about 75 min, the occurence distribution histograms for the magnitude of harmonic currents have been reported in Figure A30. Analogous results are also reported for harmonic currents phase angle in Figure A31. Table A1 reports some of the main statistics found valid for the whole BC1 charging cycle.

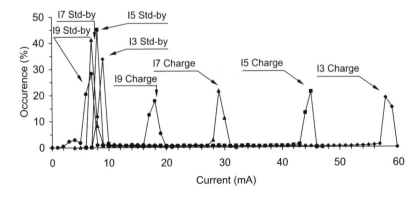

Figure A30   Main harmonic currents magnitude distribution frequency histogram

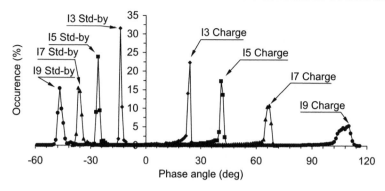

Figure A31   Main harmonic currents phase angle distribution frequency histogram

A2.5.3   Sessions BC2

The BC2 nameplate provides:

- input: AC 100–200V/160 mA/50–60 Hz,

- output: DC 6.2V/720 mA.

For session BC2a, Figure A32 reports the power demand recorded during a whole charging cycle of a Li-Ion battery. In Table A2 the main statistics of the harmonic monitoring results have been reported. For session BC2b analogous results are reported in Figure A33 and Table A3.

Figure A34 shows the cumulative frequency histograms as obtained for harmonic currents magnitude for BC2a. The figure shows a dispersion of the harmonic spectrum during the charging cycle larger than that found for BC1. The harmonic spectra obtained for BC2a and BC2b during the intervals of maximum demand are the same.

Table A1   Statistics of harmonic currents monitored in session BC1

	Magnitude (mA)			Phase angle (deg)		
	Min	Max	Std.Dev.	Min	Max	Std.Dev.
I1	8	69	26.63	−9	9	7.41
I3	8	60	22.99	−15	25	17.78
I5	7	46	16.95	−28	44	31.93
I7	5	31	10.33	−39	70	48.58
I9	1	20	5.55	−58	115	73.83
I11	0	16	4.63	−180	180	116.24
I13	2	17	5.22	−176	−43	43
I15	1	16	5.09	−146	−53	24.05
I17	1	12	3.94	−131	−58	6.04
I19	0	9	2.82	−165	−34	21.13

Table A2   Statistics of harmonic currents monitored in session BC2a

	Magnitude (mA)			Phase angle (deg)		
	Min	Max	Std.Dev.	Min	Max	Std.Dev.
I1	6	50	6.64	−5	3	0.73
I3	6	44	5.54	−22	3	2.41
I5	6	33	3.82	−37	6	4.23
I7	5	20	1.75	−53	11	6.67
I9	4	9	0.19	−68	26	11.91
I11	1	5	0.65	−84	130	49.04
I13	0	7	0.61	−179	180	10.65
I15	0	7	1.3	−153	177	11.08
I17	0	5	0.6	−180	180	13.74
I19	0	3	0.23	−180	180	157.6

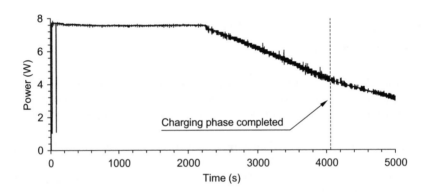

Figure A32   Demand diagram recorded for a whole charging cycle of BC2a

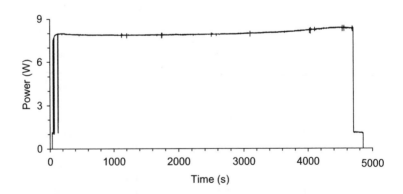

Figure A33   Demand diagram recorded for a whole charging cycle of BC2b

Table A3   Statistics of harmonic currents monitored in session BC2b

	Magnitude (mA)			Phase angle (deg)		
	Min	Max	Std.Dev.	Min	Max	Std.Dev.
I1	6	55	2.18	−7	3	0.39
I3	6	48	1.87	−24	5	1.27
I5	6	36	1.36	−41	8	2.1
I7	5	22	0.92	−58	13	3.09
I9	4	10	0.67	−75	27	4.79
I11	0	4	0.25	−135	171	13.54
I13	0	7	0.54	−177	174	13.31
I15	0	8	0.32	−180	180	10.75
I17	0	6	0.34	−180	180	60.45
I19	0	3	0.5	−180	180	19.35

A2.6   Comments

The time-varying behavior of the harmonic content as obtained for almost every nonlinear appliance is for one of the following reasons:

- load fluctuation due to different internal circuits or components involved in the actual phase of operation (e.g. desktop PC or printers, etc.),

- demand level variation due to charge status of battery (e.g. cell phone BC),

- fluctuations of the value of the individual electronic circuit components (such as capacitances, resistors, etc.) in response to possible environmental changes (e.g. the effect of a temperature increase from cold starting conditions on compact fluorescent lamps power demand),

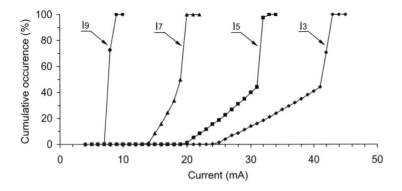

Figure A34   Harmonic currents magnitude cumulative frequency diagrams for BC2a

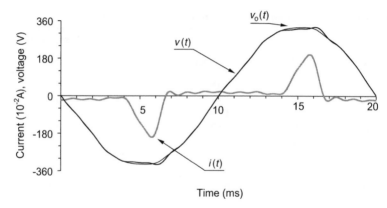

Figure A35   Voltage and current wave shapes monitored at a PC terminal supplied from a PCC with $I_{sc}/I_1 = 250$ and VTHD = 3.2%

- upstream system modifications at point of common coupling (PCC) due to the load amount connected upstream or system impedance fluctuation,

- background voltage distortion level and harmonic content daily variation at the PCC (e.g. for impact of cancellation phenomena on harmonic currents).

As an example of the effects of the even high impact of upstream system conditions and of the voltage distortion level at a PCC, Figures A35 and A36 show voltage and current wave shapes recorded at a PC terminal while it is supplied from an outlet (PCC) with varying impedance value and voltage distortion levels [7]: $I_{sc}/I_1 = 250$, VTHD = 3.2% for Figure A35; $I_{sc}/I_1 = 40$, VTHD = 11.4% for Figure A36.

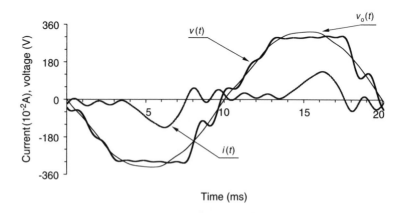

Figure A36   Voltage and current wave shapes monitored at a PC terminal supplied from a PCC with $I_{sc}/I_1 = 40$ and VTHD = 11.4%

A3   Experimental results for several loads

One important aspect of assessing the voltage quality and its impact on distribution systems is the estimation of the cumulative impact of harmonics arising from various sources. Thus, an accurate estimation of impact in distribution systems due to an ever growing number of 'micro' sources is becoming a major concern for power distribution companies.

Various approaches have been proposed in the recent past in order to estimate the net harmonic current due to multiple nonlinear sources in either probabilistic or statistical terms [10–15]. More recently, in [17], the cancellation phenomena have been identified in 'attenuation' and 'diversity', whose effects can be macroscopically well estimated by means of a correct knowledge of both the upstream system and the loading characteristics through the parameter I_{sc}/I_1 of the IEEE Standard 519-1992 [18]. In [19] both 'attenuation' and 'diversity' effects, for a large number of personal computers sharing the same point of common coupling (PCC), have been calculated for different values of the I_{sc}/I_1 ratio with a bottom-up approach.

The same approach has been adopted for further nonlinear loads via simulation based on both analytical models [19–21] and time-domain based models of the individual appliances [22–24]. A similar approach has also been followed in [25] where the positive effects of harmonic cancellation due to the combination of single- and three-phase nonlinear loads are illustrated.

However, for every approach to the problem the monitoring activity of nonlinear equipment harmonic absorption is always a preliminary step for correctly characterizing nonlinear load spectra. To this end, a simple testing configuration has been implemented in the laboratory to reproduce artificially upstream system impedance and load variation influence on harmonic quantities variability at the PCC.

The testing activity allows us to identify the parametrical models expressed by the following curves:

- attenuation curves – predicting the net harmonic current drawn by nonlinear equipment connected to a generic PCC with parametrically varying characteristics;

- impact curves – predicting the increase of the voltage distortion level at a PCC in parametrical terms of both load and upstream system characteristics.

A3.1   Experimental method

The method proposed in [26] based on a monitoring activity is carried out in the laboratory where the layout of Figure A37 is implemented. The monitored nonlinear equipment, for instance a desktop PC, is connected to a low-voltage outlet supplied from the laboratory's panel through an additional impedance made up of a variable resistor and an iron core variable inductor, whose impedance value can be adjusted.

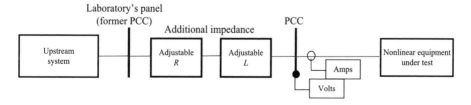

Figure A37   Laboratory tests layout

The proposed procedure provides the nonlinear equipment to be monitored during its typical operation. Every monitoring session provides several subsequent tests, each lasting a few minutes (1–3 min). Each test provides harmonic voltage and current nearly continuous recording at the PCC supplying the same load with an assigned additional impedance value. Each test, therefore, allows the measurement of harmonic quantities for certain upstream system characteristics. Both at the beginning and end of each test, no-load measurements of $VTHD$ ($VTHD_0$) are performed for verifying constancy during the test, in order to be able to calculate the impact of the nonlinear equipment in incremental terms ($\Delta VTHD = VTHD - VTHD_0$). Every test provides 2000–3000 measurement values for each harmonic quantity. The harmonic voltage and currents average values have been calculated on a time slot of four subsequent periods at 50 Hz (80 ms) and stored.

A3.2   Parametrical model identification

The laboratory test results obtained in this way allow the following two parametrical models to be identified:

$$IHD = IHD(I_{sc}/I_1, X/R, VTHD_0) \tag{A1}$$

$$VTHD = VTHD(I_{sc}/I_1, X/R, VTHD_0) \tag{A2}$$

where IHD is the current individual harmonic distortion index, $VTHD$ is the voltage total harmonic distortion index, I_{sc}/I_1 is the ratio of the IEEE Standard 519 at the PCC, X/R is the short-circuit ratio at the PCC and $VTHD_0$ is the no-load voltage total harmonic distortion at the PCC.

The models can be expressed via some curves obtained by interpolating scatter plots of the measurement results. The curves relevant to the model in Equation (A1) permit estimation of the net harmonic currents likely to be drawn by equipment supplied from a PCC with assigned characteristics. The curves can be read in terms of harmonic content decrease due to attenuation effects [19], so they can be identified as 'attenuation curves'. The curves relevant to the model in Equation (A2) give an estimation of the harmonic effect of nonlinear equipment on the voltage distortion at a PCC of given characteristics. So the curves can be defined as 'impact curves'.

A3.3   Attenuation curves

Samples of attenuation curves are reported in Figures A38 and A39. Such curves show, for the various harmonic orders, harmonic current magnitude decrease as the short-circuit ratio (I_{sc}/I_1) decreases. Although the behavior put in evidence in the curves is well known to the power systems harmonics community, the significant influence of the X/R ratio is often neglected.

In Figure A40 the variation of the attenuation effect with the X/R ratio is not negligible as derived from the monitoring activity conducted on a desktop PC.

A3.4   Impact curves

The curves reported in Figure A41 determine the variation of voltage total harmonic distortion ($VTHD$) with respect to the no-load $VTHD$ at a PCC with variable arctan(X/R) and I_{sc} values, when the PCC supplies the same type of nonlinear load (PCs). Every curve in the figure tends

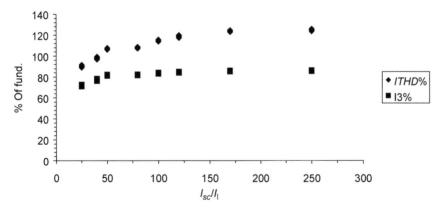

Figure A38   Attenuation of ITHD and third harmonic current for different I_{sc}/I_1

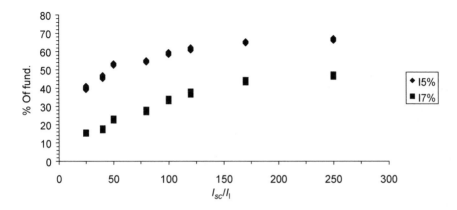

Figure A39   Attenuation of fifth and seventh harmonic current for different I_{sc}/I_1

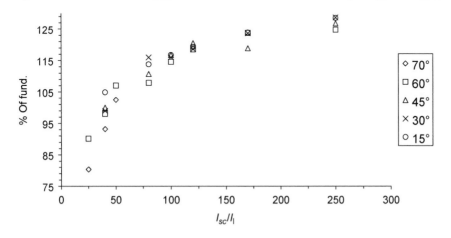

Figure A40   Attenuation of ITHD for different I_{sc}/I_1 and variousarctan (X/R) values as reported in the legend (in degrees)

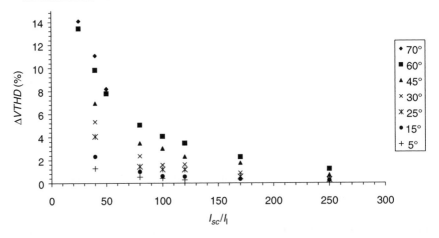

Figure A41   PC impact on VTHD% at a PCC for different I_{sc}/I_1 and arctan (X/R) values as reported in the legend (in degrees)

asymptotically to zero or, in other words, to the same *VTHD* level obtainable at the PCC in no-load conditions (background distortion level). As a matter of fact, such a value can be considered achievable for the I_{sc}/I_1 ratio tending to infinity. During the monitoring sessions on desktop PCs, whose results are reported in the Figure A42, the no-load *VTHD* was always about 2.8%.

A3.5   Comparison with analytical models of PC

The experimental results reported in Figure A41 have been compared, in Figure A42, with analogous ones as obtained from simulations, for the only case of arctan(X/R)=45°, with both analytical and time-domain models of PC.

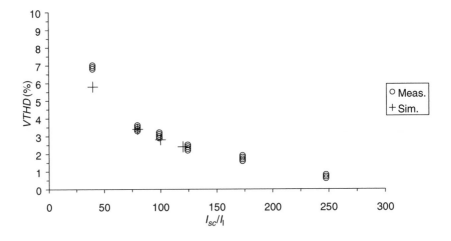

Figure A42   Comparison between simulated and measured impact of PC on the VTHD at a PCC for different I_{sc}/I_1 and arctan(X/R) = 45°

Figure A42 shows a satisfactory matching of the compared curves with the only exception of the points for the lowest I_{sc}/I_1 ratios.

A3.6   Utilization of the proposed models

The models can be used to predict the harmonic influence of nonlinear equipment absorption at a PCC of a low voltage distribution grid. For instance, the curves of Figures A38 and A41 can be used to estimate, for a desktop PC, both the net harmonic current absorption and the voltage distortion at a PCC of the grid with any specific characteristic. In fact, by adequately changing the parameter value in the diagrams, it is possible to simulate nonlinear loading conditions typical of the various low voltage distribution levels of the network. Moreover, the curves are peculiar to specific nonlinear appliance classes. The proposed curves, therefore, can be considered as ready-to-use tools for impact prediction of nonlinear load scenario evolution.

Such a use can well be explained with reference to the 'impact curves' of Figure A41. So, assume that it is necessary to predict the impact on voltage distortion of a new nonlinear loading scenario providing a doubled penetration rate of PCs (I_1''), with respect to an actual rate ($I_{1'}$). Also assume that the actual $\Delta VTHD$ is known and equal to $\Delta VTHD'$, and it is relevant to an $I_{sc}/I_{1'} = 250$ and $\arctan(X/R) = 45°$. The prediction of the new scenario impact can be approximately obtained from the curves of Figure A41. For $I_{sc}/I_{1''} = 125$ ($I_{1''} = 2 * I_{1'}$) and the same X/R value, the $\Delta VTHD''$ will be finally estimated at about 2.7%.

The values reported in the curves do not account for harmonic voltage drops in distribution circuits connecting equipment to the PCC. Therefore, such curves cannot be used directly for predicting the $VTHD$ incremental impact in points different from the PCC at which the monitoring activity of the equipment is performed. In [26], the method has been extended to further nonlinear equipment classes. Tests have been performed on electronic printers either supplied individually or combined with a desktop PC and on cellular phone battery chargers.

A4   Conclusions

This appendix has dealt with the characterization of time-varying harmonic currents from large penetration electronic equipment. The time-varying characterization of single-phase nonlinear appliances of large penetration into end-use such as a desktop PC, printer, photocopier and cell phone battery charger have been presented. The time-varying behavior of harmonics has been investigated through the use of specific measurement equipment allowing continuous harmonic monitoring. Pdfs and distribution functions have been calculated for the large amount of data made available from the monitoring activity.

Then, an experimental procedure has been presented which helps predict the mutual influence between several items of single-phase electronic equipment and the low voltage distribution network. Two parametrical models can be identified from the monitoring results. The models are expressed via some curves that have been defined: attenuation curves and impact curves. The attenuation curves allow us to estimate parametrically the effects of cancellation phenomena on net harmonic currents. The impact curves allow us to predict parametrically the nonlinear load influence on voltage harmonic distortion at PCC.

References

[1] S. R. Kaprielian and A. E. Emanuel, 'An improved real-time data acquisition method for estimation of the thermal effect of harmonics', *IEEE Trans. on Power Delivery*, **9**, 1994, 1632–1638.

[2] Probabilistic Aspects Task Force of Harmonics Working Group, 'Time-varying harmonics: part I, characterizing measured data', *IEEE Trans. on Power Delivery*, **13**, 1998, 938–944.

[3] G. Carpinelli, F. Gagliardi and. P. Verde, 'Probabilistic modelings for harmonic penetration studies in power systems', Proceedings Of the ICHPS-IV, Budapest, Hungary, October, 1990.

[4] G.T. Heydt and E. Gunther, 'Post-measurement processing of electric power quality data', *IEEE Trans. on Power Delivery*, **11**, 1996, 1853–1859.

[5] A. Capasso, R. Lamedica and A. Prudenzi, 'Experimental characterization of personal computers harmonic impact on power quality', *Computer Standards and Interfaces*, **21**, 1999, 321–333.

[6] A. Capasso, R. Lamedica and A. Prudenzi, 'Cellular phone battery chargers impact on voltage quality', IEEE PES Summer General Meeting, Seattle, Washington, USA, July 2000.

[7] R. Lamedica, C. Sorbillo and A. Prudenzi, 'The continuous harmonic monitoring of single-phase electronic appliances: desktop PC and printers', Proceedings of the IEEE ICHQP IX, Orlando, Florida, USA, October 2000.

[8] U. Grasselli, R. Lamedica and A. Prudenzi, 'Time-Varying Harmonics of Single-Phase Non-Linear Appliances', Proceedings of the IEEE PES Winter General Meeting, New York, USA, January 2002.

[9] U. Grasselli, R. Lamedica and A. Prudenzi, 'Characterization of Fluctuating Harmonics from Single-Phase Power Electronics Based Equipment', *International Journal for Computation and Mathematics in Electrical and Electronic Engineering*, **23**, 2004, 133–147.

[10] W. G. Sherman, 'Summation of harmonics with random phase angles', *Proc. IEE*, **119**, 1972, 1643–1648.

[11] N. B. Rowe, 'The summation of randomly-varying phasors or vectors with particular reference to harmonic levels', Proceedings of the IEE Conference Publication Number110, 1974, pp. 177–181.

[12] Y. Baghzouz and O. T. Tan, 'Probabilistic modeling of power system harmonics', *IEEE Trans. on Industry Applications*, **IA-23**, 1987, 173–180.

[13] W. E. Kazibwe, T. H. Ortmeyer and E. Hamman, 'Summation of probabilistic harmonic vectors', *IEEE Trans. on Power Delivery*, 1989, 621–628.

[14] L. Pierrat, 'A unified statistical approach to vectorial summation of random harmonic components', Fourth European Conference on Power Electronics and Applications, Volume III, Florence, Italy, September 1991, pp. 100–105.

[15] A. Cavallini, R. Miglio and G. C. Montanari, 'Statistical modeling of harmonic distortion at residential outlets', Proceedings of the IEEE ICHQP-VII, Las Vegas, Nevada, USA, October 1996.

[16] S. R. Kaprielian, A. E. Emanuel, R. W. Dwyer and H. Mehta, 'Predicting voltage distortion in a system with multiple random harmonic sources', IEEE PES General Winter Meeting, New York, USA, February 1993, Paper no. 94 WM 091-9 PWRD.

[17] A. Mansoor, W. M. Grady, A. H. Chowdhury and M. J. Samotyj, 'An investigation of harmonics attenuation and diversity among distributed single-phase power electronics loads', *IEEE Trans. on Power Delivery*, **10**, 1995, 467–473.

[18] IEEE Standard 519-1992, *IEEE recommended practices and requirements for harmonic control of electrical power systems*, IEEE, New York, USA, 1993.

[19] A. Mansoor, W. M. Grady, P. T. Staats, R. S. Thallam, M. T. Doyle and M. J. Samotyj, 'Predicting the net harmonic currents produced by large numbers of distributed single-phase computer loads', *IEEE Trans. on Power Delivery*, **10**, 1995, 2001–2006.

[20] F. A. Gorgette, W. M. Grady, P. Fauquembergue and K. Ahmed, 'Statistical summation of the harmonic currents produced by a large number of single-phase variable speed air conditioners', Proceedings of the IEEE ICHQP, Las Vegas, Nevada, USA, October 1996.

[21] D. J. Pileggi, C. E. Root, T. J. Gentile, A. E. Emanuel and E. M. Gulachenski, 'The effect of modern compact fluorescent lights on voltage distortion', *IEEE Trans. on Power Delivery*, **8**, 1993, 1451–1459.

[22] R. Dwyer, A. K. Khan, M. McGranaghan, Le Tang, R. K. McCluskey, R. Sung and T. Houy, 'Evaluation of harmonic impacts from compact fluorescent lights on distribution systems', *IEEE Trans. on Power Systems*, **10**, 1995, 1772–1779.

[23] A. Capasso, R. Lamedica, A. Prudenzi, P. F. Ribeiro and S. J. Ranade, 'Probabilistic assessment of harmonic distortion caused by residential load areas', Proceedings of the IEEE ICHPS-VI, Bologna, Italy, September 1994.

[24] A. Capasso, R. Lamedica and A. Prudenzi, 'Estimation of net harmonic currents due to dispersed non-linear loads within residential areas', Proceedings of the IEEE ICHQP, Athens, Greece, October, 1998.

[25] S. Hansen, P. Nielsen and F. Blaabjerg, 'Harmonic cancellation by mixing nonlinear single-phase and three-phase loads', *IEEE Trans. on Industry Applications*, **36**, 2000, 152–159.

[26] A. Prudenzi, 'A novel procedure based on lab tests for predicting single-phase power electronics-based loads harmonic impact on distribution networks', *IEEE Trans. on Power Delivery*, **19**, 2004, 702–707.

Appendix B: Sample of waveforms and decompositions

C. A. Duque, M. V. Ribeiro and P. F. Ribeiro

The waveforms presented below are from actual measurements of typical industrial, commercial and residential load signatures. A webpage will be available to readers with additional real world measurements.

Case B1: Aluminum sheet facility 88 KV

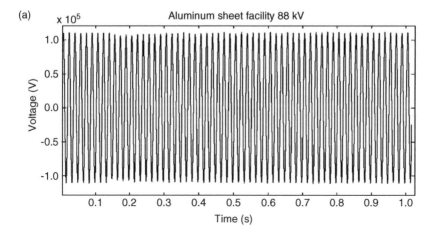

Figure B1. *(a) Voltage; (b) Current; (c) Current harmonic decomposition (odd harmonics); (d) Current harmonic decomposition (even harmonics); (e) Wavelet harmonic decomposition*

Time-Varying Waveform Distortions in Power Systems   Edited by Paulo F. Ribeiro
© 2009 John Wiley & Sons, Ltd

(b)

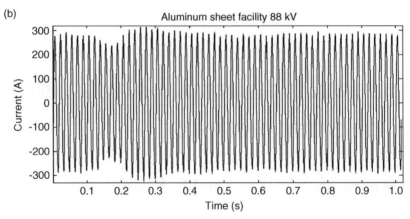

(c)

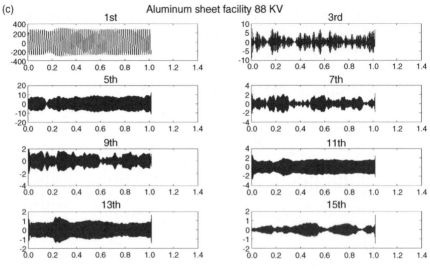

(d)

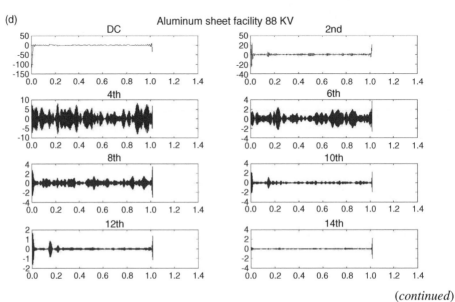

(*continued*)

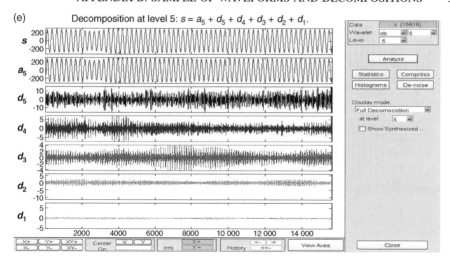

Figure B1. (continued)

Case B2: Arc furnace

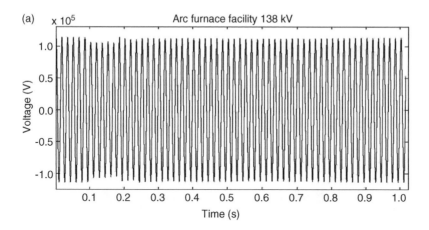

Figure B2. (a) Voltage; (b) Current; (c) Harmonic decomposition (odd); (d) Harmonic decomposition (even); (e) Reconstruction (shows the error due to the possible presence of interharmonics); (f) Wavelet decomposition

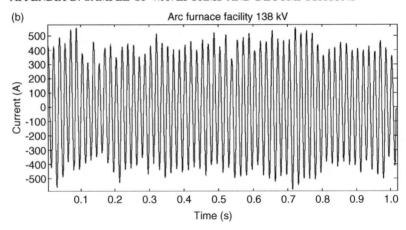

(b)

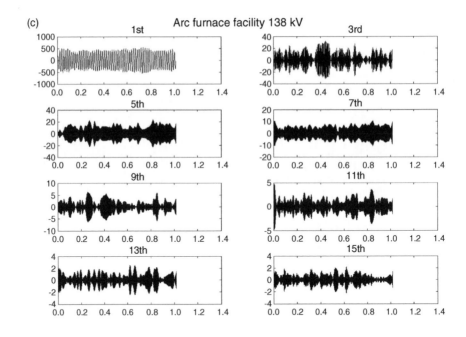

(c)

(*continued*)

(d)

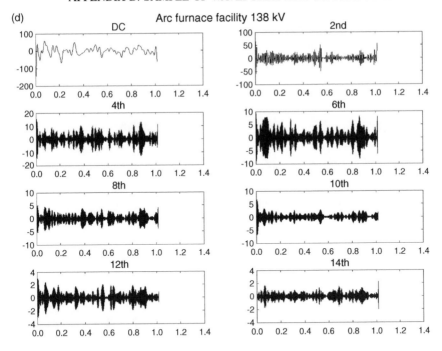

Arc furnace facility 138 kV

(e)

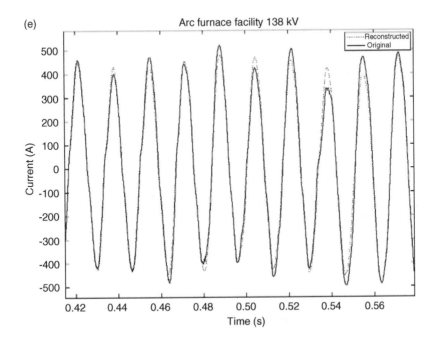

Arc furnace facility 138 kV

(continued)

(f)

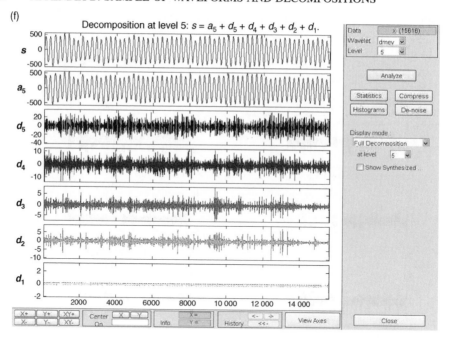

Figure B2. (continued)

Case B3: Arc furnace melting cycle

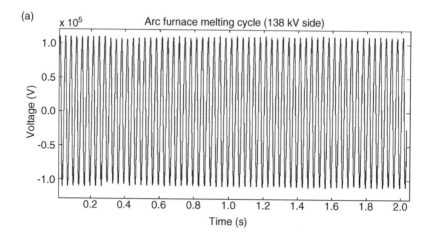

Figure B3. (a) Voltage; (b) Current; (c) Harmonic decomposition (odd); (d) Harmonic decomposition (even); (e) Reconstruction of original waveform showing error due to the presence of interharmonics; (f) Wavelet decomposition

(b)

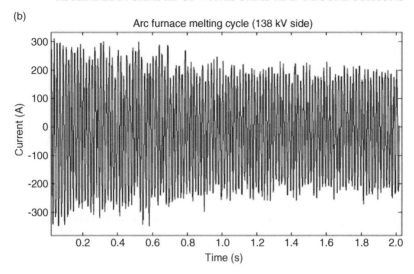

Arc furnace melting cycle (138 kV side)

(c)

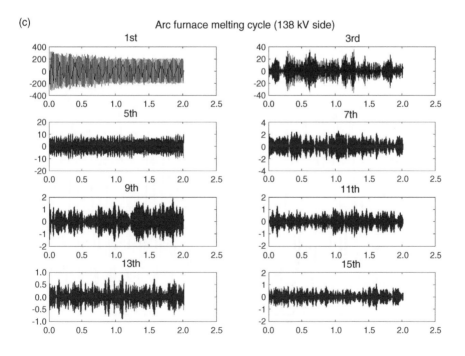

Arc furnace melting cycle (138 kV side)

(*continued*)

(d)

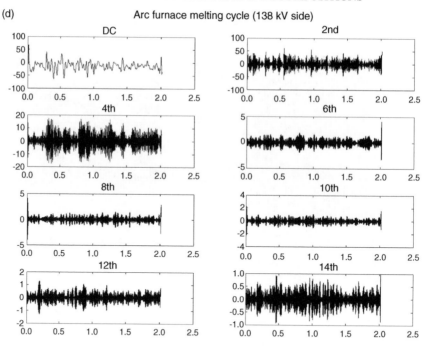

Arc furnace melting cycle (138 kV side)

(e)

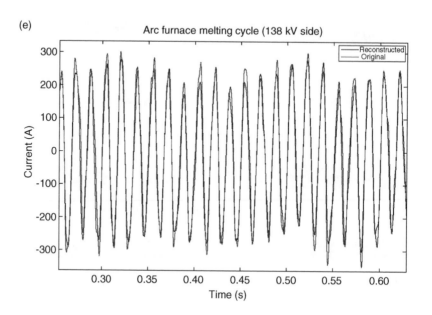

Arc furnace melting cycle (138 kV side)

(continued)

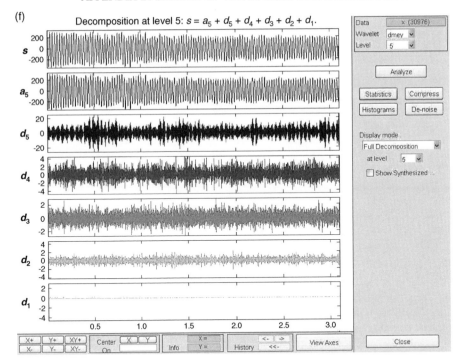

Figure B3. (continued)

Index

Time-Varying Waveform Distortions in Power Systems   Edited by Paulo F. Ribeiro
© 2009 John Wiley & Sons, Ltd